J. Wittenburg
Dynamics of Systems of Rigid Bodies

Leitfäden der angewandten Mathematik und Mechanik

Unter Mitwirkung von
Prof. Dr. E. Becker, Darmstadt
Prof. Dr. G. Hotz, Saarbrücken
Prof. Dr. P. Kall, Zürich
Prof. Dr. K. Magnus, München
Prof. Dr. E. Meister, Darmstadt
Prof. Dr. Dr. h.c. F. K. G. Odqvist, Stockholm
Prof. Dr. Dr. h.c. Dr. h.c. Dr. h.c. E. Stiefel, Zürich

herausgegeben von
Prof. Dr. Dr. h.c. H. Görtler, Freiburg

Band 33

B. G. Teubner Stuttgart

H. Windrich

Dynamics of Systems of Rigid Bodies

by Dr.-Ing. Jens Wittenburg
Professor at the Technische Universität Hannover

with 94 figures and 42 problems

B. G. Teubner Stuttgart 1977

Prof. Dr.-Ing. Jens Wittenburg

Born 1938 in Hamburg. From 1957 to 1964 study of mechanical engineering at the Technische Hochschule Hannover with final grade of Dipl.-Ing. From 1964 to 1967 studies at the University of California at Los Angeles, at the Lomonossow University Moscow and the Polytechnic Institute Leningrad. 1967 doctoral degree at the Technische Universität Hannover. From 1968 to 1969 research studies at the University of California at San Diego. 1972 habilitation, university lecturer and 1976 apl. professor at the Technische Universität Hannover.

CIP-Kurztitelaufnahme der Deutschen Bibliothek

Wittenburg, Jens
Dynamics of systems of rigid bodies.—1. Aufl.—
Stuttgart : Teubner, 1977.
(Leitfäden der angewandten Mathematik und
Mechanik ; Bd. 33)
ISBN 3-519-02337-7

Printed in Germany
Setting and Printing: Zechnersche Buchdruckerei, Speyer
Binder: F. Wochner KG, Horb

Preface

A system of rigid bodies in the sense of this book may be any finite number of rigid bodies interconnected in some arbitrary fashion by joints with ideal holonomic, nonholonomic, scleronomic and/or rheonomic constraints. Typical examples are the solar system, mechanisms in machines and living mechanisms such as the human body provided its individual members can be considered as rigid. Investigations into the dynamics of any such system require the formulation of nonlinear equations of motion, of energy expressions, kinematic relationships and other quantities. It is common practice to develop these for each system separately and to consider the labor necessary for deriving, for example, equations of motion from Lagrange's equation, as inevitable. It is the main purpose of this book to describe in detail a formalism which substantially simplifies these tasks. The formalism is general in that it provides mathematical expressions and equations which are valid for any system of rigid bodies. It is flexible in that it leaves the choice of generalized coordinates to the user. At the same time it is so explicit that its application to any particular system requires only little more than a specification of the system geometry. The book is addressed to advanced graduate students and to research workers. It tries to attract the interest of the theoretician as well as of the practitioner.

The first four out of six chapters are concerned with basic principles and with classical material. In chapter 1 the reader is made familiar with symbolic vector and tensor notation which is used throughout this book for its compact form. In order to facilitate the transition from symbolically written equations to scalar coordinate equations matrices of vector and tensor coordinates are introduced. Transformation rules for such matrices are discussed, and methods are developed for translating compound vector-tensor expressions from symbolic into scalar coordinate form. For the purpose of compact formulations of systems of symbolically written equations matrices are introduced whose elements are vectors or tensors. Generalized multiplication rules for such matrices are defined.

In chapter 2 on rigid body kinematics direction cosines, Euler angles, Bryant angles and Euler parameters are discussed. The notion of angular velocity is introduced, and kinematic differential equations are developed which relate the angular velocity to the rate of change of generalized coordinates. In chapter 3 basic principles of rigid body dynamics are discussed. The definition of, both, kinetic energy and angular momentum leads to the introduction of the inertia tensor. Formulations of the law of angular momentum for a rigid body are derived from Euler's axiom and also from d'Alembert's principle. Because of severe limitations on the length of the manuscript only those subjects are covered which are necessary for the later chapters. Other important topics such as cyclic variables or quasicoordinates, for example, had to be left out. In chapter 4 some classical problems of rigid body mechanics are treated for which closed-form solutions exist. Chapter 5 which makes up one half of the book is devoted to the presentation of a general formalism for the dynamics of systems of rigid bodies. Kinematic relationships, nonlinear equations of motion, energy expressions and other quantities are developed which are suitable for, both, numerical and nonnumerical investigations. The uniform description valid for any system of rigid bodies rests primarily on the application of concepts of graph theory (the first application to mechanics at the time of publication of ref. [1]). This mathematical tool in combination with matrix and symbolic vector and tensor notation leads to expressions which can easily be interpreted in physical terms. The usefulness of the formalism is demonstrated by means of some illustrative examples of nontrivial nature. Chapter 6 deals with phenomena which occur when a multi-body system is subject to a collision either with another system or between two of its own bodies. Instantaneous changes of velocities and internal impulses in joints between bodies caused by such collisions are determined. The investigation reveals an interesting analogy to the law of Maxwell and Betti in elastostatics.

The material presented in subsections 1, 2, 4, 6, 8 and 9 of Chap. 5.2 was developed in close cooperation with Prof. R. E. Roberson (University of Calif. at San Diego) with whom the author has a continuous exchange of ideas and results since 1965. Numerous mathematical relationships resulted from long discussions so that authorship is not claimed by any one person. It is a pleasant opportunity to express my gratitude for this fruitful cooperation. I also thank Dr. L. Lilov (Bulgarian Academy of Sciences) with whom I enjoyed close cooperation on the subject. He had a leading role in applying methods of analytical mechanics (subject of Sec. 5.2.8) and he contributed important ideas to Sec. 5.2.5. Finally, I thank the publishers for their kind patience in waiting for the completion of the manuscript.

Hannover, Spring 1977 J. Wittenburg

Dedicated to my parents

Contents

List of Symbols

$\underline{A}$	coefficient matrix in equations of motion
$\underline{A}^{ij}, \underline{G}_a$	direction cosine matrices
$\boldsymbol{b}_{ij}, \boldsymbol{b}_{ij}^*$	vectors on augmented body i
$\boldsymbol{c}_{ia}, \boldsymbol{c}_{ia}^*$	vectors locating hinge points on body i
$\underline{\boldsymbol{C}}$	matrix constructed of vectors $\boldsymbol{c}_{ia}$
$\boldsymbol{d}_{ij}, \boldsymbol{d}_{ij}^*$	vectors between hinge points on body i
$\frac{{}^{(i)}d\boldsymbol{r}}{dt}, \dot{\boldsymbol{r}}, \mathring{\boldsymbol{r}}$	time derivatives of $\boldsymbol{r}$ in the base $\underline{\boldsymbol{e}}^{(i)}$, in inertial space and in a base specified separately
$\underline{\boldsymbol{e}}, \underline{\boldsymbol{e}}^{(i)}$	vector bases, also column matrices of base vectors
E	energy
$\underline{E}$	unit matrix
$\boldsymbol{E}$	unit tensor
$\boldsymbol{F}, \boldsymbol{F}_i$	forces
$\hat{\boldsymbol{F}}, \hat{\boldsymbol{F}}_i$	impulses
$\boldsymbol{h}$	rotor angular momentum relative to carrier
$\underline{H}$	Jacobi matrix for constraint equations
$i^+(a), i^-(a)$	integer functions
$J_{\alpha\beta}$	moments and products of inertia
$\boldsymbol{J}, \boldsymbol{K}_i$	inertia tensors
$\underline{J}$	inertia matrix
$\boldsymbol{k}_{ai}$	$\partial \boldsymbol{z}_a / \partial q_{ai}$
$\boldsymbol{K}_{ij}$	tensor for multi-body system
$\underline{\boldsymbol{K}}$	matrix constructed of tensors $\boldsymbol{K}_{ij}$

L	Lagrangian
$\boldsymbol{L}$	angular momentum
m, m_i	masses
M	total system mass
$\underline{m}$	diagonal mass matrix
$\boldsymbol{M}, \boldsymbol{M}_i$	torques
$\boldsymbol{p}_{ai}$	vector along rotation axis
q_i, q_{ai}	generalized coordinates
$q_0 \cdots q_3$	Euler parameters
$\boldsymbol{r}, \boldsymbol{R}, \boldsymbol{\varrho}, \boldsymbol{z}$	radius vectors
s_i	vertex in a graph
$\underline{S}_0, \underline{S}, \underline{S}_0^*, \underline{S}^*$	submatrices of incidence matrix
T	kinetic energy
$\underline{T}$	inverse of $\underline{S}$
u_a	arc in directed graph
$\boldsymbol{v}, \boldsymbol{v}_i$	velocities
$\Delta\boldsymbol{v}, \Delta\boldsymbol{v}_i$	velocity increments during collision
$W, \delta W$	work, virtual work
$\boldsymbol{X}_a$	internal hinge force
$\hat{\boldsymbol{X}}_a$	internal hinge impulse
$\boldsymbol{Y}_a$	internal hinge couple
$\hat{\boldsymbol{Y}}_a$	internal hinge impulse couple
$\boldsymbol{z}_a$	hinge vector
$\underline{\boldsymbol{Z}}$	matrix constructed of vectors $\boldsymbol{z}_a$
$\boldsymbol{\delta\pi}, \boldsymbol{\delta\kappa}$	infinitesimal rotation vectors
$\underline{\mu}$	matrix of mass ratios
$\psi, \theta, \phi, \phi_i$	angular coordinates
$\boldsymbol{\omega}, \boldsymbol{\omega}_i, \boldsymbol{\Omega}_a$	angular velocities
$\underline{1}_n$	column matrix of n unit elements

1   Mathematical Notation

In rigid body mechanics vectors play an important role. A vectorial quantity is denoted by a boldface letter. A vector v can be represented as a linear combination of three mutually orthogonal unit vectors e_1, e_2 and e_3:

$$v = v_1 e_1 + v_2 e_2 + v_3 e_3 . \tag{1.1}$$

The unit vectors are base vectors of a vector base (also called reference base or simply base) which is denoted by the symbol $\underline{e}$. Throughout this book it is assumed that the base vectors of vector bases form right hand systems. The scalar quantities v_1, v_2 and v_3 in Eq. (1.1) are the coordinates of v in the base $\underline{e}$ or shortly the coordinates of v in $\underline{e}$. Note that the term vector is used only for the quantity v and not as an abbreviation for the coordinate triple v_α $(\alpha = 1,2,3)$ as is usually done in tensor calculus[1].

In rigid body mechanics it is usually necessary to deal with more than one vector base. The bases are then identified by a superscript in parantheses. Examples are $\underline{e}^{(1)}$, $\underline{e}^{(2)}, \underline{e}^{(s)}$ etc. The base vectors of a base $\underline{e}^{(s)}$ have the property that the scalar product of any two of them equals the Kronecker delta,

$$e_\alpha^{(s)} \cdot e_\beta^{(s)} = \delta_{\alpha\beta} \qquad \alpha, \beta = 1,2,3 . \tag{1.2}$$

The base vector $e_\alpha^{(s)}$ $(\alpha = 1,2,3)$ of a base $\underline{e}^{(s)}$ can be represented as a linear combination of the base vectors of another base $\underline{e}^{(r)}$:

$$e_\alpha^{(s)} = \sum_{\lambda=1}^{3} A_{\alpha\lambda}^{sr} e_\lambda^{(r)} \qquad \alpha = 1,2,3 . \tag{1.3}$$

$$z.B.\, A^{sr} = \begin{bmatrix} \cos\alpha & -\sin\alpha & 0 \\ \sin\alpha & \cos\alpha & 0 \\ 0 & 0 & 1 \end{bmatrix} \text{(Drehtensor)}$$

All three equations can be combined in the single matrix equation

$$\underline{e}^{(s)} = \underline{A}^{sr} \underline{e}^{(r)} . \tag{1.4}$$

$$\rightarrow \text{Spalten} \overset{\triangle}{=} \vec{e}_1, \vec{e}_2, \vec{e}_3 \text{ von } 2 \text{ im } 1\text{-System!}$$

$$A^{21} = \begin{bmatrix} : & : & : \\ : & : & : \end{bmatrix}$$

The symbol $\underline{e}^{(s)}$ which, so far, has simply been used as a name for the base is, through this equation, defined as the column matrix $[e_1^{(s)} \; e_2^{(s)} \; e_3^{(s)}]^T$ of the base vectors. The exponent T indicates transposition. The quantity $\underline{A}^{sr}$ is the (3×3) matrix composed of the vector coordinates $A_{\alpha\lambda}^{sr}$ (α is the row index).

All matrices are designated by underlining. The use of a boldface letter in $\underline{e}^{(r)}$ indicates that the elements of this matrix are vectors. Eq. (1.3) shows that the matrix product

---

[1] For different interpretations of the term vector see Lagally [2]. In some books on vector algebra the coordinates v_1, v_2 and v_3 are referred to as components. In the present book a component is understood to be itself a vector. Thus, $v_1 e_1$ in Eq. (1.1) is a component of v.

$\underline{A}^{sr}\underline{e}^{(r)}$ is evaluated by the common multiplication rule of matrix algebra although one factor is composed of scalars and the other of vectors.

The scalar product $\boldsymbol{e}_\alpha^{(s)} \cdot \boldsymbol{e}_\beta^{(r)}$ of two base vectors belonging to different bases equals the cosine of the angle between the two vectors. Because of Eqs. (1.3) and (1.2) it is also equal to $A_{\alpha\beta}^{sr}$. This identity explains the name direction cosine matrix for $\underline{A}^{sr}$. Replace in Eq. (1.2) $\boldsymbol{e}_\alpha^{(s)}$ by Eq. (1.3) and $\boldsymbol{e}_\beta^{(s)}$ by the corresponding expression with β instead of α. The equation then states that the scalar product of the rows α and β of $\underline{A}^{sr}$ equals $\delta_{\alpha\beta}$. From this follows that $\underline{A}^{sr}(\underline{A}^{sr})^{\mathrm{T}}$ is the unit matrix since every element of the product matrix represents the scalar product of two rows of $\underline{A}^{sr}$. From this identity two important properties of the matrix $\underline{A}^{sr}$ are deduced. First, the inverse of $\underline{A}^{sr}$ equals the transpose, $(\underline{A}^{sr})^{-1} = (\underline{A}^{sr})^{\mathrm{T}}$. From this follows that the inverse of Eq. (1.4) reads

$$\underline{e}^{(r)} = \underline{A}^{rs}\underline{e}^{(s)} = (\underline{A}^{sr})^{\mathrm{T}}\underline{e}^{(s)}. \tag{1.5}$$

Second, the determinant of $\underline{A}^{sr}$ is either $+1$ or -1. The case -1 occurs only if one of the two bases $\underline{e}^{(r)}$ and $\underline{e}^{(s)}$ is a right hand system and the other a left hand system. This is not the case here so that

$$\det \underline{A}^{sr} = +1. \tag{1.6}$$

The right hand side of Eq. (1.1) can be given the form of a matrix product. For this purpose the column matrix $\underline{v} = [v_1 \; v_2 \; v_3]^{\mathrm{T}}$ of the coordinates of $\boldsymbol{v}$ in the base $\underline{e}$ is introduced (a shorter name for $\underline{v}$ is coordinate matrix of $\boldsymbol{v}$ in $\underline{e}$). With this matrix Eq. (1.1) takes the form

$$\boldsymbol{v} = \underline{e}^{\mathrm{T}}\underline{v} \tag{1.7}$$

or

$$\boldsymbol{v} = \underline{v}^{\mathrm{T}}\underline{e}. \tag{1.8}$$

In two different bases $\underline{e}^{(s)}$ and $\underline{e}^{(r)}$ a vector $\boldsymbol{v}$ has different coordinate matrices. They are denoted $\underline{v}^{(s)}$ and $\underline{v}^{(r)}$, respectively. Eq. (1.7) establishes the identity

$$\underline{e}^{(s)\mathrm{T}}\underline{v}^{(s)} = \underline{e}^{(r)\mathrm{T}}\underline{v}^{(r)}.$$

On the right hand side Eq. (1.5) is substituted for $\underline{e}^{(r)}$. This yields

$$\underline{e}^{(s)\mathrm{T}}\underline{v}^{(s)} = \underline{e}^{(s)\mathrm{T}}\underline{A}^{sr}\underline{v}^{(r)}$$

or

$$\underline{v}^{(s)} = \underline{A}^{sr}\underline{v}^{(r)}. \tag{1.9}$$

This equation represents the transformation rule for vector coordinates. It states that the direction cosine matrix is also the coordinate transformation matrix. Note the mnemonic position of the superscripts s and r.

The scalar product of two vectors $\boldsymbol{a}$ and $\boldsymbol{b}$ can be written as a matrix product. Let $\underline{a}^{(r)}$ and $\underline{b}^{(r)}$ be the coordinate matrices of $\boldsymbol{a}$ and $\boldsymbol{b}$, respectively, in some vector base $\underline{e}^{(r)}$. Then, $\boldsymbol{a} \cdot \boldsymbol{b} = \underline{a}^{(r)\mathrm{T}}\underline{b}^{(r)} = \underline{b}^{(r)\mathrm{T}}\underline{a}^{(r)}$. Often the coordinate matrices of two vectors $\boldsymbol{a}$ and $\boldsymbol{b}$ are known in two different bases, say $\underline{a}^{(r)}$ in $\underline{e}^{(r)}$ and $\underline{b}^{(s)}$ in $\underline{e}^{(s)}$. Then, $\boldsymbol{a} \cdot \boldsymbol{b} = \underline{a}^{(r)\mathrm{T}}\underline{A}^{rs}\underline{b}^{(s)}$.

Consider, next, the eigenvalue problem $\underline{A}^{sr}\underline{u}^{(r)} = \lambda\underline{u}^{(r)}$ for a given direction cosine matrix $\underline{A}^{sr}$ relating two bases $\underline{e}^{(r)}$ and $\underline{e}^{(s)}$. This equation represents a special case of

Eq. (1.9) with $\underline{u}^{(s)}=\lambda\underline{u}^{(r)}$. Since the magnitude of a vector is the same in both bases the sums of squares of the elements of $\underline{u}^{(r)}$ and of $\lambda\underline{u}^{(r)}$ must be identical. From this follows that the absolute value of all (real or complex) eigenvalues λ is one. The characteristic polynomial $\det(\underline{A}^{sr}-\lambda\underline{E})$ with unit matrix $\underline{E}$ is of third order. Hence, there exists at least one real eigenvalue with $|\lambda|=1$. That this eigenvalue is $\lambda=+1$ follows from the fact that the three eigenvalues λ_1, λ_2 and λ_3 satisfy the equation $\lambda_1\lambda_2\lambda_3=\det\underline{A}^{sr}=+1$. The coordinates of the real eigenvector $\boldsymbol{u}$ which belongs to the eigenvalue $\lambda=+1$ are determined from the equation

$$\underline{A}^{sr}\underline{u}^{(r)}=\underline{u}^{(r)}. \tag{1.10}$$

This equation states that for any direction cosine matrix $\underline{A}^{sr}$ there exists (at least) one real vector $\boldsymbol{u}$ whose coordinates are the same in both bases $\underline{e}^{(r)}$ and $\underline{e}^{(s)}$. From this follows

Euler's Theorem *Two arbitrarily oriented bases $\underline{e}^{(r)}$ and $\underline{e}^{(s)}$ with common origin P can be made to coincide with one another by rotating one of them through a certain angle about an axis which is passing through P and which has the direction of the eigenvector $\boldsymbol{u}$ determined by Eq. (1.10).*

Besides vectors second-order tensors play an important role in rigid body dynamics. Such tensors are designated by boldface grotesque letters. In its most general form a tensor D is a sum of indeterminate products of two vectors each:

$$\mathsf{D}=\boldsymbol{a}_1\boldsymbol{b}_1+\boldsymbol{a}_2\boldsymbol{b}_2+\boldsymbol{a}_3\boldsymbol{b}_3+\cdots \tag{1.11}$$

A tensor is an operator. Its scalar product from the right with a vector $\boldsymbol{v}$ is defined as the vector

$$\begin{aligned}\mathsf{D}\cdot\boldsymbol{v}&=(\boldsymbol{a}_1\boldsymbol{b}_1+\boldsymbol{a}_2\boldsymbol{b}_2+\boldsymbol{a}_3\boldsymbol{b}_3+\cdots)\cdot\boldsymbol{v}\\&=\boldsymbol{a}_1\boldsymbol{b}_1\cdot\boldsymbol{v}+\boldsymbol{a}_2\boldsymbol{b}_2\cdot\boldsymbol{v}+\boldsymbol{a}_3\boldsymbol{b}_3\cdot\boldsymbol{v}+\cdots.\end{aligned} \tag{1.12}$$

Similarly, the scalar product of D from the left with $\boldsymbol{v}$ is defined as

$$\boldsymbol{v}\cdot\mathsf{D}=\boldsymbol{v}\cdot\boldsymbol{a}_1\boldsymbol{b}_1+\boldsymbol{v}\cdot\boldsymbol{a}_2\boldsymbol{b}_2+\boldsymbol{v}\cdot\boldsymbol{a}_3\boldsymbol{b}_3+\cdots.$$

If in all indeterminate products of D the order of the factors is reversed a new tensor is obtained. It is called the conjugate of D and it is denoted by the symbol $\bar{\mathsf{D}}$:

$$\begin{aligned}\mathsf{D}&=\boldsymbol{a}_1\boldsymbol{b}_1+\boldsymbol{a}_2\boldsymbol{b}_2+\boldsymbol{a}_3\boldsymbol{b}_3+\cdots\\\bar{\mathsf{D}}&=\boldsymbol{b}_1\boldsymbol{a}_1+\boldsymbol{b}_2\boldsymbol{a}_2+\boldsymbol{b}_3\boldsymbol{a}_3+\cdots.\end{aligned} \tag{1.13}$$

In vector algebra the distributive law is valid:

$$\boldsymbol{a}\boldsymbol{b}_1\cdot\boldsymbol{v}+\boldsymbol{a}\boldsymbol{b}_2\cdot\boldsymbol{v}=\boldsymbol{a}(\boldsymbol{b}_1+\boldsymbol{b}_2)\cdot\boldsymbol{v}, \qquad \boldsymbol{a}_1\boldsymbol{b}\cdot\boldsymbol{v}+\boldsymbol{a}_2\boldsymbol{b}\cdot\boldsymbol{v}=(\boldsymbol{a}_1+\boldsymbol{a}_2)\boldsymbol{b}\cdot\boldsymbol{v}.$$

Hence, the indeterminate products in a tensor are also distributive:

$$\boldsymbol{a}\boldsymbol{b}_1+\boldsymbol{a}\boldsymbol{b}_2=\boldsymbol{a}(\boldsymbol{b}_1+\boldsymbol{b}_2), \qquad \boldsymbol{a}_1\boldsymbol{b}+\boldsymbol{a}_2\boldsymbol{b}=(\boldsymbol{a}_1+\boldsymbol{a}_2)\boldsymbol{b}.$$

It is, therefore, possible to resolve all vectors on the right hand side of Eq. (1.11) in some vector base $\underline{e}$ and to regroup the resulting expression in the form

$$\boldsymbol{D} = \sum_{\alpha=1}^{3} \sum_{\beta=1}^{3} D_{\alpha\beta} \boldsymbol{e}_\alpha \boldsymbol{e}_\beta . \tag{1.14}$$

The nine scalars $D_{\alpha\beta}$ are called the coordinates of $\boldsymbol{D}$ in the base $\underline{e}$ (note that not this set of coordinates but only the quantity $\boldsymbol{D}$ is referred to as a tensor). They are combined in the (3×3) coordinate matrix $\underline{D}$. With this matrix the tensor becomes

$$\boldsymbol{D} = \underline{e}^{\mathrm{T}} \underline{D} \, \underline{e} . \tag{1.15}$$

It is a straightforward procedure to construct the matrix $\underline{D}$ from the coordinate matrices of the vectors $\boldsymbol{a}_1$, $\boldsymbol{a}_2$, $\boldsymbol{b}_1$, $\boldsymbol{b}_2$ etc. Let these latter matrices be $\underline{a}_1$, $\underline{a}_2$, $\underline{b}_1$, $\underline{b}_2$ etc. Substitution of Eqs. (1.7) and (1.8) into Eq. (1.11) yields

$$\begin{aligned} \boldsymbol{D} &= \underline{e}^{\mathrm{T}} \underline{a}_1 \underline{b}_1^{\mathrm{T}} \underline{e} + \underline{e}^{\mathrm{T}} \underline{a}_2 \underline{b}_2^{\mathrm{T}} \underline{e} + \underline{e}^{\mathrm{T}} \underline{a}_3 \underline{b}_3^{\mathrm{T}} \underline{e} + \cdots \\ &= \underline{e}^{\mathrm{T}} (\underline{a}_1 \underline{b}_1^{\mathrm{T}} + \underline{a}_2 \underline{b}_2^{\mathrm{T}} + \underline{a}_3 \underline{b}_3^{\mathrm{T}} + \cdots) \underline{e} . \end{aligned}$$

Comparison with Eq. (1.15) shows that

$$\underline{D} = \underline{a}_1 \underline{b}_1^{\mathrm{T}} + \underline{a}_2 \underline{b}_2^{\mathrm{T}} + \underline{a}_3 \underline{b}_3^{\mathrm{T}} + \cdots .$$

From this and from Eq. (1.13) follows that the coordinate matrix of the conjugate of $\boldsymbol{D}$ is the transpose of the coordinate matrix of $\boldsymbol{D}$. With Eqs. (1.14) and (1.1) the vector $\boldsymbol{D} \cdot \boldsymbol{v}$ is

$$\boldsymbol{D} \cdot \boldsymbol{v} = \sum_{\alpha=1}^{3} \sum_{\beta=1}^{3} D_{\alpha\beta} \boldsymbol{e}_\alpha \boldsymbol{e}_\beta \cdot \boldsymbol{v} = \sum_{\alpha=1}^{3} \sum_{\beta=1}^{3} D_{\alpha\beta} v_\beta \boldsymbol{e}_\alpha . \tag{1.16}$$

Its coordinate matrix in the base $\underline{e}$ is, therefore, the product $\underline{D}\,\underline{v}$ of the coordinate matrices of $\boldsymbol{D}$ and $\boldsymbol{v}$ in $\underline{e}$. The same result is obtained in a more formal way when Eqs. (1.15) and (1.7) are substituted for $\boldsymbol{D}$ and $\boldsymbol{v}$, respectively:

$$\boldsymbol{D} \cdot \boldsymbol{v} = \underline{e}^{\mathrm{T}} \underline{D} \, \underline{e} \cdot \underline{e}^{\mathrm{T}} \underline{v} . \tag{1.17}$$

In this expression a new type of matrix product appears, namely the scalar product $\underline{e} \cdot \underline{e}^{\mathrm{T}}$ of two *matrices whose elements are vectors*. Later, still another matrix product called cross product of two vectorial matrices will be met. These two products are defined as follows. Let $\underline{P}$ be an $(m \times r)$ matrix with vectors $\boldsymbol{P}_{ij}$ $(i=1 \ldots m, j=1 \ldots r)$ as elements and let $\underline{Q}$ be an $(r \times n)$ matrix with vectors $\boldsymbol{Q}_{ij}$ $(i=1 \ldots r, j=1 \ldots n)$ as elements. Then, the scalar product $\underline{P} \cdot \underline{Q}$ is a scalar $(m \times n)$ matrix with the elements

$$(\underline{P} \cdot \underline{Q})_{ij} = \sum_{k=1}^{r} \boldsymbol{P}_{ik} \cdot \boldsymbol{Q}_{kj} \qquad i=1 \ldots m, j=1 \ldots n,$$

and the cross product $\underline{P} \times \underline{Q}$ is a vectorial $(m \times n)$ matrix with the elements

$$(\underline{P} \times \underline{Q})_{ij} = \sum_{k=1}^{r} \boldsymbol{P}_{ik} \times \boldsymbol{Q}_{kj} \qquad i=1 \ldots m, j=1 \ldots n .$$

These definitions represent natural generalizations of the common matrix multiplication rule. Let us now return to Eq. (1.17). According to the definition just given the scalar product $\underline{e} \cdot \underline{e}^{\mathrm{T}}$ is the unit matrix so that $\boldsymbol{D} \cdot \boldsymbol{v} = \underline{e}^{\mathrm{T}} \underline{D}\,\underline{v}$ in agreement with Eq. (1.16).

Of particular interest is the tensor

$$\boldsymbol{E}=\boldsymbol{e}_1\boldsymbol{e}_1+\boldsymbol{e}_2\boldsymbol{e}_2+\boldsymbol{e}_3\boldsymbol{e}_3=\underline{e}^{\mathrm{T}}\underline{e} \tag{1.18}$$

whose coordinate matrix is the unit matrix. When this tensor is scalar multiplied with an arbitrary vector $\boldsymbol{v}$ the result is $\boldsymbol{v}$ itself: $\boldsymbol{E}\cdot\boldsymbol{v}=\boldsymbol{v}$ and $\boldsymbol{v}\cdot\boldsymbol{E}=\boldsymbol{v}$. For this reason $\boldsymbol{E}$ is called unit tensor.

With the help of Eq. (1.5) it is a simple matter to establish the law by which the coordinate matrix of a tensor is transformed when instead of a base $\underline{e}^{(r)}$ another base $\underline{e}^{(s)}$ is used for decomposition. Let $\underline{D}^{(r)}$ and $\underline{D}^{(s)}$ be the coordinate matrices of $\boldsymbol{D}$ in the two bases, respectively, so that by Eq. (1.15) the identity

$$\underline{e}^{(s)\mathrm{T}}\underline{D}^{(s)}\underline{e}^{(s)}=\underline{e}^{(r)\mathrm{T}}\underline{D}^{(r)}\underline{e}^{(r)}$$

holds. On the right hand side Eq. (1.5) is substituted for $\underline{e}^{(r)}$. This yields

$$\underline{e}^{(s)\mathrm{T}}\underline{D}^{(s)}\underline{e}^{(s)}=\underline{e}^{(s)\mathrm{T}}\underline{A}^{sr}\underline{D}^{(r)}\underline{A}^{rs}\underline{e}^{(s)}$$

or

$$\underline{D}^{(s)}=\underline{A}^{sr}\underline{D}^{(r)}\underline{A}^{rs}. \tag{1.19}$$

Note, here too, the mnemonic position of the superscripts.

In rigid body mechanics tensors with symmetric and with skew-symmetric coordinate matrices are met. The inertia tensor which will be defined in Chap. 3.1 and the unit tensor $\boldsymbol{E}$ have symmetric coordinate matrices. Tensors with unsymmetric coordinate matrices are found in connection with vector cross products. Consider, first, the double cross product $(\boldsymbol{a}\times\boldsymbol{b})\times\boldsymbol{v}$. It can be written in the form

$$(\boldsymbol{a}\times\boldsymbol{b})\times\boldsymbol{v}=\boldsymbol{b}\,\boldsymbol{a}\cdot\boldsymbol{v}-\boldsymbol{a}\,\boldsymbol{b}\cdot\boldsymbol{v}=(\boldsymbol{b}\,\boldsymbol{a}-\boldsymbol{a}\,\boldsymbol{b})\cdot\boldsymbol{v} \tag{1.20}$$

as scalar product of the tensor $(\boldsymbol{b}\,\boldsymbol{a}-\boldsymbol{a}\,\boldsymbol{b})$ with $\boldsymbol{v}$. If $\underline{a}$ and $\underline{b}$ are the coordinate matrices of $\boldsymbol{a}$ and $\boldsymbol{b}$, respectively, in some vector base then the coordinate matrix of the tensor in this base is the skew-symmetric matrix

$$\underline{b}\,\underline{a}^{\mathrm{T}}-\underline{a}\,\underline{b}^{\mathrm{T}}=\begin{bmatrix}0 & b_1a_2-b_2a_1 & b_1a_3-b_3a_1\\ & 0 & b_2a_3-b_3a_2\\ \text{skew-symmetric} & & 0\end{bmatrix}. \tag{1.21}$$

Also the single vector cross product $\boldsymbol{c}\times\boldsymbol{v}$ can be expressed as a scalar product of a tensor with $\boldsymbol{v}$. For this purpose two vectors $\boldsymbol{a}$ and $\boldsymbol{b}$ are constructed which satisfy the equation $\boldsymbol{a}\times\boldsymbol{b}=\boldsymbol{c}$. The tensor is then, again, $\boldsymbol{b}\,\boldsymbol{a}-\boldsymbol{a}\,\boldsymbol{b}$, and its coordinate matrix is given by Eq. (1.21). This matrix is seen to be identical with

$$\begin{bmatrix}0 & -c_3 & c_2\\ c_3 & 0 & -c_1\\ -c_2 & c_1 & 0\end{bmatrix}$$

where c_1, c_2 and c_3 are the coordinates of $\boldsymbol{c}$ in the same base in which $\underline{a}$ and $\underline{b}$ are measured. For this matrix the symbol $\tilde{\underline{c}}$ (pronounced c tilde) is introduced so that the vector $\boldsymbol{c}\times\boldsymbol{v}$ has the coordinate matrix $\tilde{\underline{c}}\,\underline{v}$. This notation simplifies the transition from symbolic vector equations to scalar coordinate equations[1]. In order to be able to change formulations of coordinate matrix equations the following basic rules will be needed. If k is a scalar then

$$(\widetilde{k\underline{a}}) = k\underline{\tilde{a}}. \tag{1.22}$$

Furthermore,

$$(\widetilde{\underline{a}+\underline{b}}) = \underline{\tilde{a}}+\underline{\tilde{b}}. \tag{1.23}$$

$$\text{From } \underline{\tilde{a}} = \underline{\tilde{b}} \text{ follows } \underline{a} = \underline{b}. \tag{1.24}$$

The identity $\boldsymbol{a}\times\boldsymbol{b} = -\boldsymbol{b}\times\boldsymbol{a}$ yields

$$\underline{\tilde{a}}\underline{b} = -\underline{\tilde{b}}\underline{a} \tag{1.25}$$

and for the special case $\boldsymbol{a} = \boldsymbol{b}$

$$\underline{\tilde{a}}\underline{a} = \underline{0}. \tag{1.26}$$

With the help of the unit tensor $\boldsymbol{E}$ the double vector cross product $\boldsymbol{a}\times(\boldsymbol{b}\times\boldsymbol{v})$ can be written in the form

$$\boldsymbol{a}\times(\boldsymbol{b}\times\boldsymbol{v}) = \boldsymbol{b}\boldsymbol{a}\cdot\boldsymbol{v} - \boldsymbol{a}\cdot\boldsymbol{b}\boldsymbol{v} = (\boldsymbol{b}\boldsymbol{a} - \boldsymbol{a}\cdot\boldsymbol{b}\boldsymbol{E})\cdot\boldsymbol{v}. \tag{1.27}$$

The corresponding coordinate equation reads $\underline{\tilde{a}}\underline{\tilde{b}}\underline{v} = (\underline{b}\underline{a}^{\mathrm{T}} - \underline{a}^{\mathrm{T}}\underline{b}\underline{E})\underline{v}$ with the unit matrix $\underline{E}$. Since this equation holds for every $\underline{v}$ the identity

$$\underline{\tilde{a}}\underline{\tilde{b}} = \underline{b}\underline{a}^{\mathrm{T}} - \underline{a}^{\mathrm{T}}\underline{b}\underline{E} \tag{1.28}$$

is valid. According to Eq. (1.20) the coordinate matrix of $(\boldsymbol{a}\times\boldsymbol{b})\times\boldsymbol{v}$ is $(\underline{b}\underline{a}^{\mathrm{T}} - \underline{a}\underline{b}^{\mathrm{T}})\underline{v}$. It can also be written in the form $(\widetilde{\underline{\tilde{a}}\underline{b}})\underline{v}$. Since both forms are identical for every $\underline{v}$ the identity

$$(\widetilde{\underline{\tilde{a}}\underline{b}}) = \underline{b}\underline{a}^{\mathrm{T}} - \underline{a}\underline{b}^{\mathrm{T}} \tag{1.29}$$

holds. Finally, the transformation rule for tensor coordinates (Eq. (1.19)) states that

$$\underline{\tilde{a}}^{(s)} = (\widetilde{\underline{A}^{sr}\underline{a}^{(r)}}) = \underline{A}^{sr}\underline{\tilde{a}}^{(r)}\underline{A}^{rs}.$$

Systems of linear vector equations can be written in a very compact form if, in addition to matrices with vectorial elements, *matrices with tensors as elements* are used. Such matrices are designated by underlined boldface grotesque letters. They have the general form

$$\underline{\boldsymbol{D}} = \begin{bmatrix} \boldsymbol{D}_{11} \cdots \boldsymbol{D}_{1r} \\ \vdots \\ \boldsymbol{D}_{m1} \quad \boldsymbol{D}_{mr} \end{bmatrix}$$

with arbitrary numbers of rows and columns. The scalar product $\underline{\boldsymbol{D}}\cdot\underline{\boldsymbol{b}}$ of the $(m\times r)$ matrix $\underline{\boldsymbol{D}}$ from the right with a vectorial matrix $\underline{\boldsymbol{b}}$ whose elements are

[1] The notation $\underline{\tilde{c}}\underline{v}$ for the coordinates of $\boldsymbol{c}\times\boldsymbol{v}$ is equivalent to the notation $\varepsilon_{\alpha\beta\gamma}c_\beta v_\gamma$ $(\alpha = 1,2,3)$ which is commonly used in tensor algebra.

$\boldsymbol{b}_{ij}$ $(i=1 \ldots r, j=1 \ldots n)$ is defined as an $(m \times n)$ matrix with the elements

$$\sum_{k=1}^{r} \underline{\boldsymbol{D}}_{ik} \cdot \boldsymbol{b}_{kj} \qquad i=1 \ldots m, j=1 \ldots n\,.$$

A similar definition holds for the scalar product of $\underline{\boldsymbol{D}}$ from the left with an $(n \times m)$ matrix with elements $\boldsymbol{b}_{ij}$ $(i=1 \ldots n, j=1 \ldots m)$. The following example illustrates the practical use of these notations. Suppose it is desired to write the scalar

$$c = \sum_{i=1}^{n} \sum_{j=1}^{n} \boldsymbol{a}_i \cdot \underline{\boldsymbol{D}}_{ij} \cdot \boldsymbol{b}_j$$

as a matrix product. This can be done in symbolic form, $c = \underline{\boldsymbol{a}}^{\mathrm{T}} \cdot \underline{\boldsymbol{D}} \cdot \underline{\boldsymbol{b}}$, with the factors

$$\underline{\boldsymbol{a}} = \begin{bmatrix} \boldsymbol{a}_1 \\ \vdots \\ \boldsymbol{a}_n \end{bmatrix}, \qquad \underline{\boldsymbol{D}} = \begin{bmatrix} \underline{\boldsymbol{D}}_{11} \ldots \underline{\boldsymbol{D}}_{1n} \\ \vdots \\ \underline{\boldsymbol{D}}_{n1} \quad \underline{\boldsymbol{D}}_{nn} \end{bmatrix}, \qquad \underline{\boldsymbol{b}} = \begin{bmatrix} \boldsymbol{b}_1 \\ \vdots \\ \boldsymbol{b}_n \end{bmatrix}.$$

When it is desired to calculate c numerically the following expression in terms of coordinate matrices is more convenient. Let $\underline{a}_i, \underline{b}_i$ and $\underline{D}_{ij}$ $(i,j=1 \ldots n)$ be the coordinate matrices of $\boldsymbol{a}_i$, $\boldsymbol{b}_i$ and $\underline{\boldsymbol{D}}_{ij}$, respectively, in some common vector base. Then,

$$c = \sum_{i=1}^{n} \sum_{j=1}^{n} \underline{a}_i^{\mathrm{T}} \underline{D}_{ij} \underline{b}_j\,.$$

This can, in turn, be written as the matrix product $c = \underline{a}^{\mathrm{T}} \underline{D}\, \underline{b}$ where

$$\underline{a} = \begin{bmatrix} \underline{a}_1 \\ \vdots \\ \underline{a}_n \end{bmatrix}, \qquad \underline{D} = \begin{bmatrix} \underline{D}_{11} \ldots \underline{D}_{1n} \\ \vdots \\ \underline{D}_{n1} \quad \underline{D}_{nn} \end{bmatrix}, \qquad \underline{b} = \begin{bmatrix} \underline{b}_1 \\ \vdots \\ \underline{b}_n \end{bmatrix}$$

are partitioned scalar matrices with the submatrices $\underline{a}_i, \underline{D}_{ij}$ and $\underline{b}_i$, respectively.

Problems

1.1 Given is the direction cosine matrix $\underline{A}^{sr}$ relating the vector bases $\underline{\boldsymbol{e}}^{(r)}$ and $\underline{\boldsymbol{e}}^{(s)}$. Express the matrix products $\underline{\boldsymbol{e}}^{(r)} \cdot \underline{\boldsymbol{e}}^{(r)\mathrm{T}}$, $\underline{\boldsymbol{e}}^{(r)\mathrm{T}} \cdot \underline{\boldsymbol{e}}^{(r)}$, $\underline{\boldsymbol{e}}^{(r)} \times \underline{\boldsymbol{e}}^{(r)\mathrm{T}}$, $\underline{\boldsymbol{e}}^{(r)\mathrm{T}} \times \underline{\boldsymbol{e}}^{(r)}$, $\underline{\boldsymbol{e}}^{(s)} \cdot \underline{\boldsymbol{e}}^{(r)\mathrm{T}}$, $\underline{\boldsymbol{e}}^{(r)} \cdot \underline{\boldsymbol{e}}^{(s)\mathrm{T}}$ and $\underline{\boldsymbol{e}}^{(s)\mathrm{T}} \cdot \underline{\boldsymbol{e}}^{(r)}$ in terms of $\underline{A}^{sr}$ or of elements of $\underline{A}^{sr}$.

1.2 Let $\underline{\boldsymbol{a}}$ and $\underline{\boldsymbol{b}}$ be vectorial matrices and let $\underline{c}$ be a scalar matrix of such dimensions that the products $\underline{\boldsymbol{a}} \cdot \underline{c}\, \underline{\boldsymbol{b}}$ and $\underline{\boldsymbol{a}} \times \underline{c}\, \underline{\boldsymbol{b}}$ exist. Show that the former product is identical with $\underline{\boldsymbol{a}}\, \underline{c} \cdot \underline{\boldsymbol{b}}$ and the latter with $\underline{\boldsymbol{a}}\, \underline{c} \times \underline{\boldsymbol{b}}$.

1.3 $\underline{\boldsymbol{e}}^{(r)}$ and $\underline{\boldsymbol{e}}^{(s)} = \underline{A}^{sr} \underline{\boldsymbol{e}}^{(r)}$ are two vector bases, and $\boldsymbol{a}$, $\boldsymbol{b}$ and $\boldsymbol{c}$ are vectors whose coordinate matrices $\underline{a}^{(r)}$ and $\underline{b}^{(r)}$ in $\underline{\boldsymbol{e}}^{(r)}$ and $\underline{c}^{(s)}$ in $\underline{\boldsymbol{e}}^{(s)}$, respectively, are given. Furthermore, $\boldsymbol{D}$ is a tensor with the coordinate matrix $\underline{D}^{(s)}$ in $\underline{\boldsymbol{e}}^{(s)}$. Formulate in terms of $\underline{A}^{sr}$ and of the given coordinate matrices the scalars $\boldsymbol{a} \cdot \boldsymbol{b} \times \boldsymbol{c}$, $\boldsymbol{a} \times \boldsymbol{b} \cdot \boldsymbol{b} \times \boldsymbol{c}$, $\boldsymbol{c} \cdot \boldsymbol{D} \cdot \boldsymbol{a}$ and $\boldsymbol{c} \cdot \boldsymbol{b} \times \boldsymbol{D} \cdot \boldsymbol{c}$ as well as the coordinate matrices in $\underline{\boldsymbol{e}}^{(r)}$ of the vectors $\boldsymbol{a} \times \boldsymbol{b}$, $\boldsymbol{a} \times \boldsymbol{c}$, $\boldsymbol{a} \times (\boldsymbol{c} \times \boldsymbol{b})$, $\boldsymbol{c} \times \boldsymbol{D} \cdot \boldsymbol{a}$ and $\boldsymbol{a} \times [(\boldsymbol{D} \cdot \boldsymbol{b}) \times \boldsymbol{c}]$.

1.4 Rewrite the vector equations

$$\boldsymbol{a}_1 = \boldsymbol{b} \times (\boldsymbol{v}_1 \times \boldsymbol{b} + \boldsymbol{v}_2 \times \boldsymbol{c}) + \boldsymbol{d} \times \boldsymbol{v}_2$$

$$\boldsymbol{a}_2 = \boldsymbol{c} \times (\boldsymbol{v}_1 \times \boldsymbol{b} + \boldsymbol{v}_2 \times \boldsymbol{c}) - \boldsymbol{d} \times \boldsymbol{v}_1$$

in the form

$$\begin{bmatrix} \boldsymbol{a}_1 \\ \boldsymbol{a}_2 \end{bmatrix} = \begin{bmatrix} \boldsymbol{D}_{11} & \boldsymbol{D}_{12} \\ \boldsymbol{D}_{21} & \boldsymbol{D}_{22} \end{bmatrix} \cdot \begin{bmatrix} \boldsymbol{v}_1 \\ \boldsymbol{v}_2 \end{bmatrix}$$

with explicit expressions for the tensors $\boldsymbol{D}_{ij}$ $(i,j=1,2)$. How are $\boldsymbol{D}_{12}$ and $\boldsymbol{D}_{21}$ related to one another? In some vector base the vectors in the original equations have the coordinate matrices $\underline{a}_1, \underline{a}_2, \underline{v}_1, \underline{v}_2, \underline{b}, \underline{c}$ and $\underline{d}$, respectively. Write down the coordinate matrix equation

$$\begin{bmatrix} \underline{a}_1 \\ \underline{a}_2 \end{bmatrix} = \begin{bmatrix} \underline{D}_{11} & \underline{D}_{12} \\ \underline{D}_{21} & \underline{D}_{22} \end{bmatrix} \begin{bmatrix} \underline{v}_1 \\ \underline{v}_2 \end{bmatrix}$$

giving explicit expressions for the (3×3) submatrices $\underline{D}_{ij}$ $(i,j=1,2)$. What can be said about the (6×6) matrix on the right hand side?

2 Rigid Body Kinematics

2.1 Generalized coordinates for the angular orientation of a rigid body

In order to specify the angular orientation of a rigid body in a vector base $\underline{e}^{(1)}$ it is sufficient to specify the angular orientation of a vector base $\underline{e}^{(2)}$ which is rigidly attached to the body. This can be done, for instance, by means of the direction cosine matrix:

$$\underline{e}^{(2)} = \underline{A}^{21} \underline{e}^{(1)}.$$

The nine elements of this matrix are generalized coordinates which describe the angular orientation of the body in the base $\underline{e}^{(1)}$. Between these coordinates there exist six constraint equations of the form

$$\sum_{\lambda=1}^{3} A_{\alpha\lambda}^{21} A_{\beta\lambda}^{21} = \delta_{\alpha\beta} \qquad \alpha, \beta = 1, 2, 3. \tag{2.1}$$

It is often inconvenient to work with nine coordinates and six constraint equations. There are several useful systems of three coordinates without constraint equations and of four coordinates with one constraint equation which can be used as alternatives to direction cosines. In the following subsections generalized coordinates known as Euler angles, Bryant angles and Euler parameters will be discussed.

2.1.1 Euler angles $\left(\vec{e}_3^{(1)} - \vec{e}_1^{(2)''} - \vec{e}_3^{(2)'} \right)$ entspr. $\left(\psi - \theta - \phi \right)$

The angular orientation of the body-fixed base $\underline{e}^{(2)}$ is thought to be the result of three successive rotations. Before the first rotation the base $\underline{e}^{(2)}$ coincides with the base $\underline{e}^{(1)}$. The first rotation is carried out about the axis $e_3^{(1)}$ through an angle ψ. It carries the base from its original orientation to an orientation denoted $\underline{e}^{(2)''}$ (Fig. 2.1). The second

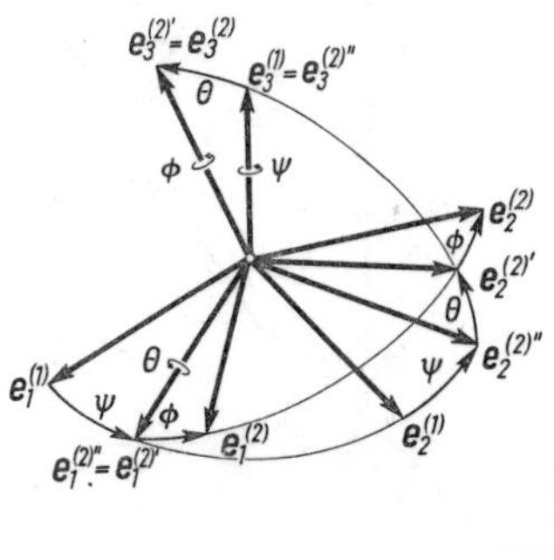

Fig. 2.1
Euler angles ψ, θ, ϕ

rotation through the angle θ about the axis $e_1^{(2)''}$ results in the orientation denoted $\underline{e}^{(2)'}$. The third rotation through the angle ϕ about the axis $e_3^{(2)'}$ produces the final orientation of the base. It is denoted $\underline{e}^{(2)}$ in Fig. 2.1. A characteristic property of Euler angles is that each rotation is carried out about a base vector of the body-fixed base in a position which is the result of all previous rotations. A further characteristic is the sequence (3,1,3) of indices of rotation axes. The desired presentation of the transformation matrix $\underline{A}^{21}$ in terms of ψ, θ and ϕ is found from the transformation equations for the individual rotations which are according to Fig. 2.1

$$\underline{e}^{(2)''} = \underline{A}^{\psi}\underline{e}^{(1)}, \qquad \underline{e}^{(2)'} = \underline{A}^{\theta}\underline{e}^{(2)''}, \qquad \underline{e}^{(2)} = \underline{A}^{\phi}\underline{e}^{(2)'}$$

with

$$\underline{A}^{\psi} = \begin{bmatrix} \cos\psi & \sin\psi & 0 \\ -\sin\psi & \cos\psi & 0 \\ 0 & 0 & 1 \end{bmatrix}, \quad \underline{A}^{\theta} = \begin{bmatrix} 1 & 0 & 0 \\ 0 & \cos\theta & \sin\theta \\ 0 & -\sin\theta & \cos\theta \end{bmatrix}, \quad \underline{A}^{\phi} = \begin{bmatrix} \cos\phi & \sin\phi & 0 \\ -\sin\phi & \cos\phi & 0 \\ 0 & 0 & 1 \end{bmatrix}$$

It follows that $\underline{A}^{21} = \underline{A}^{\phi}\underline{A}^{\theta}\underline{A}^{\psi}$ or explicitly with the abbreviations c_ψ, c_θ and c_ϕ for $\cos\psi$, $\cos\theta$ and $\cos\phi$ and s_ψ, s_θ and s_ϕ for $\sin\psi$, $\sin\theta$ and $\sin\phi$, respectively,

$$\underline{A}^{21} = \begin{bmatrix} c_\psi c_\phi - s_\psi c_\theta s_\phi & s_\psi c_\phi + c_\psi c_\theta s_\phi & s_\theta s_\phi \\ -c_\psi s_\phi - s_\psi c_\theta c_\phi & -s_\psi s_\phi + c_\psi c_\theta c_\phi & s_\theta c_\phi \\ s_\psi s_\theta & -c_\psi s_\theta & c_\theta \end{bmatrix}. \tag{2.2}$$

The advantage of having only three coordinates without any constraint equation is paid for with the disadvantage that the direction cosines are complicated circular functions. There is still another problem. Fig. 2.1 shows that in the case $\theta = n\pi\ (n=0,1,\ldots)$ the axes of the first and third rotation coincide so that ψ and ϕ cannot be distinguished. Euler angles can be illustrated by means of a rigid body in a two-gimbal cardanic suspension. The bases $\underline{e}^{(1)}$ and $\underline{e}^{(2)}$ are attached to the material base and to the

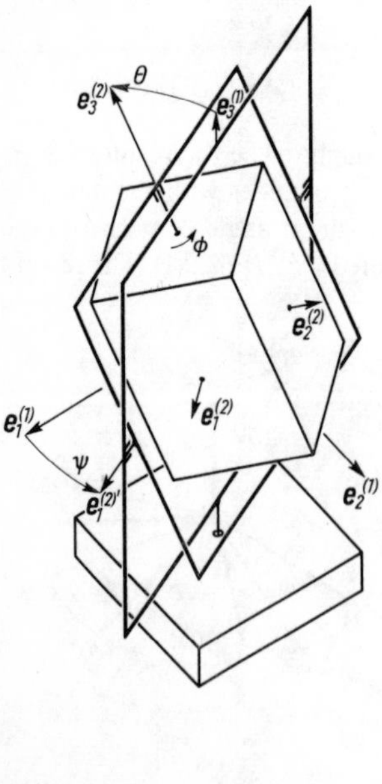

Fig. 2.2
Euler angles in a cardan suspension

suspended body, respectively, as shown in Fig. 2.2. The angles ψ, θ and ϕ are in this order the rotation angle of the outer gimbal relative to the material base, of the inner gimbal relative to the outer gimbal and of the body relative to the inner gimbal. For $\theta = n\pi$ $(n=0,1,\ldots)$ the two gimbals coincide (gimbal lock). With this device all three angles can be adjusted independently since the auxiliary vector bases $\underline{e}^{(2)''}$ and $\underline{e}^{(2)'}$ are materially realized by the gimbals. Euler angles owe their practical importance to the fact that there are many technical systems in which a rigid body is moving in such a way that θ is exactly or approximately constant and that, both, ψ and ϕ are exactly or approximately proportional to time, i.e. $\dot{\psi} \approx$ const and $\dot{\phi} \approx$ const. The use of Euler angles is advantageous also whenever there exist two axes of particular physical significance, one fixed in the base $\underline{e}^{(1)}$ and the other fixed on the body. In such cases the base vectors $e_3^{(1)}$ and $e_3^{(2)}$ are given these directions so that θ is the angle between the two axes (an example of this kind will be treated in Sec. 4.1.4). The use of Euler angles as generalized coordinates is not limited to such cases, however.

It is sometimes necessary to calculate Euler angles which correspond to a given matrix $\underline{A}^{21}$. For this purpose the following formulas are deduced from Eq. (2.2).

$$\begin{aligned}
\cos\theta &= A_{33}^{21}\ , & \sin\theta &= \varepsilon\sqrt{1-\cos^2\theta}, \quad \varepsilon = +1 \text{ or } -1 \\
\cos\psi &= -\frac{A_{32}^{21}}{\sin\theta}, & \sin\psi &= \frac{A_{31}^{21}}{\sin\theta} \\
\cos\phi &= \frac{A_{23}^{21}}{\sin\theta}, & \sin\phi &= \frac{A_{13}^{21}}{\sin\theta}\ .
\end{aligned} \tag{2.3}$$

The formulas show that numerical difficulties are to be expected for values of θ which are close to the critical values $n\pi$ $(n=0,1,\ldots)$.

2.1.2 Bryant angles

$(\vec{e}_1^{\,(1)} - \vec{e}_2^{\,(2)''} - \vec{e}_3^{\,(2)'})$ entspr. $(\phi_1 - \phi_2 - \phi_3)$

These angles are also referred to as Cardan angles. The angular orientation of the body-fixed base $\underline{e}^{(2)}$ is, again, represented as the result of a sequence of three rotations at the beginning of which the base coincides with the reference base $\underline{e}^{(1)}$. The first rotation through an angle ϕ_1 is carried out about the axis $e_1^{(1)}$. It results in the auxiliary base $\underline{e}^{(2)''}$ (Fig. 2.3). The second rotation through an angle ϕ_2 about the axis $e_2^{(2)''}$

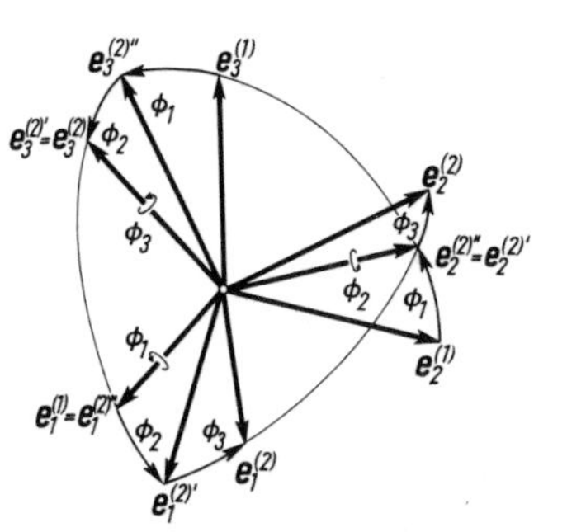

Fig. 2.3 Bryant angles ϕ_1, ϕ_2, ϕ_3

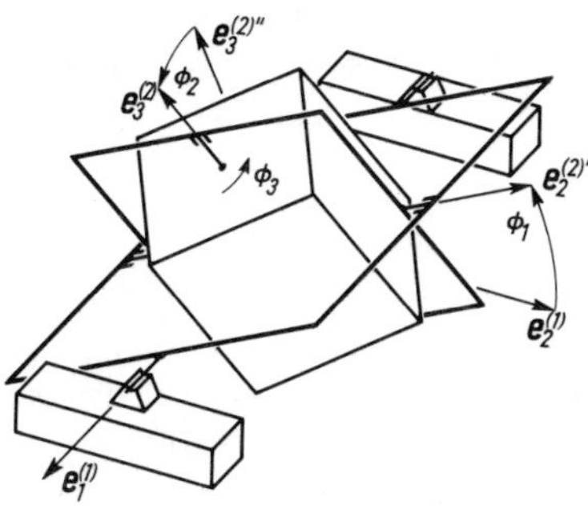

Fig. 2.4
Bryant angles in a cardan suspension

produces the base $\underline{e}^{(2)'}$. The third rotation through an angle ϕ_3 about the axis $e_3^{(2)'}$ gives the body-fixed base its final orientation denoted $\underline{e}^{(2)}$ in Fig. 2.3. The transformation equations for the individual rotations are

$$\underline{e}^{(2)''} = \underline{A}^1 \underline{e}^{(1)}, \qquad \underline{e}^{(2)'} = \underline{A}^2 \underline{e}^{(2)''}, \qquad \underline{e}^{(2)} = \underline{A}^3 \underline{e}^{(2)'}$$

with

$$\underline{A}^1 = \begin{bmatrix} 1 & 0 & 0 \\ 0 & \cos\phi_1 & \sin\phi_1 \\ 0 & -\sin\phi_1 & \cos\phi_1 \end{bmatrix}, \quad \underline{A}^2 = \begin{bmatrix} \cos\phi_2 & 0 & -\sin\phi_2 \\ 0 & 1 & 0 \\ \sin\phi_2 & 0 & \cos\phi_2 \end{bmatrix},$$

$$\underline{A}^3 = \begin{bmatrix} \cos\phi_3 & \sin\phi_3 & 0 \\ -\sin\phi_3 & \cos\phi_3 & 0 \\ 0 & 0 & 1 \end{bmatrix}. \tag{2.4}$$

The direction cosine matrix relating $\underline{e}^{(1)}$ and $\underline{e}^{(2)}$ is the product $\underline{A}^{21} = \underline{A}^3 \underline{A}^2 \underline{A}^1$ or explicitly with the abbreviations $c_\alpha = \cos\phi_\alpha$, $s_\alpha = \sin\phi_\alpha$ $(\alpha = 1,2,3)$

$$\underline{A}^{21} = \begin{bmatrix} c_2 c_3 & c_1 s_3 + s_1 s_2 c_3 & s_1 s_3 - c_1 s_2 c_3 \\ -c_2 s_3 & c_1 c_3 - s_1 s_2 s_3 & s_1 c_3 + c_1 s_2 s_3 \\ s_2 & -s_1 c_2 & c_1 c_2 \end{bmatrix}. \tag{2.5}$$

The only significant difference as compared with Euler angles is the sequence (1,2,3) of indices of rotation axes. Bryant angles can also be illustrated by means of a rigid body in a two-gimbal cardanic suspension. The bases $\underline{e}^{(1)}$ and $\underline{e}^{(2)}$ are attached to the material base and to the body, respectively, as shown in Fig. 2.4. The angles ϕ_1, ϕ_2 and ϕ_3 are in this order the rotation angle of the outer gimbal relative to the material base, of the inner gimbal relative to the outer gimbal and of the body relative to the inner gimbal. For $\phi_2 = 0$ the three rotation axes are mutually orthogonal. As with Euler angles there exists a critical case, namely the case $\phi_2 = \pi/2 + n\pi$ $(n = 0, 1 \ldots)$ in which the planes of the two gimbals coincide so that the rotation axes of ϕ_1 and ϕ_3 become identical. In contrast to Euler angles no mathematical difficulties arise if all three angles ϕ_1, ϕ_2 and ϕ_3 are close to zero. For this reason Bryant angles are particularly useful in cases where a body is moving in such a way that the body-fixed base $\underline{e}^{(2)}$ deviates only little from $\underline{e}^{(1)}$. For sufficiently small angles the linear approximation $\sin\phi_\alpha \approx \phi_\alpha$, $\cos\phi_\alpha \approx 1$ $(\alpha = 1,2,3)$ yields

$$\underline{A}^{21} \approx \begin{bmatrix} 1 & \phi_3 & -\phi_2 \\ -\phi_3 & 1 & \phi_1 \\ \phi_2 & -\phi_1 & 1 \end{bmatrix}. \tag{2.6}$$

This expression suggests defining a vector $\boldsymbol{\phi}$ with coordinates ϕ_1, ϕ_2 and ϕ_3 and writing $\underline{A}^{21} \approx \underline{E} - \tilde{\underline{\phi}}$. Note that it makes no difference whether ϕ_1, ϕ_2 and ϕ_3 are interpreted as coordinates of $\boldsymbol{\phi}$ in the base $\underline{e}^{(1)}$ or in the base $\underline{e}^{(2)}$ or along the axes $e_1^{(1)}$, $e_2^{(2)''}$ and $e_3^{(2)'}$, respectively. This can be shown as follows. If $\underline{\phi} = [\phi_1 \;\; \phi_2 \;\; \phi_3]^T$ designates the coordinate matrix in $\underline{e}^{(2)}$ then the coordinate matrix in the base $\underline{e}^{(1)}$ is in the linear approximation $\underline{A}^{21}\underline{\phi} \approx (\underline{E} - \tilde{\underline{\phi}})\underline{\phi}$. Because of the identity $\tilde{\underline{\phi}}\underline{\phi} = \underline{0}$ (cf. Eq. (1.26)) this is identical with ϕ. The result of these considerations is that within linear approximations small rotation angles can be added like vectors.

It is sometimes necessary to calculate Bryant angles which correspond to a given direction cosine matrix $\underline{A}^{21}$. This can be done with the help of the formulas derived from Eq. (2.5)

$$\begin{aligned} \sin\phi_2 &= A_{31}^{21}, & \cos\phi_2 &= \varepsilon\sqrt{1-\sin^2\phi_2}, \quad \varepsilon=+1 \text{ or } -1 \\ \sin\phi_1 &= -\frac{A_{32}^{21}}{\cos\phi_2}, & \cos\phi_1 &= \frac{A_{33}^{21}}{\cos\phi_2} \\ \sin\phi_3 &= -\frac{A_{21}^{21}}{\cos\phi_2}, & \cos\phi_3 &= \frac{A_{11}^{21}}{\cos\phi_2}. \end{aligned} \tag{2.7}$$

2.1.3 Euler parameters

The angular orientation of the body-fixed base $\underline{e}^{(2)}$ is considered to be the result of a single rotation at the beginning of which the base coincides with the reference base $\underline{e}^{(1)}$. The rotation is carried out clockwise through an angle χ about an axis which is defined by a unit vector $\boldsymbol{u}$. Euler's theorem (see Chap. 1) states that for any angular orientation of the base $\underline{e}^{(2)}$ to be described a real angle χ and a unit vector $\boldsymbol{u}$ with real coordinates exist. These coordinates are the same in both bases $\underline{e}^{(1)}$ and $\underline{e}^{(2)}$. The following considerations are based upon the assumption that χ as well as $\boldsymbol{u}$ are given and that the direction cosine matrix $\underline{A}^{21}$ relating $\underline{e}^{(1)}$ and $\underline{e}^{(2)}$ is to be expressed as a function of these quantities. The function is found from a comparison of the coordinate matrices in the bases $\underline{e}^{(1)}$ and $\underline{e}^{(2)}$ of a body-fixed vector. In Fig. 2.5 this vector is shown in its positions $\boldsymbol{r}^*$ and $\boldsymbol{r}$ before and after the rotation, respectively. In both positions the vector lies on a circular cone whose axis is defined by the unit vector $\boldsymbol{u}$. Let $\underline{r}^{(1)}$ and $\underline{r}^{(2)}$ be the coordinate matrices of $\boldsymbol{r}$ in $\underline{e}^{(1)}$ and $\underline{e}^{(2)}$, respectively. Then, $\underline{r}^{(1)}=\underline{A}^{12}\underline{r}^{(2)}$ where $\underline{A}^{12}$ is the transpose of the matrix $\underline{A}^{21}$ under consideration. The coordinate matrix $\underline{r}^{(2)}$ is identical with the coordinate matrix $\underline{r}^{*(1)}$ of $\boldsymbol{r}^*$ in the base $\underline{e}^{(1)}$ since before the rotation $\underline{e}^{(2)}$ coincides with $\underline{e}^{(1)}$ and the body-fixed vector coincides with $\boldsymbol{r}^*$. Therefore,

$$\underline{r}^{(1)}=\underline{A}^{12}\underline{r}^{*(1)}. \tag{2.8}$$

According to Fig. 2.5 $\boldsymbol{r}$ and $\boldsymbol{r}^*$ are related by the equation $\boldsymbol{r}=\boldsymbol{r}^*+(1-\cos\chi)\boldsymbol{b}+\sin\chi\,\boldsymbol{a}$ or recognizing that $\boldsymbol{a}$ equals $\boldsymbol{u}\times\boldsymbol{r}^*$ and that $\boldsymbol{b}$ equals $\boldsymbol{u}\times\boldsymbol{a}$

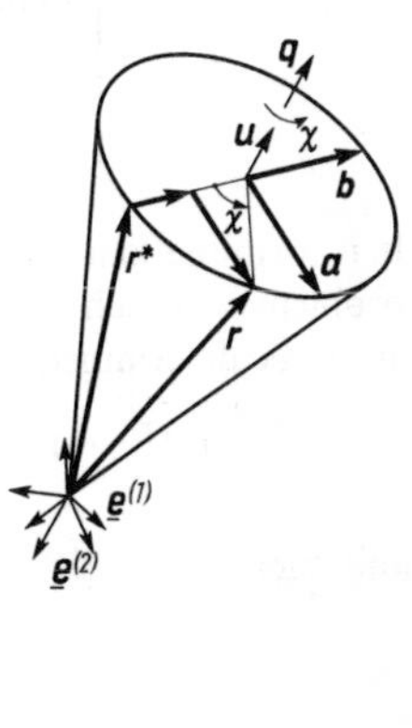

Fig. 2.5
Rotation of a body-fixed vector $\boldsymbol{r}$ about an axial unit vector $\boldsymbol{u}$

$$\boldsymbol{r}=\boldsymbol{r}^*+(1-\cos\chi)\boldsymbol{u}\times(\boldsymbol{u}\times\boldsymbol{r}^*)+\sin\chi\,\boldsymbol{u}\times\boldsymbol{r}^*. \tag{2.9}$$

By means of the relationships

$$1-\cos\chi=2\sin^2\frac{\chi}{2}, \qquad \sin\chi=2\sin\frac{\chi}{2}\cos\frac{\chi}{2} \tag{2.10}$$

the semi rotation angle is introduced. Now the new quantities

$$q_0=\cos\frac{\chi}{2} \quad\text{and}\quad \boldsymbol{q}=\boldsymbol{u}\sin\frac{\chi}{2} \tag{2.11}$$

are defined. The vector $\boldsymbol{q}$ has equal coordinates in both bases $\underline{\boldsymbol{e}}^{(1)}$ and $\underline{\boldsymbol{e}}^{(2)}$ since $\boldsymbol{u}$ has this property. The coordinates of $\boldsymbol{q}$ are denoted q_1, q_2 and q_3. It is the four scalars q_0, q_1, q_2 and q_3 which are called Euler parameters. They satisfy the constraint equation

$$q_0^2+\boldsymbol{q}\cdot\boldsymbol{q}=\cos^2\frac{\chi}{2}+\boldsymbol{u}\cdot\boldsymbol{u}\sin^2\frac{\chi}{2}=1.$$

Alternative formulations are

$$\begin{aligned} & q_0^2+q_1^2+q_2^2+q_3^2=1 \\ \text{or}\quad & q_0^2+\underline{q}^{\mathrm{T}}\underline{q}=1. \end{aligned} \tag{2.12}$$

Mathematically speaking Euler parameters represent normalized quaternions. Together with Eqs. (2.11) and (2.10) Eq. (2.9) becomes ↳s. Hamel, S. 116

$$\boldsymbol{r}=\boldsymbol{r}^*+2\boldsymbol{q}\times(\boldsymbol{q}\times\boldsymbol{r}^*)+2q_0\boldsymbol{q}\times\boldsymbol{r}^*.$$

In the base $\underline{\boldsymbol{e}}^{(1)}$ this yields the coordinate equation

$$\underline{r}^{(1)}=(\underline{\underline{E}}+2\underline{\underline{\tilde{q}}}\,\underline{\underline{\tilde{q}}}+2q_0\underline{\underline{\tilde{q}}})\underline{r}^{*(1)}.$$

Comparison with Eq. (2.8) shows that the term in brackets is the desired expression for the matrix $\underline{\underline{A}}^{12}$. Applying Eq. (1.28) to the product $\underline{\underline{\tilde{q}}}\,\underline{\underline{\tilde{q}}}$ and using Eq. (2.12) the transpose $\underline{\underline{A}}^{21}$ becomes

$$\underline{\underline{A}}^{21}=(2q_0^2-1)\underline{\underline{E}}+2(\underline{q}\,\underline{q}^{\mathrm{T}}-q_0\underline{\underline{\tilde{q}}}) \tag{2.13}$$

or explicitly

$$\underline{\underline{A}}^{21}=\begin{bmatrix} 2(q_0^2+q_1^2)-1 & 2(q_1q_2+q_0q_3) & 2(q_1q_3-q_0q_2) \\ 2(q_1q_2-q_0q_3) & 2(q_0^2+q_2^2)-1 & 2(q_2q_3+q_0q_1) \\ 2(q_1q_3+q_0q_2) & 2(q_2q_3-q_0q_1) & 2(q_0^2+q_3^2)-1 \end{bmatrix}. \tag{2.14}$$

It is a simple matter to derive from this expression explicit formulas for the inverse problem in which the matrix $\underline{\underline{A}}^{21}$ is given and the corresponding Euler parameters are to be determined. For the trace of $\underline{\underline{A}}^{21}$ the relationship is found

$$\frac{\operatorname{tr}\underline{\underline{A}}^{21}+1}{2}=q_0^2+q_1^2+q_2^2+q_3^2+2q_0^2-1=2q_0^2$$

and, therefore,

$$q_0^2 = \frac{\operatorname{tr} \underline{A}^{21} + 1}{4}. \tag{2.15}$$

Substitution of this expression into the formula for the diagonal elements, $A_{ii}^{21} = 2(q_0^2 + q_i^2) - 1$, results in

$$q_i^2 = \frac{A_{ii}^{21}}{2} - \frac{\operatorname{tr} \underline{A}^{21} - 1}{4} \qquad i = 1, 2, 3. \tag{2.16}$$

In contrast to Euler and Bryant angles (and to any other set of three generalized coordinates) there is no critical case in which the right hand sides of these inverse formulas are singular.

Problems

2.1 The angular orientation of a body is described in terms of Euler angles. It is desired to convert ψ, θ and ϕ into equivalent Euler parameters. How can this be done?

2.2 Show that the rotation angle χ and the coordinates of the unit vector $\boldsymbol{u}$ of Fig. 2.5 are determined by the equations

$$\cos\chi = \frac{\operatorname{tr} \underline{A}^{21} - 1}{2}, \qquad u_i^2 = \frac{A_{ii}^{21} - \cos\chi}{1 - \cos\chi} \qquad i = 1, 2, 3.$$

2.3 A reference base $\underline{e}^{(1)}$ and a body-fixed base $\underline{e}^{(2)}$ are initially coincident. The base $\underline{e}^{(2)}$ is then subject to a sequence of three rotations. It is rotated, first, through an angle ϕ_1 about the axis $\boldsymbol{e}_1^{(1)}$, then through an angle ϕ_2 about $\boldsymbol{e}_2^{(1)}$ and, finally, through an angle ϕ_3 about $\boldsymbol{e}_3^{(1)}$. Note that in contrast to Bryant angles all three rotations are carried out about base vectors of the reference base $\underline{e}^{(1)}$. The matrix $\underline{A}^{21}$ relating the final orientation of $\underline{e}^{(2)}$ to $\underline{e}^{(1)}$ is a function of ϕ_1, ϕ_2 and ϕ_3. Find this function and evaluate it numerically for the three sets of angles $(\pi/2, \pi/2, \pi/2)$, $(0, \pi/2, 0)$ and (π, π, π). Check the results experimentally.

2.2 The notion of angular velocity

Let $\underline{e}^{(1)}$ be some arbitrarily moving base. Relative to this base a rigid body is in arbitrary motion. Fixed on this body there is a base $\underline{e}^{(2)}$ with origin P. Furthermore, a point Q is considered which is moving relative to the body. With the notations of Fig. 2.6

$$\boldsymbol{z} = \boldsymbol{z}_P + \boldsymbol{r}. \tag{2.17}$$

The goal of the present investigation is an expression for the velocity of Q relative to the base $\underline{e}^{(1)}$ in terms of the velocity of Q relative to the body, of the velocity of P relative to

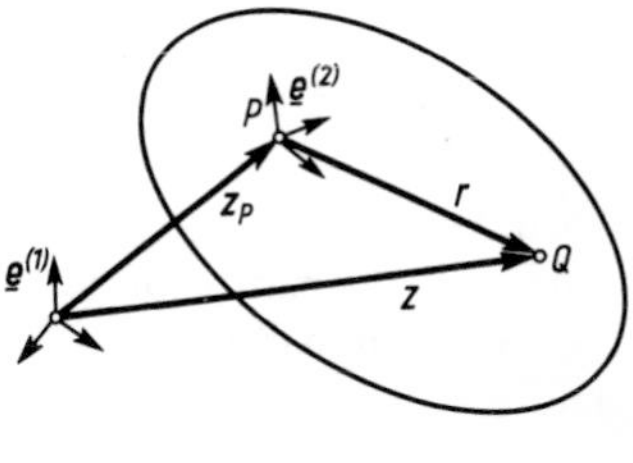

Fig. 2.6
Radius vectors of two points P (body-fixed) and Q (not body-fixed)

$\underline{e}^{(1)}$ and of some as yet unknown quantity which accounts for changes of the body angular orientation in the base $\underline{e}^{(1)}$. Velocities are represented as time derivatives of radius vectors. Since a point of a body has, in general, different velocities relative to different vector bases it must be specified in which base a radius vector is differentiated with respect to time. The same is true also for vectors which are not radius vectors. One frequently used notation for the time derivative of a vector c in a base $\underline{e}^{(s)}$ is $^{(s)}dc/dt$. This derivative is calculated from the equation

$$\frac{^{(s)}d}{dt}c = \sum_{\alpha=1}^{3} \frac{d}{dt} c_\alpha^{(s)} e_\alpha^{(s)}. \tag{2.18}$$

In words: The coordinates of $^{(s)}dc/dt$ in the base $\underline{e}^{(s)}$ are found by calculating the coordinates $c_\alpha^{(s)}$ $(\alpha=1,2,3)$ of c in $e^{(s)}$ and by differentiating them with respect to time. The time derivative $dc_\alpha^{(s)}/dt$ of the scalar $c_\alpha^{(s)}$ is unambiguous. $(= unzweideutig)$

After these preparatory remarks Eq. (2.17) is considered, again. The vector r is expressed in terms of its coordinates in $\underline{e}^{(2)}$, $r = \sum_{\alpha=1}^{3} r_\alpha^{(2)} e_\alpha^{(2)}$. This is substituted into Eq. (2.17), and then the entire equation is differentiated with respect to time in the base $\underline{e}^{(1)}$:

$$\frac{^{(1)}d}{dt}z = \frac{^{(1)}d}{dt}z_P + \sum_{\alpha=1}^{3} \frac{d}{dt} r_\alpha^{(2)} e_\alpha^{(2)} + \sum_{\alpha=1}^{3} r_\alpha^{(2)} \frac{^{(1)}d}{dt} e_\alpha^{(2)}. \tag{2.19}$$

The derivative on the left hand side is the velocity of Q relative to $\underline{e}^{(1)}$. It is called v. Similarly, the first term on the right hand side is the velocity v_P of the point P relative to $\underline{e}^{(1)}$. The second term on the right represents (according to Eq. (2.18)) the velocity $^{(2)}dr/dt$ of the point Q relative to $\underline{e}^{(2)}$. It is abbreviated v_{rel}. In the last term the derivatives $^{(1)}de_\alpha^{(2)}/dt$ can be calculated from Eq. (2.18). Since the coordinates of $e_\alpha^{(2)}$ in the base $\underline{e}^{(1)}$ are the direction cosines $A_{\alpha\beta}^{21}$ $(\beta=1,2,3)$ we get

$$\frac{^{(1)}d}{dt} e_\alpha^{(2)} = \sum_{\beta=1}^{3} \frac{d}{dt} A_{\alpha\beta}^{21} e_\beta^{(1)} \qquad \alpha=1,2,3. \tag{2.20}$$

This expression leads to very complicated formulas. Therefore, another approach is chosen. The derivatives can also be represented as linear combinations of base vectors of $\underline{e}^{(2)}$ with as yet unknown coefficients $c_{\alpha\beta}$:

$$\frac{^{(1)}d}{dt} e_\alpha^{(2)} = \sum_{\beta=1}^{3} c_{\alpha\beta} e_\beta^{(2)} \qquad \alpha=1,2,3. \tag{2.21}$$

The base vectors satisfy the relationship $e_\alpha^{(2)} \cdot e_\gamma^{(2)} = \delta_{\alpha\gamma}$ $(\alpha,\gamma=1,2,3)$. Differentiation of this relationship with respect to time in the base $e^{(1)}$ leads to

$$\frac{^{(1)}d}{dt} e_\alpha^{(2)} \cdot e_\gamma^{(2)} + e_\alpha^{(2)} \cdot \frac{^{(1)}d}{dt} e_\gamma^{(2)} = 0 \qquad \alpha,\gamma=1,2,3.$$

When Eq. (2.21) is substituted for the derivatives one obtains $c_{\alpha\gamma}+c_{\gamma\alpha}=0$. This means that the matrix formed by the coefficients $c_{\alpha\beta}$ $(\alpha,\beta=1,2,3)$ is skew-symmetric. Its three nonzero elements are given the new names $\omega_1=c_{23}=-c_{32}$, $\omega_2=-c_{13}=c_{31}$ and $\omega_3=c_{12}=-c_{21}$. These three quantities are interpreted as coordinates of a vector $\boldsymbol{\omega}$ in the body-fixed base $\underline{e}^{(2)}$. Eq. (2.21) then becomes

$$\begin{aligned}
\frac{{}^{(1)}\mathrm{d}}{\mathrm{d}t}\boldsymbol{e}_1^{(2)} &= \omega_3\boldsymbol{e}_2^{(2)}-\omega_2\boldsymbol{e}_3^{(2)}=\boldsymbol{\omega}\times\boldsymbol{e}_1^{(2)}\\
\frac{{}^{(1)}\mathrm{d}}{\mathrm{d}t}\boldsymbol{e}_2^{(2)} &= -\omega_3\boldsymbol{e}_1^{(2)}+\omega_1\boldsymbol{e}_3^{(2)}=\boldsymbol{\omega}\times\boldsymbol{e}_2^{(2)}\\
\frac{{}^{(1)}\mathrm{d}}{\mathrm{d}t}\boldsymbol{e}_3^{(2)} &= \omega_2\boldsymbol{e}_1^{(2)}-\omega_1\boldsymbol{e}_2^{(2)}=\boldsymbol{\omega}\times\boldsymbol{e}_3^{(2)}\,.
\end{aligned}$$

With this the third term in Eq. (2.19) takes the simple form $\boldsymbol{\omega}\times\boldsymbol{r}$, and the basic relationship is obtained

$$\boldsymbol{v}=\boldsymbol{v}_P+\boldsymbol{v}_{\mathrm{rel}}+\boldsymbol{\omega}\times\boldsymbol{r}\,. \tag{2.22}$$

The vector $\boldsymbol{\omega}$ is referred to as angular velocity of the body relative to the base $\underline{e}^{(1)}$. It has some important properties two of which will now be discussed. From the identity of the right hand side expressions of Eqs. (2.20) and (2.21) follows that the coefficients $c_{\alpha\beta}$ $(\alpha,\beta=1,2,3)$ and, hence, also $\boldsymbol{\omega}$ depend on direction cosines and on time derivatives of direction cosines only. This means that $\boldsymbol{\omega}$ (in contrast to $\boldsymbol{v}_P$) is independent of the choice of the body-fixed point P in Fig. 2.6 because the direction cosine matrix $\underline{A}^{21}$ is independent of it.

In order to find another important property of $\boldsymbol{\omega}$ it is necessary to investigate, first, the relationship which exists between the time derivatives of one and the same vector in two different vector bases. Let $\underline{e}^{(1)}$ and $\underline{e}^{(2)}$ be two vector bases which move relative to each other in an arbitrary way and let $\boldsymbol{c}$ be some vector (not necessarily a radius vector). In the base $\underline{e}^{(2)}$ the vector $\boldsymbol{c}$ has coordinates $c_\alpha^{(2)}$ $(\alpha=1,2,3)$ so that $\boldsymbol{c}=\sum_{\alpha=1}^{3}c_\alpha^{(2)}\boldsymbol{e}_\alpha^{(2)}$. This equation is differentiated with respect to time in the base $\underline{e}^{(1)}$:

$$\frac{{}^{(1)}\mathrm{d}}{\mathrm{d}t}\boldsymbol{c}=\sum_{\alpha=1}^{3}\frac{\mathrm{d}}{\mathrm{d}t}c_\alpha^{(2)}\boldsymbol{e}_\alpha^{(2)}+\sum_{\alpha=1}^{3}c_\alpha^{(2)}\frac{{}^{(1)}\mathrm{d}}{\mathrm{d}t}\boldsymbol{e}_\alpha^{(2)}\,.$$

The terms on the right hand side have the same form as the last two terms in Eq. (2.19). Using the same arguments as before one gets

$$\frac{{}^{(1)}\mathrm{d}}{\mathrm{d}t}\boldsymbol{c}=\frac{{}^{(2)}\mathrm{d}}{\mathrm{d}t}\boldsymbol{c}+\boldsymbol{\omega}\times\boldsymbol{c}\,. \tag{2.23}$$

This is the desired general relationship between the time derivatives of an arbitrary vector in two different bases. The vector $\boldsymbol{\omega}$ represents the angular velocity of the base $\underline{e}^{(2)}$ relative to $\underline{e}^{(1)}$. For the vector $\boldsymbol{\omega}$ itself the equation has the special form

$$\frac{{}^{(1)}\mathrm{d}}{\mathrm{d}t}\boldsymbol{\omega}=\frac{{}^{(2)}\mathrm{d}}{\mathrm{d}t}\boldsymbol{\omega}\,. \tag{2.24}$$

This is the second important property of the angular velocity which is mentioned here. We now return to Eq. (2.22). If only body-fixed points Q are considered the equation takes the special form

$$\boldsymbol{v} = \boldsymbol{v}_P + \boldsymbol{\omega} \times \boldsymbol{r}\,. \tag{2.25}$$

It describes the velocity distribution of a rigid body. All points along the straight line parallel to $\boldsymbol{\omega}$ and passing through P have the same velocity $\boldsymbol{v}_P$ since for these points $\boldsymbol{\omega} \times \boldsymbol{r} = \boldsymbol{0}$. The velocity distribution can, therefore, be interpreted as the result of a superposition of two separate motions. One is a pure translation with the velocity $\boldsymbol{v}_P$ of the point P, and the other is a pure rotation with the angular velocity $\boldsymbol{\omega}$ about an axis which has the direction of $\boldsymbol{\omega}$ and which passes through P. This interpretation holds for any arbitrarily chosen body-fixed point P. It becomes particularly simple if a point P^* is chosen whose velocity $\boldsymbol{v}_P^*$ has the same direction as $\boldsymbol{\omega}$. The pure translation then has the direction of the axis of the pure rotation. Thus, the velocity distribution of the body is the same as that of a screw. It remains to be shown that in the case $\boldsymbol{\omega} \neq \boldsymbol{0}$ there exists a unique screw axis (in the trivial case $\boldsymbol{\omega} = \boldsymbol{0}$ the motion is a pure translation). It is assumed that the velocity $\boldsymbol{v}_P$ is known for some point P of the body. Let $\boldsymbol{\varrho}$ be the vector from P to a point on the screw axis. Then, $\boldsymbol{v}_P^* = \boldsymbol{v}_P + \boldsymbol{\omega} \times \boldsymbol{\varrho}$ is the velocity of this point and by definition of the screw axis it is parallel to $\boldsymbol{\omega}$ so that cross multiplication with $\boldsymbol{\omega}$ yields $\boldsymbol{0} = \boldsymbol{\omega} \times \boldsymbol{v}_P + \boldsymbol{\omega} \cdot \boldsymbol{\varrho}\,\boldsymbol{\omega} - \omega^2 \boldsymbol{\varrho}$. Since the direction of the screw axis is known it suffices to determine a single point on the axis. We choose the particular point P^* for which $\boldsymbol{\omega} \cdot \boldsymbol{\varrho}^*$ is zero. The radius vector $\boldsymbol{\varrho}^*$ for this point is $\boldsymbol{\varrho}^* = \boldsymbol{\omega} \times \boldsymbol{v}_P / \omega^2$. It is, indeed, uniquely defined if $\boldsymbol{\omega}$ is different from zero.

In general, the location of the screw axis in the body varies with time. Of particular interest is the case in which one point of the body is fixed in the reference base $\underline{e}^{(1)}$. The screw axis then degenerates to an instantaneous axis of rotation which is always passing through the fixed point. In the course of time this axis with the direction of $\boldsymbol{\omega}(t)$ is sweeping out two cones one of which is fixed in the body and the other in the base $\underline{e}^{(1)}$ (Fig. 2.7). At time t the two cones share the instantaneous axis as a

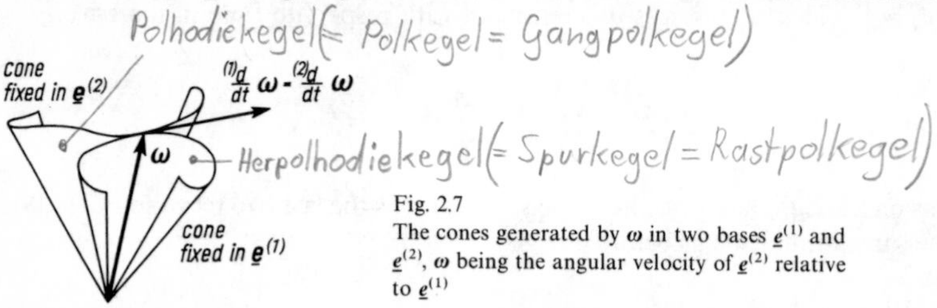

Fig. 2.7
The cones generated by $\boldsymbol{\omega}$ in two bases $\underline{e}^{(1)}$ and $\underline{e}^{(2)}$, $\boldsymbol{\omega}$ being the angular velocity of $\underline{e}^{(2)}$ relative to $\underline{e}^{(1)}$

common generating line. That the cones also have a common tangential plane at time t is a consequence of Eq. (2.24) which states that $\boldsymbol{\omega}(t)$ is sweeping out both cones with equal rates of change. Summarizing all these facts *the general motion of a rigid body relative to a base* $\underline{e}^{(1)}$ *with a point fixed in this base can be interpreted as rolling motion without slipping of a body-fixed cone on a cone fixed in the base* $\underline{e}^{(1)}$.

Problems

2.4 A body has an angular velocity $\boldsymbol{\omega} \neq \boldsymbol{0}$, and a point P of the body has a velocity $\boldsymbol{v}_P \neq \boldsymbol{0}$, both measured relative to the same reference base. What is the condition for the existence of body points with zero velocity and where are these points located?

2.5 On a rigid body three non-collinear points are defined by their radius vectors $\boldsymbol{r}_1, \boldsymbol{r}_2$ and $\boldsymbol{r}_3$. The velocities $\boldsymbol{v}_1$, $\boldsymbol{v}_2$ and $\boldsymbol{v}_3$ of these points relative to a base $\underline{e}^{(1)}$ are known. Show that the angular velocity of the body relative to the same base is

$$\boldsymbol{\omega}=2\frac{\boldsymbol{v}_1\times\boldsymbol{v}_2+\boldsymbol{v}_2\times\boldsymbol{v}_3+\boldsymbol{v}_3\times\boldsymbol{v}_1}{\boldsymbol{v}_1\cdot(\boldsymbol{r}_2-\boldsymbol{r}_3)+\boldsymbol{v}_2\cdot(\boldsymbol{r}_3-\boldsymbol{r}_1)+\boldsymbol{v}_3\cdot(\boldsymbol{r}_1-\boldsymbol{r}_2)}.$$

2.3 Relationships between the angular velocity of a body and generalized coordinates describing the angular orientation of the body

The angular velocity of a body cannot, in general, be represented as the time derivative of another vector (this is possible only in the trivial case where the direction of $\boldsymbol{\omega}$ in the body is constant). The coordinates ω_1, ω_2 and ω_3 of $\boldsymbol{\omega}$ in a body-fixed vector base do not, therefore, represent generalized velocities in the sense of analytic mechanics. From this follows that generalized coordinates for the angular orientation of a body cannot be determined from $\omega_\alpha(t)$ $(\alpha=1,2,3)$ by simple integration. Instead, differential equations must be solved in which $\omega_\alpha(t)$ $(\alpha=1,2,3)$ appear as variable coefficients. These equations will now be formulated for direction cosines, Euler angles, Bryant angles and Euler parameters as generalized coordinates.

2.3.1 Direction cosines

Let $\boldsymbol{\omega}$ be the angular velocity of a body-fixed base $\underline{e}^{(2)}$ relative to another base $\underline{e}^{(1)}$ and let $\boldsymbol{r}$ be a body-fixed vector with a constant coordinate matrix $\underline{r}^{(2)}$ in $\underline{e}^{(2)}$. The time varying coordinate matrix of $\boldsymbol{r}$ in $\underline{e}^{(1)}$ is then $\underline{r}^{(1)}(t)=\underline{A}^{12}(t)\underline{r}^{(2)}$ where $\underline{A}^{12}(t)$ is the direction cosine matrix relating the two bases. The time derivative of $\underline{r}^{(1)}(t)$ is

$$\dot{\underline{r}}^{(1)}=\dot{\underline{A}}^{12}\underline{r}^{(2)} \tag{2.26}$$

(here and in the remainder of this chapter time derivatives of scalars are designated by a dot). The same quantity is obtained by decomposing the vector ${}^{(1)}\mathrm{d}\boldsymbol{r}/\mathrm{d}t=\boldsymbol{\omega}\times\boldsymbol{r}$ in the base $\underline{e}^{(1)}$. This gives $\dot{\underline{r}}^{(1)}=\underline{A}^{12}\tilde{\underline{\omega}}^{(2)}\underline{r}^{(2)}$. Comparison with Eq. (2.26) yields $\dot{\underline{A}}^{12}\underline{r}^{(2)}=\underline{A}^{12}\tilde{\underline{\omega}}^{(2)}\underline{r}^{(2)}$. Since this holds for any coordinate matrix $\underline{r}^{(2)}$ the factors in front must be identical. Omitting the superscript of $\underline{\omega}^{(2)}$ and taking the transpose of either side one obtains

$$\dot{\underline{A}}^{21}=-\tilde{\underline{\omega}}\underline{A}^{21}. \tag{2.27}$$

These are, in matrix form, the desired differential equations for the nine direction cosines. They are known as Poisson's equations. For the individual elements of $\underline{A}^{21}$ they read

$$\dot{A}^{21}_{11}=\omega_3 A^{21}_{21}-\omega_2 A^{21}_{31}\quad\text{etc.}$$

Because of the six constraint equations (2.1) only three differential equations need be integrated.

2.3.2 Euler angles

From Fig. 2.1 the angular velocity $\boldsymbol{\omega}$ of the base $\underline{e}^{(2)}$ relative to $\underline{e}^{(1)}$ is seen to be

$$\boldsymbol{\omega}=\dot{\psi}\,e_3^{(1)}+\dot{\theta}\,e_1^{(2)\prime}+\dot{\phi}\,e_3^{(2)}.$$

Decomposition in $\underline{e}^{(2)}$ yields the coordinate equations

$$\begin{bmatrix}\omega_1\\ \omega_2\\ \omega_3\end{bmatrix}=\begin{bmatrix}\sin\theta\sin\phi & \cos\phi & 0\\ \sin\theta\cos\phi & -\sin\phi & 0\\ \cos\theta & 0 & 1\end{bmatrix}\begin{bmatrix}\dot{\psi}\\ \dot{\theta}\\ \dot{\phi}\end{bmatrix}. \tag{2.28}$$

Their explicit solutions for $\dot{\psi}, \dot{\theta}$ and $\dot{\phi}$ read

$$\begin{bmatrix}\dot{\psi}\\ \dot{\theta}\\ \dot{\phi}\end{bmatrix}=\begin{bmatrix}\dfrac{\sin\phi}{\sin\theta} & \dfrac{\cos\phi}{\sin\theta} & 0\\ \cos\phi & -\sin\phi & 0\\ -\sin\phi\,\mathrm{ctan}\,\theta & -\cos\phi\,\mathrm{ctan}\,\theta & 1\end{bmatrix}\begin{bmatrix}\omega_1\\ \omega_2\\ \omega_3\end{bmatrix}. \tag{2.29}$$

These are the desired kinematic differential equations. They show, again, that numerical problems will arise if θ is close to the critical values $n\pi$ ($n=0,1,\ldots$).

2.3.3 Bryant angles

Fig. 2.3 yields

$$\boldsymbol{\omega}=\dot{\phi}_1\,e_1^{(1)}+\dot{\phi}_2\,e_2^{(2)\prime}+\dot{\phi}_3\,e_3^{(2)}. \tag{2.30}$$

Decomposition in $\underline{e}^{(2)}$ leads to the coordinate equations

$$\begin{bmatrix}\omega_1\\ \omega_2\\ \omega_3\end{bmatrix}=\begin{bmatrix}\cos\phi_2\cos\phi_3 & \sin\phi_3 & 0\\ -\cos\phi_2\sin\phi_3 & \cos\phi_3 & 0\\ \sin\phi_2 & 0 & 1\end{bmatrix}\begin{bmatrix}\dot{\phi}_1\\ \dot{\phi}_2\\ \dot{\phi}_3\end{bmatrix}. \tag{2.31}$$

Their solutions for $\dot{\phi}_1, \dot{\phi}_2$ and $\dot{\phi}_3$ read

$$\begin{bmatrix}\dot{\phi}_1\\ \dot{\phi}_2\\ \dot{\phi}_3\end{bmatrix}=\begin{bmatrix}\dfrac{\cos\phi_3}{\cos\phi_2} & -\dfrac{\sin\phi_3}{\cos\phi_2} & 0\\ \sin\phi_3 & \cos\phi_3 & 0\\ -\cos\phi_3\tan\phi_2 & \sin\phi_3\tan\phi_2 & 1\end{bmatrix}\begin{bmatrix}\omega_1\\ \omega_2\\ \omega_3\end{bmatrix}. \tag{2.32}$$

These are the kinematic differential equations for Bryant angles. They fail numerically in the vicinity of the critical values $\phi_2=\pi/2+n\pi$ ($n=0,1,\ldots$). In connection with Eq. (2.6) it has been shown that for small angles ϕ_1, ϕ_2 and ϕ_3 the orientation of the body in the base $\underline{e}^{(1)}$ can, within linear approximations, be characterized by a rotation vector $\boldsymbol{\phi}$ whose coordinates are ϕ_1, ϕ_2 and ϕ_3. From Eq. (2.31) follow the linear approximations $\omega_1\approx\dot{\phi}_1$, $\omega_2\approx\dot{\phi}_2$ and $\omega_3\approx\dot{\phi}_3$. The angular orientation of the body is, therefore, found by simple integration:

$$\phi_\alpha(t) \approx \int \omega_\alpha(t)\,\mathrm{d}t \qquad \alpha=1,2,3\ (\phi_1,\phi_2,\phi_3 \text{ small}).$$

These approximation formulas for small angular displacements are often used in technical problems.

2.3.4 Euler parameters

In Poisson's equations (2.27) which can be written in the form $\underline{\tilde{\omega}}=-\underline{\dot{A}}^{21}\underline{A}^{12}$ the matrices $\underline{A}^{12}$ and $\underline{\dot{A}}^{21}$ are replaced by the transpose of the expression in Eq. (2.13) and by its time derivative, respectively:

$$\underline{A}^{12}=(2q_0^2-1)\underline{E}+2(\underline{q}\underline{q}^{\mathrm{T}}+q_0\underline{\tilde{q}})$$
$$\underline{\dot{A}}^{21}=2(2q_0\dot{q}_0\underline{E}+\underline{\dot{q}}\underline{q}^{\mathrm{T}}+\underline{q}\underline{\dot{q}}^{\mathrm{T}}-\dot{q}_0\underline{\tilde{q}}-q_0\underline{\tilde{\dot{q}}}).$$

This yields

$$\begin{aligned}-\frac{\underline{\tilde{\omega}}}{2} &= 2q_0\dot{q}_0(2q_0^2-1)\underline{E}+4q_0\dot{q}_0(\underline{q}\underline{q}^{\mathrm{T}}+q_0\underline{\tilde{q}})+\\ &\quad+(2q_0^2-1)(\underline{\dot{q}}\underline{q}^{\mathrm{T}}+\underline{q}\underline{\dot{q}}^{\mathrm{T}}-\dot{q}_0\underline{\tilde{q}}-q_0\underline{\tilde{\dot{q}}})+\\ &\quad+2(\underline{\dot{q}}\underline{q}^{\mathrm{T}}+\underline{q}\underline{\dot{q}}^{\mathrm{T}}-\dot{q}_0\underline{\tilde{q}}-q_0\underline{\tilde{\dot{q}}})(\underline{q}\underline{q}^{\mathrm{T}}+q_0\underline{\tilde{q}}).\end{aligned} \tag{2.33}$$

In simplifying this expression the constraint equation (2.12) and its time derivative will be used:

$$\underline{q}^{\mathrm{T}}\underline{q}=1-q_0^2, \qquad \underline{\dot{q}}^{\mathrm{T}}\underline{q}=\underline{q}^{\mathrm{T}}\underline{\dot{q}}=-q_0\dot{q}_0\,.$$

Also Eqs. (1.22) to (1.29) will be applied. The product of the last two expressions in brackets in Eq. (2.33) contains, among others, the following terms

$$\begin{aligned}&\underline{\dot{q}}\underline{q}^{\mathrm{T}}\underline{q}\underline{q}^{\mathrm{T}}=(1-q_0^2)\underline{\dot{q}}\underline{q}^{\mathrm{T}}, &&\underline{\dot{q}}\underline{q}^{\mathrm{T}}q_0\underline{\tilde{q}}=\underline{0}\\ &\underline{q}\underline{\dot{q}}^{\mathrm{T}}\underline{q}\underline{q}^{\mathrm{T}}=-q_0\dot{q}_0\underline{q}\underline{q}^{\mathrm{T}}, &&\dot{q}_0\underline{\tilde{q}}\underline{q}\underline{q}^{\mathrm{T}}=\underline{0}\\ &\dot{q}_0\underline{\tilde{q}}q_0\underline{\tilde{q}}=\dot{q}_0q_0[\underline{q}\underline{q}^{\mathrm{T}}-(1-q_0^2)\underline{E}], &&q_0^2\underline{\tilde{\dot{q}}}\underline{\tilde{q}}=q_0^2(\underline{q}\underline{\dot{q}}^{\mathrm{T}}+q_0\dot{q}_0\underline{E}).\end{aligned}$$

With them Eq. (2.33) can be rewritten in the form

$$-\frac{\underline{\tilde{\omega}}}{2}=\underline{\dot{q}}\underline{q}^{\mathrm{T}}-\underline{q}\underline{\dot{q}}^{\mathrm{T}}+\dot{q}_0\underline{\tilde{q}}+q_0\underline{\tilde{\dot{q}}}+2q_0[(q_0\dot{q}_0\underline{E}+\underline{q}\underline{\dot{q}}^{\mathrm{T}})\underline{\tilde{q}}\underline{\tilde{\dot{q}}}(q_0^2\underline{E}+\underline{q}\underline{q}^{\mathrm{T}})].$$

Druckfehler!
$\mapsto -\underline{\tilde{\dot{q}}}(q_0^2\underline{E}+\underline{q}\underline{q}^{\mathrm{T}})]$

In this equation the identities

$$q_0\dot{q}_0\underline{E}+\underline{q}\underline{\dot{q}}^{\mathrm{T}}=\underline{\tilde{\dot{q}}}\underline{\tilde{q}}, \qquad q_0^2\underline{E}+\underline{q}\underline{q}^{\mathrm{T}}=\underline{E}+\underline{\tilde{q}}\underline{\tilde{q}}$$

are used. They reduce the expression in square brackets to $-\underline{\tilde{\dot{q}}}$. This yields

$$-\frac{\underline{\tilde{\omega}}}{2}=\underline{\dot{q}}\underline{q}^{\mathrm{T}}-\underline{q}\underline{\dot{q}}^{\mathrm{T}}+\dot{q}_0\underline{\tilde{q}}-q_0\underline{\tilde{\dot{q}}}$$

or with Eq. (1.29)

$$\underline{\tilde{\omega}}=-2[(\widetilde{\underline{\tilde{q}}\underline{\dot{q}}})+\dot{q}_0\underline{\tilde{q}}-q_0\underline{\tilde{\dot{q}}}]$$

so that $\quad \underline{\omega}=-2(\underline{\tilde{q}}\underline{\dot{q}}+\dot{q}_0\underline{q}-q_0\underline{\dot{q}})$.

Thus, the coordinates of $\boldsymbol{\omega}$ in the base $\underline{e}^{(2)}$ are linear combinations of q_i as well as of

$\dot{q}_i\,(i=0\ldots 3)$. In explicit form the equations represent the last three rows of the matrix equation

$$\begin{bmatrix} 0 \\ \omega_1 \\ \omega_2 \\ \omega_3 \end{bmatrix} = 2 \begin{bmatrix} q_0 & q_1 & q_2 & q_3 \\ -q_1 & q_0 & q_3 & -q_2 \\ -q_2 & -q_3 & q_0 & q_1 \\ -q_3 & q_2 & -q_1 & q_0 \end{bmatrix} \begin{bmatrix} \dot{q}_0 \\ \dot{q}_1 \\ \dot{q}_2 \\ \dot{q}_3 \end{bmatrix}.$$

The first row is the time derivative of the constraint equation. It is added in order to produce a square coefficient matrix. Because of the constraint equation this matrix is orthogonal and normalized to unity. Its inverse equals, therefore, its transpose. Solving for $\dot{q}_i\,(i=0\ldots 3)$ and rearranging the right hand side results in

$$\begin{bmatrix} \dot{q}_0 \\ \dot{q}_1 \\ \dot{q}_2 \\ \dot{q}_3 \end{bmatrix} = \frac{1}{2} \left[\begin{array}{c|c} 0 & -\underline{\omega}^{\mathrm{T}} \\ \hline \underline{\omega} & -\underline{\tilde{\omega}} \end{array}\right] \begin{bmatrix} q_0 \\ q_1 \\ q_2 \\ q_3 \end{bmatrix}. \tag{2.34}$$

These are the desired kinematic differential equations for Euler parameters. In numerical calculations all four equations are integrated. The constraint equation is used for correcting round-off errors. When after a few integration steps values for $q_i\ (i=0\ldots 3)$ are obtained which do not strictly satisfy the constraint equation then the calculation is continued not with q_i but with the corrected values

$$q_i^* = q_i \left(\sum_{j=0}^{3} q_j^2 \right)^{-1/2} \qquad (i=0\ldots 3).$$

Problems

2.6 Develop the above correction formula for $q_i^*(i=0\ldots 3)$ from the condition that the sum of squares of the corrections, i.e. the sum $\sum_{i=0}^{3} (q_i^* - q_i)^2$, is a minimum.

2.7 A rigid body is suspended in two gimbals as shown in Fig. 2.8. In the outer gimbal the two axes are offset from 90° by an angle α and in the inner gimbal by an angle β. Let ϕ_1, ϕ_2 and ϕ_3 be defined like Bryant angles, i.e. as rotation angles of the outer gimbal about $\underline{e}_1^{(1)}$, of the inner gimbal relative to the outer gimbal and of the body relative to the inner gimbal, respectively. For $\phi_1=\phi_2=\phi_3=0$ the planes of the gimbals are perpendicular to one another and, furthermore, the base vectors $\underline{e}_1^{(1)}$ and $\underline{e}_2^{(1)}$ of the reference base as well as the body-fixed base vector $\underline{e}_1^{(2)}$ lie in the plane of the outer gimbal. Develop an expression for the direction cosine matrix $\underline{A}^{21}$ and kinematic differential equations similar to Eq. (2.32).

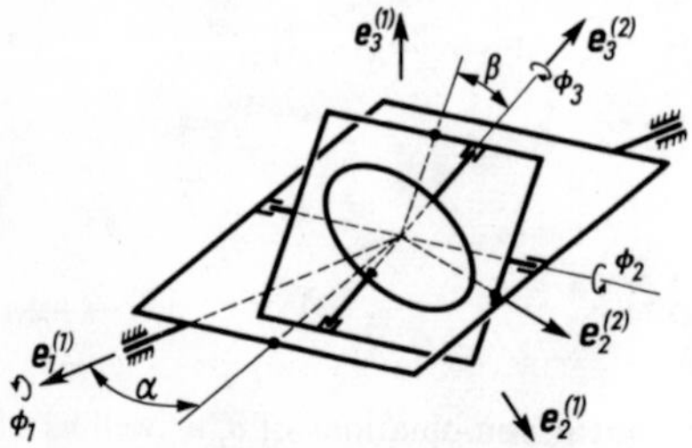

Fig. 2.8
Two-gimbal suspension with non-orthogonal gimbal axes

3 Basic Principles of Rigid Body Dynamics

The two most important physical quantities in rigid body dynamics are kinetic energy and angular momentum. Both lead directly to the definition of the inertia tensor.

3.1 Kinetic energy

The kinetic energy is a scalar quantity. For a point mass of mass m it is defined as $T = m\dot{z}^2/2$ where $\dot{z}$ is the absolute velocity of m, i.e. its velocity relative to an inertial reference base. Throughout this chapter a dot over a vector designates differentiation with respect to time in an inertial base. For a rigid body as for any extended body the kinetic energy is the integral

$$T = \frac{1}{2} \int_m \dot{z}^2 \, \mathrm{d}m .$$

Let P be an arbitrary point fixed on the body (Fig. 3.1). The absolute velocity $\dot{z}$ of a mass particle $\mathrm{d}m$ is according to Eq. (2.25) $\dot{z} = v_P + \omega \times r$ where $v_P = \dot{z}_P$ is the absolute velocity of the reference point P, ω the absolute angular velocity of the body and r

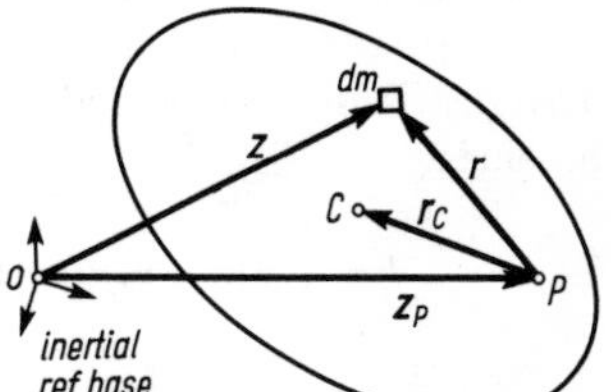

Fig. 3.1
Radius vectors of a mass particle $\mathrm{d}m$ of a rigid body. Center of mass C and body-fixed reference point P

the radius vector from P to the mass particle. The point C in Fig. 3.1 at the radius vector $r_C = \overrightarrow{PC}$ indicates the body center of mass. Evaluation of the integral yields

$$T = \frac{1}{2} v_P^2 m + v_P \cdot (\omega \times r_C) m + \frac{1}{2} \int_m (\omega \times r)^2 \, \mathrm{d}m . \tag{3.1}$$

This expression becomes particularly simple if either the body-fixed point P is also fixed in inertial space or the center of mass C is used as reference point P. In the former case $v_P = 0$ so that the first two terms equal zero. In the latter case $r_C = 0$ so that the central term vanishes. The first term is then called kinetic energy of translation T_{trans} and the third term kinetic energy of rotation T_{rot}. For the integrand in the third term the identity holds [1)]

$$(\boldsymbol{\omega}\times\boldsymbol{r})^2=\boldsymbol{\omega}\cdot[\boldsymbol{r}\times(\boldsymbol{\omega}\times\boldsymbol{r})].$$

In this expression tensor notation is introduced (cf. Eq. (1.27)):

$$\boldsymbol{r}\times(\boldsymbol{\omega}\times\boldsymbol{r})=(r^2\boldsymbol{E}-\boldsymbol{r}\,\boldsymbol{r})\cdot\boldsymbol{\omega}.$$

With this

$$\int_m(\boldsymbol{\omega}\times\boldsymbol{r})^2\,\mathrm{d}m=\boldsymbol{\omega}\cdot\boldsymbol{J}^P\cdot\boldsymbol{\omega} \tag{3.2}$$

where $\boldsymbol{J}^P$ is the tensor

$$\boldsymbol{J}^P=\int_m(r^2\boldsymbol{E}-\boldsymbol{r}\,\boldsymbol{r})\,\mathrm{d}m. \tag{3.3}$$

It is called inertia tensor of the body with respect to P. In a body-fixed base $\underline{e}^{(2)}$ in which $\boldsymbol{r}$ has the coordinate matrix $\underline{r}$ $\boldsymbol{J}^P$ has the coordinate matrix

$$\underline{J}^P=\int_m(\underline{r}^{\mathrm{T}}\underline{r}\,\underline{E}-\underline{r}\,\underline{r}^{\mathrm{T}})\,\mathrm{d}m$$

or explicitly

$$\underline{J}^P=\begin{bmatrix}\int_m(r_2^2+r_3^2)\mathrm{d}m & -\int_m r_1r_2\,\mathrm{d}m & -\int_m r_1r_3\,\mathrm{d}m\\ & \int_m(r_3^2+r_1^2)\mathrm{d}m & -\int_m r_2r_3\,\mathrm{d}m\\ \text{symmetric} & & \int_m(r_1^2+r_2^2)\mathrm{d}m\end{bmatrix}=\begin{bmatrix}J_{11} & -J_{12} & -J_{13}\\ -J_{12} & J_{22} & -J_{23}\\ -J_{13} & -J_{23} & J_{33}\end{bmatrix}. \tag{3.4}$$

The integrals J_{11}, J_{22} and J_{33} along the diagonal are called moments of inertia and the integrals J_{12}, J_{13} and J_{23} products of inertia, both with respect to P and to the base $\underline{e}^{(2)}$. The symmetric matrix $\underline{J}^P$ itself is called inertia matrix of the body with respect to P and to $\underline{e}^{(2)}$. It is a geometric quantity which is determined by the mass distribution of the body. With Eq. (3.2) the kinetic energy expression in Eq. (3.1) becomes

$$T=\frac{1}{2}\,\boldsymbol{v}_P^2m+\boldsymbol{v}_P\cdot(\boldsymbol{\omega}\times\boldsymbol{r}_C)m+\frac{1}{2}\,\boldsymbol{\omega}\cdot\boldsymbol{J}^P\cdot\boldsymbol{\omega}.$$

3.2 Angular momentum

The absolute angular momentum—also referred to as absolute moment of momentum—of a point mass m having an absolute velocity $\dot{\boldsymbol{z}}$ is a vector. For its definition the specification of a reference point is required. Let O be a point fixed in inertial space. The absolute angular momentum with respect to O is $\boldsymbol{L}^O=\boldsymbol{z}\times\dot{\boldsymbol{z}}\,m$ where $\boldsymbol{z}$ is the radius vector from O to the point mass. The expression explains the name moment of momentum since $\dot{\boldsymbol{z}}\,m$ is the linear momentum of the point mass. For

[1)] In mixed products the symbols of dot and cross multiplication can be interchanged so that $\boldsymbol{\omega}\times\boldsymbol{r}\cdot\boldsymbol{c}=\boldsymbol{\omega}\cdot\boldsymbol{r}\times\boldsymbol{c}$. Here, $\boldsymbol{c}$ equals $\boldsymbol{\omega}\times\boldsymbol{r}$.

a rigid body as for any extended body the absolute angular momentum with respect to O is the integral

$$\boldsymbol{L}^O = \int_m \boldsymbol{z} \times \dot{\boldsymbol{z}}\, \mathrm{d}m\,. \tag{3.5}$$

In evaluating this integral the quantities defined in Fig. 3.1 are used again (center of mass C, arbitrary body-fixed point P). With

$$\boldsymbol{z} = \boldsymbol{z}_P + \boldsymbol{r} \quad \text{and} \quad \dot{\boldsymbol{z}} = \boldsymbol{v}_P + \boldsymbol{\omega} \times \boldsymbol{r}$$

the integral becomes

$$\boldsymbol{L}^O = \int_m (\boldsymbol{z}_P + \boldsymbol{r}) \times (\boldsymbol{v}_P + \boldsymbol{\omega} \times \boldsymbol{r})\mathrm{d}m = \boldsymbol{z}_P \times (\boldsymbol{v}_P + \boldsymbol{\omega} \times \boldsymbol{r}_C)m + \boldsymbol{r}_C \times \boldsymbol{v}_P m + \int_m \boldsymbol{r} \times (\boldsymbol{\omega} \times \boldsymbol{r})\mathrm{d}m\,. \tag{3.6}$$

The expression $\boldsymbol{v}_P + \boldsymbol{\omega} \times \boldsymbol{r}_C$ in the first brackets is the absolute velocity of the center of mass C so that $(\boldsymbol{v}_P + \boldsymbol{\omega} \times \boldsymbol{r}_C)m$ represents the absolute linear momentum of the body. The expression for $\boldsymbol{L}^O$ becomes particularly simple if either the body-fixed point P is also fixed in inertial space or the center of mass C is used as reference point P. In the former case $\boldsymbol{v}_P = \boldsymbol{0}$ and in the latter $\boldsymbol{r}_C = \boldsymbol{0}$. In both cases the central term in Eq. (3.6) vanishes. The first term then represents the angular momentum with respect to O due to translation of the body center of mass and the last term represents the angular momentum caused by a rotation of the body. Using tensor notation this last term becomes

$$\int_m \boldsymbol{r} \times (\boldsymbol{\omega} \times \boldsymbol{r})\mathrm{d}m = \int_m (r^2 \boldsymbol{E} - \boldsymbol{r}\boldsymbol{r})\mathrm{d}m \cdot \boldsymbol{\omega} = \boldsymbol{J}^P \cdot \boldsymbol{\omega}\,. \tag{3.7}$$

This introduces, again, the inertia tensor of the body with respect to P. The angular momentum $\boldsymbol{L}^O$ is with this expression

$$\boldsymbol{L}^O = \boldsymbol{z}_P \times (\boldsymbol{v}_P + \boldsymbol{\omega} \times \boldsymbol{r}_C)m + \boldsymbol{r}_C \times \boldsymbol{v}_P m + \boldsymbol{J}^P \cdot \boldsymbol{\omega}\,. \tag{3.8}$$

Problem

3.1 Give an expression for the absolute angular momentum $\boldsymbol{L}^C$ of a rigid body with respect to its center of mass C. For comparison consider a body one point P of which is fixed in inertial space. Use Eq. (3.8) to find the absolute angular momentum $\boldsymbol{L}^P$ with respect to this point.

3.3 Properties of moments and products of inertia

3.3.1 Transition to another reference point without change of the reference base

Given the inertia tensor $\boldsymbol{J}^P$ of a body with respect to a point P what is the inertia tensor $\boldsymbol{J}^{P^*}$ with respect to another point P^*? For solving this problem it is sufficient to establish the relationship between $\boldsymbol{J}^P$ and the central inertia tensor $\boldsymbol{J}^C$, i.e. the inertia tensor with respect to the body center of mass C. Let $\boldsymbol{\varrho}$ be the radius vector of a mass particle measured from C (Fig. 3.2). According to the general definition the central inertia tensor is

$$\boldsymbol{J}^C = \int_m (\varrho^2 \boldsymbol{E} - \boldsymbol{\varrho}\boldsymbol{\varrho})\mathrm{d}m\,,$$

and the inertia tensor with respect to P is (with $\boldsymbol{r}=\boldsymbol{r}_C+\boldsymbol{\varrho}$ and $\int_m \boldsymbol{\varrho}\,\mathrm{d}m=\mathbf{0}$)

$$\begin{aligned}\boldsymbol{J}^P &= \int_m [(\boldsymbol{r}_C+\boldsymbol{\varrho})^2\boldsymbol{E}-(\boldsymbol{r}_C+\boldsymbol{\varrho})(\boldsymbol{r}_C+\boldsymbol{\varrho})]\,\mathrm{d}m\\ &= \int_m [(r_C^2+\varrho^2)\boldsymbol{E}-(\boldsymbol{r}_C\boldsymbol{r}_C+\boldsymbol{\varrho}\boldsymbol{\varrho})]\,\mathrm{d}m=\boldsymbol{J}^C+(r_C^2\boldsymbol{E}-\boldsymbol{r}_C\boldsymbol{r}_C)m.\end{aligned}$$

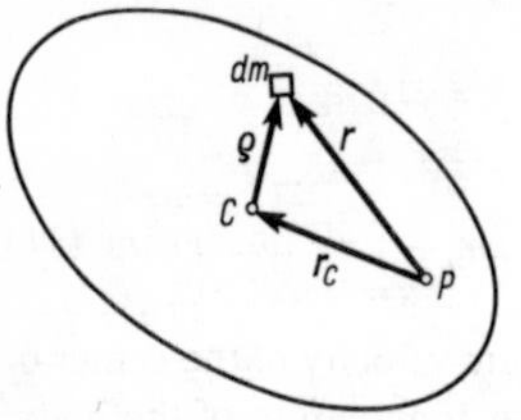

Fig. 3.2
Radius vectors of a mass particle dm. Center of mass C and body-fixed point P

This is the desired relationship between $\boldsymbol{J}^P$ and $\boldsymbol{J}^C$. Resolved in a body-fixed reference base it yields the coordinate equation in matrix form

$$\underline{J}^P=\underline{J}^C+(\underline{r}_C^T\underline{r}_C\underline{E}-\underline{r}_C\underline{r}_C^T)m$$

and for the single moments and products of inertia

$$\left.\begin{aligned}J_{\alpha\alpha}^P&=J_{\alpha\alpha}^C+(r_{C\beta}^2+r_{C\gamma}^2)m\\ J_{\alpha\beta}^P&=J_{\alpha\beta}^C+r_{C\alpha}r_{C\beta}m\end{aligned}\right\}\quad \alpha,\beta,\gamma \text{ different}. \tag{3.9}$$

These formulas are known as Huygens-Steiner formulas.

3.3.2 Transition to another reference base without change of the reference point

Let $\underline{e}^{(2)}$ and $\underline{e}^{(3)}=\underline{A}^{32}\underline{e}^{(2)}$ be two vector bases fixed on one and the same body and let, furthermore, P be a point on this body. Given the inertia matrix $\underline{J}^{(2)P}$ of the body with respect to P in the base $\underline{e}^{(2)}$ what is the inertia matrix $\underline{J}^{(3)P}$ in the base $\underline{e}^{(3)}$ with respect to the same point P? The answer is provided by the transformation equation (1.19) for tensor coordinates:

$$\underline{J}^{(3)P}=\underline{A}^{32}\underline{J}^{(2)P}\underline{A}^{23}. \tag{3.10}$$

From this matrix equation transformation formulas for all moments and products of inertia are found by multiplying out the product on the right hand side. These formulas are, however, rather complicated so that it is preferable to memorize only the matrix equation as a whole.

3.3.3 Principal axes and principal moments of inertia

Suppose that the inertia matrix $\underline{J}^{(2)P}$ is known for a body with respect to a certain reference point P and for a certain body-fixed base $\underline{e}^{(2)}$ and that it is not a diagonal matrix. Does another body-fixed base $\underline{e}^{(3)}$ exist for which the inertia matrix $\underline{J}^{(3)P}$

(with respect to the same point P) is diagonal? If so, how are the diagonal elements of $\underline{J}^{(3)P}$ and the transformation matrix $\underline{A}^{23}$ relating $\underline{e}^{(2)}$ and $\underline{e}^{(3)}$ determined from $\underline{J}^{(2)P}$? The answer to these questions is found as follows. The unknowns $\underline{J}^{(3)P}$ and $\underline{A}^{23}$ are related to $\underline{J}^{(2)P}$ by Eq. (3.10). Omitting the superscript P this equation can be written in the form $\underline{J}^{(2)}\underline{A}^{23}=\underline{A}^{23}\underline{J}^{(3)}$. Let J_1, J_2 and J_3 be the unknown diagonal elements of $\underline{J}^{(3)}$ and let $\underline{A}_\alpha^{23}$ ($\alpha=1,2,3$) be the α-th column of $\underline{A}^{23}$, i.e. the coordinate matrix of the base vector $e_\alpha^{(3)}$ in the base $\underline{e}^{(2)}$. The transformation equation is then equivalent to $\underline{J}^{(2)}\underline{A}_\alpha^{23}=J_\alpha\underline{A}_\alpha^{23}$ ($\alpha=1,2,3$). Each of these three equations represents the same eigenvalue problem

$$(\underline{J}^{(2)}-J_\alpha\underline{E})\underline{A}_\alpha^{23}=\underline{0}. \tag{3.11}$$

The unknowns J_1, J_2 and J_3 are the eigenvalues. They are the solutions of the cubic equation

$$\det(\underline{J}^{(2)}-J_\alpha\underline{E})=0. \tag{3.12}$$

The unknown column matrices $\underline{A}_\alpha^{23}$ ($\alpha=1,2,3$) are the corresponding eigenvectors. These results not only answer the question how $\underline{J}^{(3)}$ and $\underline{A}^{23}$ are determined from $\underline{J}^{(2)}$. They also show that for any inertia matrix $\underline{J}^{(2)}$ there exists a real base $\underline{e}^{(3)}$ in which the inertia matrix $\underline{J}^{(3)}$ is diagonal and real. This follows from the fact that a symmetric matrix has real eigenvalues and eigenvectors and that, in addition, the eigenvectors are mutually orthogonal (see Gantmacher [3]).

The eigenvalues J_1, J_2 and J_3 are called principal moments of inertia (with respect to P), and the base vectors $e_\alpha^{(3)}$ ($\alpha=1,2,3$) determine the directions of what is called principal axes of inertia (with respect to P). In determining these principal axes it must be distinguished whether all three eigenvalues are different from one another or whether there exists a double or a triple root of Eq. (3.12). In the case of three different eigenvalues each of the three coefficient matrices $(\underline{J}^{(2)}-J_\alpha\underline{E})$ in Eq. (3.11) has defect one. Each equation then determines uniquely the direction of one principal axis of inertia. The elements of $\underline{A}_\alpha^{23}$ are found if the constraint equation $\sum_{\beta=1}^{3}(A_{\alpha\beta}^{23})^2=1$ is taken into account. In the case of a double eigenvalue $J_1=J_2\neq J_3$ the principal axis which corresponds to J_3 is determined uniquely as before. For the eigenvalue J_1, however, the coefficient matrix $(\underline{J}^{(2)}-J_\alpha\underline{E})$ in Eq. (3.11) has defect two so that the equation defines only a plane. This is the plane spanned by the two principal axes which correspond to J_1 and $J_2=J_1$. Any two mutually perpendicular axes in this plane (and passing through P) can serve as principal axes of inertia since for any such axis the moment of inertia has magnitude J_1. In the case of a triple eigenvalue $J_1=J_2=J_3$ the original matrix $\underline{J}^{(2)}$ is already diagonal. All axes passing through P are then principal axes of inertia.

3.3.4 Invariants and inequalities for moments and products of inertia

In connection with stability investigations and with other problems it is sometimes necessary to determine the sign of expressions which are composed of moments and products of inertia. In such cases the knowledge of invariants of the inertia matrix and of inequalities involving moments and products of inertia is helpful. Eq. (3.12) represents a cubic equation for the principal moments of inertia. Since these moments are independent of the orientation of the vector base in which the inertia matrix $\underline{J}^{(2)}$ is measured, the coefficients of the cubic must also be independent. Omitting the superscript (2) this yields the invariants

$$\left.\begin{aligned} \operatorname{tr}\underline{J}&=J_{11}+J_{22}+J_{33} &&=J_1+J_2+J_3\\ (J_{11}J_{22}-J_{12}^2)+(J_{22}J_{33}-J_{23}^2)+(J_{33}J_{11}-J_{31}^2) &&&=J_1J_2+J_2J_3+J_3J_1\\ \det\underline{J}=J_{11}J_{22}J_{33}-J_{11}J_{23}^2-J_{22}J_{31}^2-J_{33}J_{12}^2-2J_{12}J_{23}J_{31}&&&=J_1J_2J_3. \end{aligned}\right\} \quad (3.13)$$

The roots of the cubic are positive or—in the case of a physically unrealizable, infinitely thin rod—zero. Therefore, Hurwitz's criterion is satisfied. It yields Sylvester's inequalities: $J_{\alpha\alpha}\geqslant 0$ (this follows already from the definition of moment of inertia), $\det\underline{J}\geqslant 0$ (this is implied by Eq. (3.13)) and

$$J_{\alpha\alpha}J_{\beta\beta}-J_{\alpha\beta}^2\geqslant 0 \qquad \alpha,\beta=1,2,3;\quad \alpha\neq\beta.$$

This last inequality is not contained in Eq. (3.13).

From the definitions of moments and products of inertia in Eq. (3.4) follows that for α, β and γ being any permutation of 1, 2 and 3

$$J_{\alpha\alpha}+J_{\beta\beta}=\int_m (r_\alpha^2+r_\beta^2+2r_\gamma^2)\,dm=J_{\gamma\gamma}+2\int_m r_\gamma^2\,dm$$

and, hence,

$$J_{\alpha\alpha}+J_{\beta\beta}\geqslant J_{\gamma\gamma}.$$

The equality sign requires $r_\gamma\equiv 0$, i.e. a physically unrealizable body in the form of an infinitely thin disc. Also from Eq. (3.4) in combination with the inequality $(r_\beta\pm r_\gamma)^2\geqslant 0$, i.e. with $r_\beta^2+r_\gamma^2\geqslant 2|r_\beta r_\gamma|$ follows

$$\int_m (r_\beta^2+r_\gamma^2)\,dm\geqslant 2\int |r_\beta r_\gamma|\,dm\geqslant 2\left|\int_m r_\beta r_\gamma\,dm\right|$$

or $$J_{\alpha\alpha}\geqslant 2|J_{\beta\gamma}| \qquad \alpha,\beta,\gamma \text{ different}. \quad (3.14)$$

Problems

3.2 Under which conditions is the equality sign valid in Eq. (3.14)?

3.3 Eq. (3.10) for the transformation of moments and products of inertia is particularly simple if the two vector bases related by the matrix $\underline{A}^{23}$ have one base vector in common, say $e_3^{(2)}=e_3^{(3)}$. Show that in this special case Eq. (3.10) can be interpreted geometrically by what is known as Mohr's circle.

3.4 In Fig. 3.3 a homogeneous, solid tetrahedron of density ϱ and of side lengths l, $l/2$ and $l/2$ is shown. Calculate, first, from triple integrals the moments and products of inertia in the

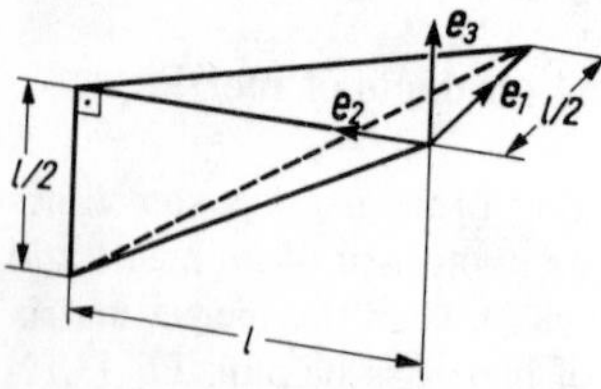

Fig. 3.3
Tetrahedron with three mutually orthogonal edges

base $\underline{e}$ with respect to the origin of this base. Then, determine the location of the body center of mass and calculate the central moments and products of inertia in the same base. From these quantities, finally, determine the central principal moments and principal axes of inertia.

3.4 The law of moment of momentum

Newton's second axiom for translational motions finds its complement in the law of moment of momentum as the basic law governing rotational motions. In symbolic form this law reads

$$\dot{\boldsymbol{L}}^O = \boldsymbol{M}^O \tag{3.15}$$

and in words: The absolute time derivative (i.e. the time derivative in an inertial reference base) of the absolute angular momentum with respect to a reference point O fixed in inertial space equals the resultant torque with respect to the same reference point. This law was first formulated as an axiom by Euler[1]. It is valid for any material system. It cannot be derived from Newton's axioms without making use of further assumptions. In the special case where the system is a single rigid body the nature of these assumptions can be shown as follows. Let the body be interpreted as a finite set of point masses m_i $(i=1\ldots n)$ which are kept at constant distances from one another by internal forces. The location of m_i in inertial space is described by the radius vector $\boldsymbol{z}_i$ which originates at O (Fig. 3.4). Newton's second axiom for a single point mass reads

$$\ddot{\boldsymbol{z}}_i m_i = \boldsymbol{F}_i + \sum_{j=1}^{n} \boldsymbol{F}_{ij} \qquad i=1\ldots n$$

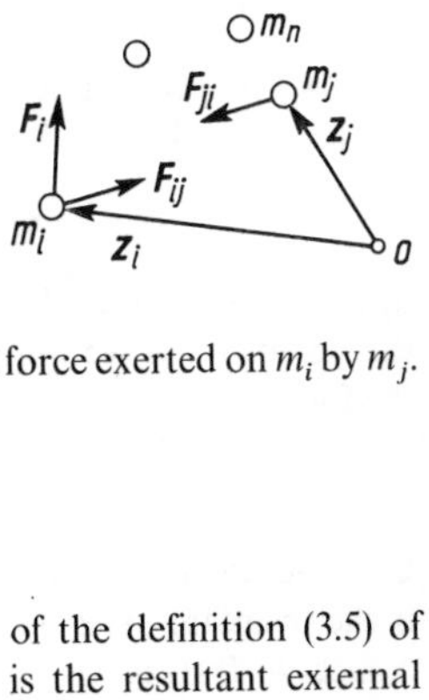

Fig. 3.4
A system of point masses as model of a rigid body. Point O is fixed in inertial space

where $\boldsymbol{F}_i$ is the resultant external force on m_i and $\boldsymbol{F}_{ij}$ the internal force exerted on m_i by m_j. Cross multiplication with $\boldsymbol{z}_i$ and summation over i yield

$$\sum_{i=1}^{n} \boldsymbol{z}_i \times \ddot{\boldsymbol{z}}_i m_i = \sum_{i=1}^{n} \boldsymbol{z}_i \times \boldsymbol{F}_i + \sum_{i=1}^{n} \sum_{j=1}^{n} \boldsymbol{z}_i \times \boldsymbol{F}_{ij}.$$

The expression on the left hand side represents $\dot{\boldsymbol{L}}^O$ in view of the definition (3.5) of angular momentum. The first sum on the right hand side is the resultant external torque $\boldsymbol{M}^O$. For the internal forces Newton's third axiom $\boldsymbol{F}_{ij} = -\boldsymbol{F}_{ji}$ $(i,j=1\ldots n)$ is valid. Hence,

$$\dot{\boldsymbol{L}}^O = \boldsymbol{M}^O + \frac{1}{2} \sum_{i=1}^{n} \sum_{j=1}^{n} (\boldsymbol{z}_i - \boldsymbol{z}_j) \times \boldsymbol{F}_{ij}.$$

This equation is identical with the law of moment of momentum in the form of Eq. (3.15) only if the internal forces do not cause a resultant torque. This condition is fulfilled if for all $i,j=1\ldots n$ $\boldsymbol{F}_{ij}$ and $\boldsymbol{F}_{ji}$ have one and the same line of action. This is

[1] For the history of this law see Truesdell [4].

the assumption which goes beyond Newton's axioms. It is not a trivial assumption since about the internal forces nothing is known except that they keep the point masses at constant distances from one another.

A special form of the law of moment of momentum for a rigid body is obtained when $\dot{\boldsymbol{L}}^O$ is expressed in terms of the inertia tensor, of the angular velocity and of the angular acceleration of the body. The simplest way to do this is to differentiate the expression for $\boldsymbol{L}^O$ in Eq. (3.8) and to substitute this into Eq. (3.15). For reasons which will become clear later it is preferable to go back to the definition of moment of momentum and to write

$$\dot{\boldsymbol{L}}^O = \int_m \boldsymbol{z} \times \ddot{\boldsymbol{z}}\,\mathrm{d}m\,. \tag{3.16}$$

In this integral $\boldsymbol{z}$ and $\ddot{\boldsymbol{z}}$ are expressed in terms of the quantities shown in Fig. 3.1:

$$\boldsymbol{z} = \boldsymbol{z}_P + \boldsymbol{r}\,, \qquad \ddot{\boldsymbol{z}} = \ddot{\boldsymbol{z}}_P + \dot{\boldsymbol{\omega}} \times \boldsymbol{r} + \boldsymbol{\omega} \times (\boldsymbol{\omega} \times \boldsymbol{r})\,. \tag{3.17}$$

The vector $\ddot{\boldsymbol{z}}_P$ is the absolute acceleration of the body-fixed point P. With these expressions

$$\begin{aligned}\dot{\boldsymbol{L}}^O &= \int_m (\boldsymbol{z}_P + \boldsymbol{r}) \times [\ddot{\boldsymbol{z}}_P + \dot{\boldsymbol{\omega}} \times \boldsymbol{r} + \boldsymbol{\omega} \times (\boldsymbol{\omega} \times \boldsymbol{r})]\,\mathrm{d}m \\ &= m\{\boldsymbol{z}_P \times [\ddot{\boldsymbol{z}}_P + \dot{\boldsymbol{\omega}} \times \boldsymbol{r}_C + \boldsymbol{\omega} \times (\boldsymbol{\omega} \times \boldsymbol{r}_C)] + \boldsymbol{r}_C \times \ddot{\boldsymbol{z}}_P\} + \\ &\quad + \int_m \boldsymbol{r} \times (\dot{\boldsymbol{\omega}} \times \boldsymbol{r})\,\mathrm{d}m + \int_m \boldsymbol{r} \times [\boldsymbol{\omega} \times (\boldsymbol{\omega} \times \boldsymbol{r})]\,\mathrm{d}m\,.\end{aligned}$$

In view of Eq. (3.17) the expression in square brackets in the leading term is interpreted as absolute acceleration $\ddot{\boldsymbol{z}}_C$ of the body center of mass. The first integral equals $\boldsymbol{J}^P \cdot \dot{\boldsymbol{\omega}}$ (cf. Eq. (3.7)). Under the second integral

$$\boldsymbol{r} \times [\boldsymbol{\omega} \times (\boldsymbol{\omega} \times \boldsymbol{r})] = \boldsymbol{\omega} \times [\boldsymbol{r} \times (\boldsymbol{\omega} \times \boldsymbol{r})]$$

(this can be verified by expanding the double-cross products on either side). With this identity the second integral becomes $\boldsymbol{\omega} \times \boldsymbol{J}^P \cdot \boldsymbol{\omega}$. The equation for $\dot{\boldsymbol{L}}^O$ now takes the form

$$\dot{\boldsymbol{L}}^O = m(\boldsymbol{z}_P \times \ddot{\boldsymbol{z}}_C + \boldsymbol{r}_C \times \ddot{\boldsymbol{z}}_P) + \boldsymbol{J}^P \cdot \dot{\boldsymbol{\omega}} + \boldsymbol{\omega} \times \boldsymbol{J}^P \cdot \boldsymbol{\omega}\,. \tag{3.18}$$

Consider next the resultant torque $\boldsymbol{M}^O$ on the body with respect to the point O. If $\boldsymbol{F}$ is the resultant external force on the body and $\boldsymbol{M}^P$ the resultant external torque with respect to P then $\boldsymbol{M}^O = \boldsymbol{M}^P + \boldsymbol{z}_P \times \boldsymbol{F}$. Furthermore, according to Newton's law,

$$m\ddot{\boldsymbol{z}}_C = \boldsymbol{F}\,. \tag{3.19}$$

When these two expressions together with Eq. (3.18) are substituted into Eq. (3.15) the law of moment of momentum for a rigid body is obtained in the final form

$$m\boldsymbol{r}_C \times \ddot{\boldsymbol{z}}_P + \boldsymbol{J}^P \cdot \dot{\boldsymbol{\omega}} + \boldsymbol{\omega} \times \boldsymbol{J}^P \cdot \boldsymbol{\omega} = \boldsymbol{M}^P\,. \tag{3.20}$$

It is obvious that Eq. (3.17) as well as all subsequent statements remain valid if the reference point P in Fig. 3.1 is not fixed on the body but moves relative to it. The only change that has to be made is to interpret $\ddot{\boldsymbol{z}}_P$ as absolute acceleration not of P but

→ koinzidierender Punkt!

of the body-fixed point which momentarily coincides with P. The simplicity of this argument was the reason not to differentiate Eq. (3.8) but to start from Eq. (3.16). The law of moment of momentum takes its simplest form

$$\boldsymbol{J}^P \cdot \dot{\boldsymbol{\omega}} + \boldsymbol{\omega} \times \boldsymbol{J}^P \cdot \boldsymbol{\omega} = \boldsymbol{M}^P \tag{3.21}$$

if as reference point P either the body center of mass is chosen ($\boldsymbol{r}_C = \boldsymbol{0}$) or a point (if it exists) for which $\ddot{\boldsymbol{z}}_P = \boldsymbol{0}$ or a point (if it exists) for which $\boldsymbol{r}_C$ and $\ddot{\boldsymbol{z}}_P$ are parallel to one another. The case $\ddot{\boldsymbol{z}}_P = \boldsymbol{0}$ applies when a body-fixed point P is also fixed in inertial space. From now on the superscript P will be omitted. The coordinate formulation of Eq. (3.21) in a body-fixed base is then

$$\underline{J}\,\underline{\dot{\omega}} + \underline{\tilde{\omega}}\,\underline{J}\,\underline{\omega} = \underline{M}\,.$$

Using, in particular, principal axes of inertia as directions for the base vectors this matrix equation is equivalent to

$$\begin{aligned} J_1 \dot{\omega}_1 - (J_2 - J_3)\omega_2 \omega_3 &= M_1 \\ J_2 \dot{\omega}_2 - (J_3 - J_1)\omega_3 \omega_1 &= M_2 \\ J_3 \dot{\omega}_3 - (J_1 - J_2)\omega_1 \omega_2 &= M_3\,. \end{aligned} \tag{3.22}$$

These are Euler's equations of motion for a single rigid body. They can be integrated in closed form in a few special cases only. Mathematical problems arise for two reasons. One is the nonlinearity of the left hand side of the equations. The other is the generally complicated form of the right hand side expressions. Three types of problems can be distinguished. In the first and simplest case the torque coordinates M_1, M_2 and M_3 are known functions of ω_1, ω_2, ω_3 and t (and possibly of $\dot{\omega}_1$, $\dot{\omega}_2$ and $\dot{\omega}_3$). Physically this means that the source of the torque $\boldsymbol{M}$ is rotating together with the body. A typical example is the torque caused by the reaction of a rocket engine which is mounted on a missile and which moves relative to the missile according to some prescribed function of time. In such cases the rigid body is said to be self-excited. All problems not being of this type can be subdivided into two classes. To the first class belong problems in which M_1, M_2 and M_3 depend not only on ω_1, ω_2, ω_3 and t but also on generalized coordinates which describe the angular orientation of the body in some external reference base. To give an example gravity acting on a body which is suspended as a pendulum causes a torque whose coordinates are functions of the direction of the vertical in the principal axes system. The dependence of M_1, M_2 and M_3 on such generalized coordinates causes a mathematical coupling between Euler's equations and the kinematic differential equations describing the angular orientation of the body (Eq. (2.27) or (2.29) or (2.32) or (2.34) depending on the choice of generalized coordinates). Still more complicated are problems in which M_1, M_2 and M_3 depend also on the location and velocity of the body center of mass. This dependency provides a coupling with Newton's law (Eq. (3.19)). Examples for this most general case are motions of airplanes and ships.

Problem

3.5 In Fig. 3.5 an inhomogeneous circular cylinder of radius R, mass m and moment of inertia J^C about an axis through the center of mass C is shown. The center of mass is located at

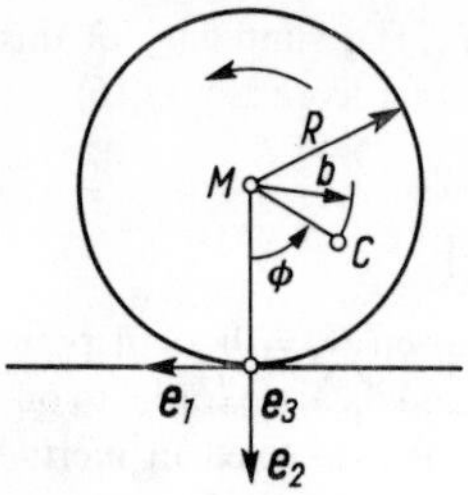

the radius b. The cylinder is rolling without slipping on a horizontal plane. Formulate the equation of motion for the angular coordinate ϕ by using as reference point P in Eq. (3.20) (i) the center of mass C, (ii) the geometric center M and (iii) the point of contact with the plane.

Fig. 3.5
An unbalanced rolling cylinder with center of mass C

3.5 D'Alembert's principle applied to a rigid body

D'Alembert's principle in the general form

$$\int_m \delta z \cdot (\mathrm{d}\boldsymbol{F} - \ddot{z}\,\mathrm{d}m) + \delta W = 0$$

is valid for any material system (axiom). The integral is taken over the total system mass. The vector $\boldsymbol{z}$ is the radius vector of the mass particle $\mathrm{d}m$ measured from a point which is fixed in inertial space. The second derivative $\ddot{\boldsymbol{z}}$ is the absolute acceleration of the mass particle, and $\delta \boldsymbol{z}$ is the variation of $\boldsymbol{z}$, i.e. any arbitrary, infinitesimally small displacement which is compatible with all constraints of the system. The quantity $\mathrm{d}\boldsymbol{F}$ represents the external force acting on the mass element and δW the total virtual work done by internal forces during the variation of the system location. When d'Alembert's principle is applied to a rigid body it is useful to introduce a reference point P which may be either fixed on the body or moving relative to it. Using the symbols of Fig. 3.1 we have

$$\boldsymbol{z} = \boldsymbol{z}_P + \boldsymbol{r}. \tag{3.23}$$

As in Eq. (3.17) $\ddot{\boldsymbol{z}}_P$ is the absolute acceleration not of P but of the body-fixed point which momentarily coincides with P, and $\ddot{\boldsymbol{r}}$ is the absolute second time derivative of the body-fixed vector from this point to the mass particle. Likewise, the variation $\delta \boldsymbol{z}$ is the sum $\delta \boldsymbol{z}_P + \delta \boldsymbol{r}$ where $\delta \boldsymbol{r}$ is the variation of the body-fixed vector $\boldsymbol{r}$. According to Euler's theorem any (finite or infinitesimally small) change of angular orientation of a rigid body can be represented as rotation through a certain angle about a certain body-fixed axis. The variation $\delta \boldsymbol{r}$ can, therefore, be expressed in the form

$$\delta \boldsymbol{r} = \boldsymbol{\delta\pi} \times \boldsymbol{r} \tag{3.24}$$

where the vector $\boldsymbol{\delta\pi}$ has the direction of an axis of rotation and the magnitude of a rotation angle about this axis. The virtual work δW is zero since in a rigid body there are no displacements of particles relative to one another along which internal forces could do work. Substituting this and Eqs. (3.23) and (3.24) into d'Alembert's principle one obtains

$$\int_m (\delta \boldsymbol{z}_P + \boldsymbol{\delta\pi} \times \boldsymbol{r}) \cdot [\mathrm{d}\boldsymbol{F} - (\ddot{\boldsymbol{z}}_P + \ddot{\boldsymbol{r}})\mathrm{d}m] = 0.$$

The symbols for dot and cross multiplication can be interchanged. Writing $\int_m \mathrm{d}\boldsymbol{F} = \boldsymbol{F}$

and $\int_m \boldsymbol{r}\,\mathrm{d}m = \boldsymbol{r}_C m$ one gets

$$\delta \boldsymbol{z}_P \cdot [\boldsymbol{F} - (\ddot{\boldsymbol{z}}_P + \ddot{\boldsymbol{r}}_C)m] + \boldsymbol{\delta\pi} \cdot [\int_m \boldsymbol{r} \times \mathrm{d}\boldsymbol{F} - \boldsymbol{r}_C \times \ddot{\boldsymbol{z}}_P m - \int_m \boldsymbol{r} \times \ddot{\boldsymbol{r}}\,\mathrm{d}m] = 0.$$

The sum $\ddot{\boldsymbol{z}}_P + \ddot{\boldsymbol{r}}_C$ represents the absolute acceleration $\ddot{\boldsymbol{z}}_C$ of the body center of mass. The first integral is the resultant external torque $\boldsymbol{M}^P$ with respect to P. The second integral is the absolute time derivative of the integral $\int_m \boldsymbol{r} \times \dot{\boldsymbol{r}}\,\mathrm{d}m$, hence, according to Eq. (3.7), of $\boldsymbol{J}^P \cdot \boldsymbol{\omega}$. The last equation can, therefore, be rewritten as

$$\delta \boldsymbol{z}_P \cdot (\boldsymbol{F} - \ddot{\boldsymbol{z}}_C m) + \boldsymbol{\delta\pi} \cdot (\boldsymbol{M}^P - \boldsymbol{r}_C \times \ddot{\boldsymbol{z}}_P m - \boldsymbol{J}^P \cdot \dot{\boldsymbol{\omega}} - \boldsymbol{\omega} \times \boldsymbol{J}^P \cdot \boldsymbol{\omega}) = 0. \tag{3.25}$$

If the body is unconstrained, $\delta \boldsymbol{z}_P$ and $\boldsymbol{\delta\pi}$ are independent variations. D'Alembert's principle then yields

$$\ddot{\boldsymbol{z}}_C m = \boldsymbol{0}$$

and $$\boldsymbol{r}_C \times \ddot{\boldsymbol{z}}_P m + \boldsymbol{J}^P \cdot \dot{\boldsymbol{\omega}} + \boldsymbol{\omega} \times \boldsymbol{J}^P \cdot \boldsymbol{\omega} = \boldsymbol{M}^P.$$

These two equations represent Newton's second axiom and Euler's axiom in the general form of Eq. (3.20), respectively. Thus, it has been shown that for a rigid body d'Alembert's principle is equivalent to these two axioms. This is not the only useful result, however. In Sec. 5.2.8 it will be seen that equations of motion for complicated systems can be developed conveniently on the basis of Eq. (3.25) in cases where the variations $\delta \boldsymbol{z}_P$ and $\boldsymbol{\delta\pi}$ are not independent (see also Problem 3.6).

Problem

3.6 Two straight lines forming an angle α are frictionless guides for two pegs P_1 and P_2 which are fixed on a rigid body of mass m and central moment of inertia J^C (Fig. 3.6). The body center of mass C is constrained to move in the plane of the guides. The body is subject to

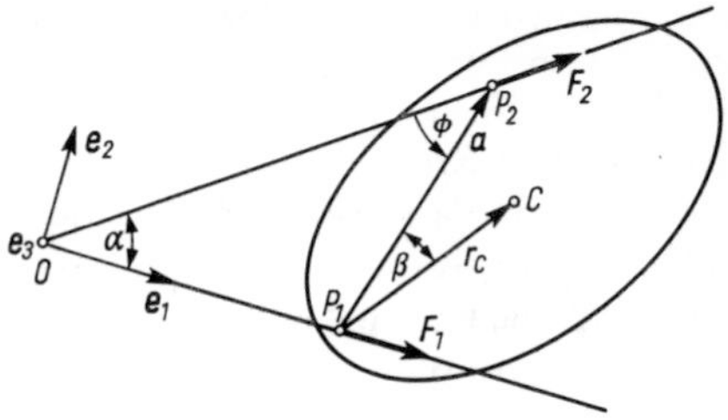

Fig. 3.6
Plane motion of a rigid body with center of mass C. The body-fixed pegs P_1 and P_2 move along straight rigid guides

external forces $\boldsymbol{F}_1$ and $\boldsymbol{F}_2$ which are applied to the pegs in the direction of the guides. Use Eq. (3.25) to formulate an equation of motion for the variable ϕ and with parameters α, β, $a = |\boldsymbol{a}|$, $r_C = |\boldsymbol{r}_C|$ and J^C. Let P_1 be the reference point for $\boldsymbol{J}^P$ and $\boldsymbol{M}^P$.

4 Classical Problems of Rigid Body Mechanics

In this chapter some of the rare rigid body problems are considered in which the equations of motion can be integrated in closed form. With the exception of the gyrostat in Chap. 4.7 all problems are treated in similar forms in other books on rigid body mechanics.

4.1 The unsymmetric torque-free rigid body

In the absence of external torques the equations of motion in the form of Eq. (3.21) and (3.22) read

$$\boldsymbol{J}\cdot\dot{\boldsymbol{\omega}}+\boldsymbol{\omega}\times\boldsymbol{J}\cdot\boldsymbol{\omega}=\boldsymbol{0} \tag{4.1}$$

and

$$\left.\begin{aligned} J_1\dot{\omega}_1-(J_2-J_3)\omega_2\omega_3&=0\\ J_2\dot{\omega}_2-(J_3-J_1)\omega_3\omega_1&=0\\ J_3\dot{\omega}_3-(J_1-J_2)\omega_1\omega_2&=0, \end{aligned}\right\} \tag{4.2}$$

numerische Bearbeitung nach Runge-Kutta vgl. Collatz, S. 75

respectively. Reference point of the moments of inertia is the body center of mass. These equations apply, for instance, to celestial bodies which are isolated from any external torque. They describe also the motion of a body which is supported without friction at its center of mass (provided torques caused by air resistence can be neglected). Such a suspension is approximately realized by a system of gimbals of the kind shown in Fig. 2.2. Strictly speaking this system consists of three kinematically coupled bodies. The influence of the gimbals can, however, often be neglected (in Chap. 4.5 this influence will be the subject of investigation).

Eq. (4.1) has two algebraic first integrals which are obtained through scalar multiplication by $\boldsymbol{\omega}$ and by $\boldsymbol{J}\cdot\boldsymbol{\omega}$, respectively:

$\vec{\omega}\times J\vec{\omega}\perp\vec{\omega}$!

$$\boldsymbol{\omega}\cdot(\boldsymbol{J}\cdot\dot{\boldsymbol{\omega}}+\boldsymbol{\omega}\times\boldsymbol{J}\cdot\boldsymbol{\omega})=\boldsymbol{\omega}\cdot\boldsymbol{J}\cdot\dot{\boldsymbol{\omega}}=\frac{1}{2}\frac{\mathrm{d}}{\mathrm{d}t}[\boldsymbol{\omega}\cdot\boldsymbol{J}\cdot\boldsymbol{\omega}]=0$$

$$\boldsymbol{J}\cdot\boldsymbol{\omega}\cdot(\boldsymbol{J}\cdot\dot{\boldsymbol{\omega}}+\boldsymbol{\omega}\times\boldsymbol{J}\cdot\boldsymbol{\omega})=\boldsymbol{J}\cdot\boldsymbol{\omega}\cdot\boldsymbol{J}\cdot\dot{\boldsymbol{\omega}}=\frac{1}{2}\frac{\mathrm{d}}{\mathrm{d}t}(\boldsymbol{J}\cdot\boldsymbol{\omega})^2=0.$$

From this follows

$$\boldsymbol{\omega}\cdot\boldsymbol{J}\cdot\boldsymbol{\omega}=2T=\text{const}$$
$$(\boldsymbol{J}\cdot\boldsymbol{\omega})^2=L^2=\text{const}$$

or in terms of coordinates in the principal axes frame of reference

$$\sum_{\alpha=1}^{3} J_\alpha \omega_\alpha^2 = 2T \tag{4.3}$$

$$\sum_{\alpha=1}^{3} J_\alpha^2 \omega_\alpha^2 = L^2 = 2DT. \tag{4.4}$$

The quantities T and L respresent the kinetic energy of rotation and the magnitude of the absolute angular momentum, respectively. In Eq. (4.4) a parameter D with the physical dimension of a moment of inertia has been introduced. The use of $2T$ and $2DT$ instead of $2T$ and L^2 as parameters simplifies subsequent formulations. Only the general case of three different principal moments of inertia will be considered. Without loss of generality it can be assumed that the moments of inertia are arranged in the order

$$J_3 < J_2 < J_1 .$$

The body-fixed base vector in the principal axis associated with J_α is called $\boldsymbol{e}_\alpha$ ($\alpha = 1, 2, 3$). Eqs. (4.3) and (4.4) define two ellipsoids which are fixed on the body and which are both geometric locus of the angular velocity vector $\boldsymbol{\omega}$. The vector is, therefore, confined to the line of intersection of the ellipsoids. These lines are called polhodes.

4.1.1 Polhodes and permanent rotations

An investigation of the geometric properties of the polhodes contributes to an understanding of the dynamic behavior of the torque-free rigid body. It is useful to think of the energy ellipsoid as given and to imagine that the angular momentum ellipsoid is "blown up" by increasing the parameter D so that on the invariable energy ellipsoid the family of all physically realizable polhodes is generated. This family corresponds to a certain interval of D values for which the angular momentum ellipsoid lies neither entirely inside nor entirely outside the energy ellipsoid. The minimum and maximum D values are found by multiplying Eq. (4.3) with J_1 and with J_3, respectively, and by subtracting both equations separately from Eq. (4.4). In the resulting equations

$$J_2(J_1 - J_2)\omega_2^2 + J_3(J_1 - J_3)\omega_3^2 = 2T(J_1 - D), \tag{4.5}$$

$$J_1(J_1 - J_3)\omega_1^2 + J_2(J_2 - J_3)\omega_2^2 = 2T(D - J_3) \tag{4.6}$$

the left hand side expressions are non-negative so that the inequalities

$$J_3 \leqslant D \leqslant J_1$$

must be satisfied. Thus, J_3 and J_1 are the extreme values of D for which the equations of motion have real solutions.

Of particular interest are degenerate polhodes which consist of singular points. In such points the ellipsoids have a common tangential plane. Each singular point marks a solution $\boldsymbol{\omega} \equiv \boldsymbol{\omega}^* = \text{const}$ of the equations of motion. This particular state of motion is referred to as permanent rotation. From Eq. (4.1) it is seen that a solution $\boldsymbol{\omega} \equiv \boldsymbol{\omega}^* = \text{const}$ is possible only if either $\boldsymbol{\omega}^*$ equals zero (in this trivial case the body is not rotating. Both ellipsoids degenerate into a single point) or $\boldsymbol{\omega}^*$ and the angular momentum $\boldsymbol{L}^* = \boldsymbol{J} \cdot \boldsymbol{\omega}^*$ are parallel to one another. In matrix form this latter condition yields the coordinate equation $\underline{J}\,\underline{\omega}^* = \lambda \underline{\omega}^*$ with an unknown factor λ, i.e. the eigenvalue problem

$$(\underline{J} - \lambda \underline{E})\underline{\omega}^* = \underline{0}. \tag{4.7}$$

This equation is identical in form with Eq. (3.11) which led to principal moments and principal axes of inertia. From this identity follows that the eigenvectors of Eq. (4.7), i.e. the axes of permanent rotation are identical with the principal axes of inertia. It is now possible to specify the particular values $D = D^*$ which cause the angular momentum ellipsoid to touch the energy ellipsoid in singular points. In a state of permanent rotation with an angular velocity of magnitude ω^* about the principal axis $\boldsymbol{e}_\alpha$ $(\alpha = 1, 2, 3)$ the integrals of motion are $2T = J_\alpha \omega^{*2}$ and $2D^* T = J_\alpha^2 \omega_\alpha^{*2}$. From this follows

$$D^* = J_\alpha \, .$$

Consider, again, the entire family of polhodes. A clear picture can be obtained if the polhodes are seen in projections along principal axes. The projection along $\boldsymbol{e}_\alpha$ $(\alpha = 1, 2, 3)$ requires the elimination of the coordinate ω_α from Eqs. (4.3) and (4.4). For the projections along $\boldsymbol{e}_1$ and $\boldsymbol{e}_3$ this has been done already. The resulting equations are Eqs. (4.5) and (4.6). In a similar manner the projection along $\boldsymbol{e}_2$ yields

$$J_1(J_1 - J_2)\omega_1^2 - J_3(J_2 - J_3)\omega_3^2 = 2T(D - J_2) \, . \tag{4.8}$$

Because of the inequalities $J_3 < J_2 < J_1$ and $J_3 \leqslant D \leqslant J_1$ Eqs. (4.5) and (4.6) represent families of ellipses whereas Eq. (4.8) represents a family of hyperbolas. The asymptotes of the hyperbolas correspond to $D = J_2$. In Figs. 4.1a to c all three projections of polhodes for one and the same set of D values are illustrated. Heavy lines represent the contour ellipses of the energy ellipsoid. Of the polhode ellipses and hyperbolas only those parts are relevant which lie inside these contour ellipses. All three projections together produce an image of the three-dimensional pattern of polhodes. A perspective view is shown in Fig. 4.1d. The results just obtained can be summarized as follows. To each of the parameter values $D = D^* = J_\alpha$ $(\alpha = 1, 2, 3)$ there belongs one axis of permanent rotation which coincides with the principal axis $\boldsymbol{e}_\alpha$. To $D = D^* = J_2$ there belong, in addition, two particular polhodes which intersect one another in the points of permanent rotation about the axis $\boldsymbol{e}_2$ and which separate all other polhodes into four families. Two families corresponding to $D < J_2$ envelop the axis $\boldsymbol{e}_3$ and the others corresponding to $D > J_2$ envelop the axis $\boldsymbol{e}_1$. The separating polhodes are called separatrices.

4.1.2 Poinsot's geometric interpretation of the motion

So far the integrals of motion have been used for characterizing the geometric locus of the angular velocity vector on the body. The integrals can also be used for an interpretation of the motion of the body relative to inertial space. This interpretation is due to Poinsot. The energy equation states that the scalar product of angular velocity $\boldsymbol{\omega}$ and angular momentum $\boldsymbol{L} = \boldsymbol{J} \cdot \boldsymbol{\omega}$ is constant. Since the magnitude and direction of $\boldsymbol{L}$ are also constant it follows that the projection of $\boldsymbol{\omega}$ onto this invariable

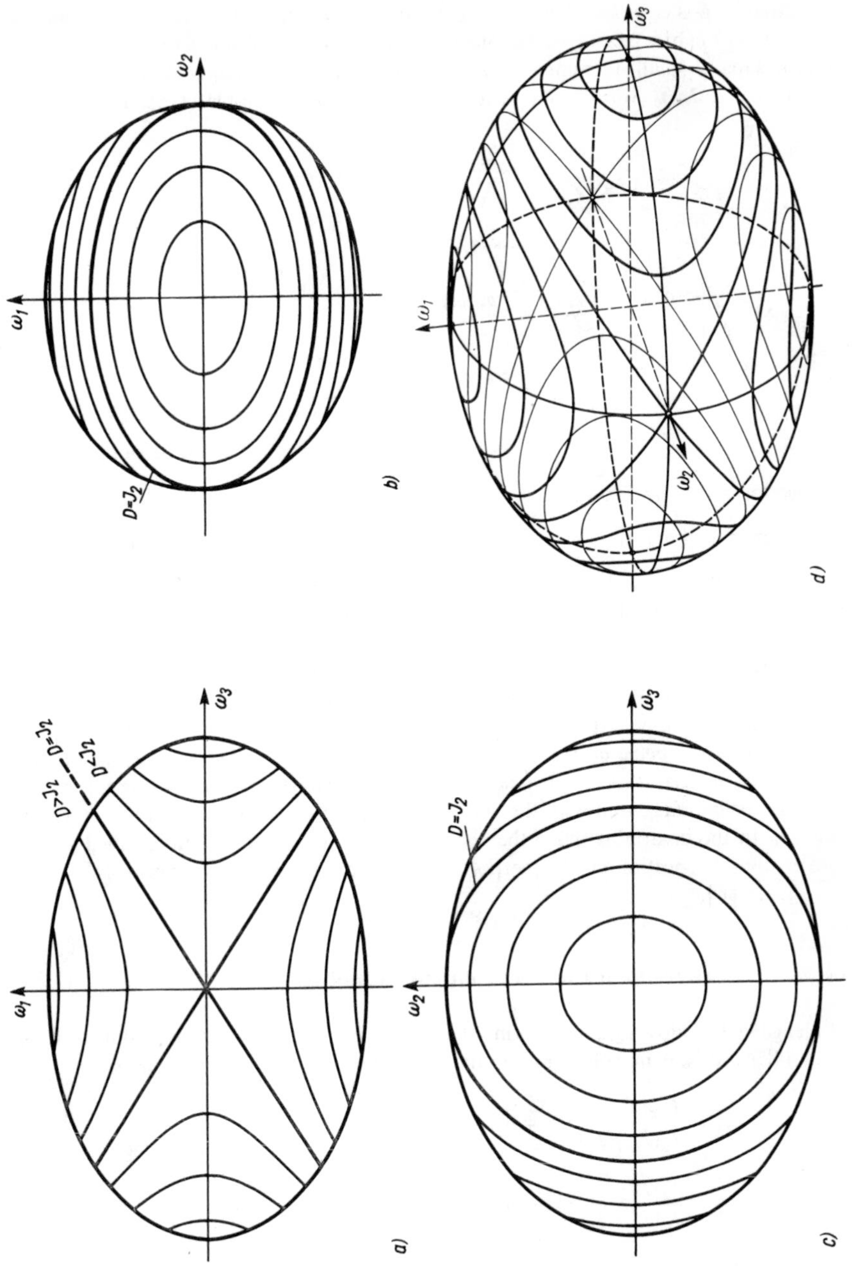

Fig. 4.1 Polhodes on the energy ellipsoid of an unsymmetric rigid body seen in projections along principal axes (a), b), c)) and in perspective (d)). The figures are based on the parameters $J_1=7$, $J_2=5$, $J_3=3$, $D=3$, 3.3, 3.9, 4.5, 5, 5.5, 6.1, 6.7 and 7 (same physical unit for all quantities)

direction of $\boldsymbol{L}$ is constant. Fig. 4.2a illustrates this situation. The vector $\boldsymbol{\omega}$ is confined to an invariable plane perpendicular to $\boldsymbol{L}$. From this follows that any (finite or infinitesimally small) increment $\Delta\boldsymbol{\omega}$ between two arbitrary moments of time is perpendicular to $\boldsymbol{L}$: $\boldsymbol{L}\cdot\Delta\boldsymbol{\omega}=0$. This equation defines the invariable plane. The vector $\boldsymbol{\omega}$

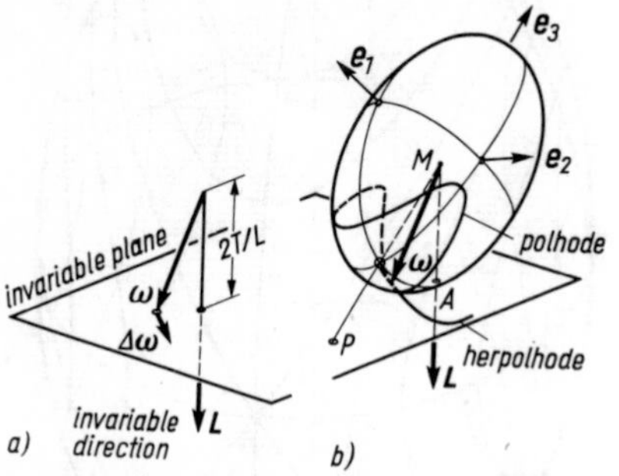

Fig. 4.2
Poinsot's interpretation of the motion of a torque-free rigid body

is also located on the energy ellipsoid defined by Eq. (4.3). The total differential of this equation has the form $\sum_{\alpha=1}^{3} J_\alpha\omega_\alpha \mathrm{d}\omega_\alpha = \boldsymbol{L}\cdot\mathrm{d}\boldsymbol{\omega}=0$. It defines the tangential plane of the energy ellipsoid at the point with coordinates ω_1, ω_2 and ω_3. Comparison with the equation for the invariable plane reveals that both planes are parallel. Morever, since both are geometric locus of $\boldsymbol{\omega}$ the planes coincide with one another. This situation is illustrated in Fig. 4.2b. The point of contact between the ellipsoid and the invariable plane is located on the instantaneous axis of rotation. Poinsot's interpretation of the motion can be summarized as follows. *The body is moving as if its energy ellipsoid were rolling without slipping on the invariable plane the geometric center M of the ellipsoid being fixed in inertial space a distance* $\overline{AM}=2T/L$ *above this plane.* During this rolling motion the contact point with the invariable plane traces a polhode on the energy ellipsoid. In the invariable plane the contact point traces another curve called a herpolhode. Properties of the herpolhodes are discussed by Grammel [5] and Magnus [6].

4.1.3 The solution of Euler's equations of motion

Euler gave the following closed-form solution of the dynamic equations of motion. First, Eqs. (4.5) and (4.6) are solved for ω_1 and ω_3, respectively, as functions of ω_2:

$$\omega_3^2 = \frac{J_2(J_1-J_2)}{J_3(J_1-J_3)}(b^2-\omega_2^2).$$

$$\omega_1^2 = \frac{J_2(J_2-J_3)}{J_1(J_1-J_3)}(a^2-\omega_2^2), \tag{4.9}$$

In these expressions a^2 and b^2 are the non-negative constants

$$a^2 = \frac{2T(D-J_3)}{J_2(J_2-J_3)}, \qquad b^2 = \frac{2T(J_1-D)}{J_2(J_1-J_2)}$$

which satisfy the relationship

$$a^2-b^2=\frac{2T(J_1-J_3)(D-J_2)}{J_2(J_1-J_2)(J_2-J_3)}.$$

Substitution of Eq. (4.9) into the second of Eqs. (4.2) and separation of the variables yield

$$\int\frac{d\omega_2}{\sqrt{(a^2-\omega_2^2)(b^2-\omega_2^2)}}=s_2(t-t_0)\sqrt{\frac{(J_1-J_2)(J_2-J_3)}{J_1J_3}} \tag{4.10}$$

where s_2 is an abbreviation for the resultant sign of the two square roots. This sign will be determined later. The integral on the left hand side is an elliptic integral of the first kind. In reducing it to a Legendre normal form three cases have to be distinguished:

a) $a^2<b^2$ or $D<J_2$

b) $a^2=b^2$ or $D=J_2$

c) $a^2>b^2$ or $D>J_2$.

Cases a) and c) correspond to polhodes which envelop the axes $\boldsymbol{e}_3$ and $\boldsymbol{e}_1$, respectively, and case b) corresponds to the separatrices (see Fig. 4.1). Consider, first, case a). Eq. (4.10) can be rewritten in the form

$$\int\frac{d\omega_2/a}{\sqrt{(1-\omega_2^2/a^2)(1-\omega_2^2/b^2)}}=s_2b(t-t_0)\sqrt{\frac{(J_1-J_2)(J_2-J_3)}{J_1J_3}}$$

or

$$\int\frac{dx}{\sqrt{(1-x^2)(1-k^2x^2)}}=s_2\tau$$

with the abbreviations

$$x=\frac{\omega_2}{a},\qquad k=\frac{a}{b},\qquad \tau=(t-t_0)\sqrt{\frac{2T(J_1-D)(J_2-J_3)}{J_1J_2J_3}}. \tag{4.11}$$

The solution has the form $x=s_2\,\mathrm{sn}\,\tau$ or

$$\omega_2=s_2\sqrt{\frac{2T(D-J_3)}{J_2(J_2-J_3)}}\,\mathrm{sn}\,\tau \tag{4.12}$$

(for elliptic integrals and Jacobian elliptic functions see Tölke [7]). When this is substituted into Eq. (4.9) and use is made of the addition theorems $\mathrm{sn}^2\tau+\mathrm{cn}^2\tau=1$ and $\mathrm{dn}^2\tau+k^2\mathrm{sn}^2\tau=1$ solutions for ω_1 and ω_3 are obtained in the form

$$\omega_1=s_1\sqrt{\frac{2T(D-J_3)}{J_1(J_1-J_3)}}\,\mathrm{cn}\,\tau,\qquad \omega_3=s_3\sqrt{\frac{2T(J_1-D)}{J_3(J_1-J_3)}}\,\mathrm{dn}\,\tau. \tag{4.13}$$

The quantities s_1 and s_3 are as yet undetermined signs of the respective square roots. The missing relationship between s_1, s_2 and s_3 is found when Eqs. (4.12) and (4.13) are substituted back into the second of Eqs. (4.2). Taking into account the relationship $d\,\mathrm{sn}\,\tau/d\tau=\mathrm{cn}\,\tau\,\mathrm{dn}\,\tau$ one gets $s_2=-s_1s_3$ or

$$s_1 s_2 s_3 = -1 . \tag{4.14}$$

Altogether four combinations of signs satisfying this relationship are possible. This result is in accordance with the fact that for each value of the parameter D two separate polhodes exist and that on each of them ω_2 passes through zero in two different points (see Fig. 4.1 d).

Case c) for $D>J_2$ is treated in a similar manner. The results read

$$\omega_1 = s_1 \sqrt{\frac{2T(D-J_3)}{J_1(J_1-J_3)}}\,\mathrm{dn}\,\tau, \qquad \omega_2 = s_2 \sqrt{\frac{2T(J_1-D)}{J_2(J_1-J_2)}}\,\mathrm{sn}\,\tau$$

$$\omega_3 = s_3 \sqrt{\frac{2T(J_1-D)}{J_3(J_1-J_3)}}\,\mathrm{cn}\,\tau$$

with the modulus $k=b/a$ and the argument

$$\tau = (t-t_0)\sqrt{\frac{2T(J_1-J_2)(D-J_3)}{J_1 J_2 J_3}} .$$

The signs s_1, s_2 and s_3 satisfy Eq. (4.14), again.

It is unnecessary to integrate case b) since the solutions for the cases a) and c) converge both in the limit $D \to J_2$ toward the same result

$$\omega_1 = s_1 \sqrt{\frac{2T(J_2-J_3)}{J_1(J_1-J_3)}}\,\frac{1}{\cosh\tau}, \qquad \omega_2 = s_2 \sqrt{\frac{2T}{J_2}}\,\tanh\tau$$

$$\omega_3 = s_3 \sqrt{\frac{2T(J_1-J_2)}{J_3(J_1-J_3)}}\,\frac{1}{\cosh\tau}$$

which, therefore, represents the solution for case b). These formulas show that the motion of $\boldsymbol{\omega}$ along a separatrix is aperiodic. For $t\to\infty$, i.e. for $\tau\to\infty$ ω_1 and ω_3 tend toward zero and ω_2 approaches asymptotically the value $s_2\sqrt{2T/J_2}$. This represents a permanent rotation about the axis $\boldsymbol{e}_2$.

Problem

4.1 Determine from the second equation of Eq. (4.2) the sense of direction in which $\boldsymbol{\omega}$ traces the polhodes in Fig. 4.1d.

4.1.4 The solution of the kinematic differential equations

The last part of the problem is to show how the body is moving in inertial space. For this purpose Euler angles are used as generalized coordinates. According to Fig. 2.1 the intermediate angle θ is measured between two axes one of which is fixed in the base $\underline{e}^{(1)}$ (here inertial space) and the other is fixed on the body. It is convenient to choose as axis fixed in inertial space the direction of the angular momentum $\boldsymbol{L}$ since this is the only significant axis. On the body the axis $\boldsymbol{e}_3$ is an appropriate choice for polhodes which envelop this axis, i.e. for $D<J_2$. In the case $D>J_2$ the polhodes envelop the axis $\boldsymbol{e}_1$. Therefore, this axis will be chosen. Consider first the case $D<J_2$ which is

illustrated in Fig. 4.2. In this figure θ is the angle PMA and ψ is measured in the invariable plane between the straight line $\overline{PA}$ and some inertially fixed reference line through A. It is unnecessary to solve kinematic differential equations of the form of Eq. (2.29). The problem can be simplified substantially if use is made of the fact that the angular momentum $\boldsymbol{L}$ has constant magnitude and direction in inertial space. In the principal axes frame $\boldsymbol{L}$ has coordinates $L_\alpha = J_\alpha \omega_\alpha$ ($\alpha = 1,2,3$). These coordinates can also be expressed as functions of Euler angles and of time derivatives of Euler angles. Using the notation of Fig. 2.1 $\boldsymbol{L}$ has the direction of $\boldsymbol{e}_3^{(1)}$. Its coordinates in the principal axes frame are, therefore, found in the third column of the direction cosine matrix $\underline{A}^{21}$ in Eq. (2.2). The results are the equations

$$J_1\omega_1 = L\sin\theta\sin\phi, \qquad J_2\omega_2 = L\sin\theta\cos\phi, \qquad J_3\omega_3 = L\cos\theta. \tag{4.15}$$

They yield without integration

$$\cos\theta = \frac{J_3}{L}\omega_3, \qquad \tan\phi = \frac{J_1\omega_1}{J_2\omega_2}.$$

Only the angle ψ is not found directly. Its time derivative is according to Eq. (2.28)

$$\dot{\psi} = \frac{\omega_3 - \dot{\phi}}{\cos\theta} = \frac{L}{J_3}\left(1 - \frac{\dot{\phi}}{\omega_3}\right). \tag{4.16}$$

The expressions for $\cos\theta$ and $\tan\phi$ become with Eqs. (4.12), (4.13) and (4.14)

$$\cos\theta = s_3\sqrt{\frac{J_3(J_1 - D)}{D(J_1 - J_3)}}\,\mathrm{dn}\,\tau, \qquad \tan\phi = -s_3\sqrt{\frac{J_1(J_2 - J_3)}{J_2(J_1 - J_3)}}\,\frac{\mathrm{cn}\,\tau}{\mathrm{sn}\,\tau} \tag{4.17}$$

From this follows by differentiation

$$\dot{\phi} = s_3\cos^2\phi\sqrt{\frac{J_1(J_2 - J_3)}{J_2(J_1 - J_3)}}\,\frac{\mathrm{dn}\,\tau}{\mathrm{sn}^2\tau}\,\frac{\mathrm{d}\tau}{\mathrm{d}t}.$$

Expressing $\cos^2\phi$ through $\tan^2\phi$ and $\mathrm{dn}\,\tau$ through ω_3 this can be given the form

$$\dot{\phi} = \omega_3\left(\frac{J_2}{J_2 - J_3}\,\mathrm{sn}^2\tau + \frac{J_1}{J_1 - J_3}\,\mathrm{cn}^2\tau\right)^{-1}.$$

Eq. (4.16) for $\dot{\psi}$ takes with this the form

$$\dot{\psi} = \frac{L}{J_3}\,\frac{(J_3/(J_2 - J_3))\,\mathrm{sn}^2\tau + (J_3/(J_1 - J_3))\,\mathrm{cn}^2\tau}{(J_2/(J_2 - J_3))\,\mathrm{sn}^2\tau + (J_1/(J_1 - J_3))\,\mathrm{cn}^2\tau}. \tag{4.18}$$

The results show that θ, $\dot{\phi}$ and $\dot{\psi}$ are periodic functions of time. The sign of $\dot{\phi}$ equals that of ω_3, and $\dot{\psi}$ is always positive (clockwise rotation about $\boldsymbol{L}$). If the function $\psi(t)$ is desired Eq. (4.18) is preferably rewritten in the form

$$\psi - \psi_0 = a\tau + b\int\frac{\mathrm{d}\tau}{\mathrm{sn}^2\tau + c^2}$$

with new constants a, b and c. This represents the normal form of an elliptic integral of the third kind (see Tölke [7]).

The solution for $\theta(t)$ can be used to answer the question whether the orientation of the axis $\boldsymbol{e}_3$ in inertial space during a permanent rotation about this axis is stable or not. For reasons of symmetry of the polhodes on the energy ellipsoid only the case $s_3=+1$ need be considered. The elliptic function $\operatorname{dn}\tau$ has the lower bound $\sqrt{1-k^2}$ where k is the modulus given by Eq. (4.11). This yields for θ the inequality

$$\cos^2\theta \geqslant (1-k^2)\frac{J_3(J_1-D)}{D(J_1-J_3)} = \frac{J_3(J_2-D)}{D(J_2-J_3)}.$$

For motions close to a permanent rotation about the axis $\boldsymbol{e}_3$ D is slightly larger than J_3. Setting $D=J_3+\delta$ with $\delta \ll J_3$ one obtains the inequality

$$\sin^2\theta \leqslant \frac{J_2}{J_3(J_2-J_3)}\delta.$$

This indicates that by choosing appropriate initial conditions $\theta(t)$ can be kept smaller than any given arbitrarily small angle. Thus, the orientation in inertial space of the axis of permanent rotation is stable.

Arguments similar to those just used also lead to solutions in the case $D>J_2$. As was previously stated the angle θ is now measured between $\boldsymbol{L}$ and $\boldsymbol{e}_1$. Starting from the equations

$$J_1\omega_1 = L\cos\theta, \qquad J_2\omega_2 = L\sin\theta\sin\phi, \qquad J_3\omega_3 = L\sin\theta\cos\phi$$

it is easy to find results of the form

$$\cos\theta = s_1\sqrt{\frac{J_1(D-J_3)}{D(J_1-J_3)}}\operatorname{dn}\tau \qquad \tan\phi = -s_1\sqrt{\frac{J_2(J_1-J_3)}{J_3(J_1-J_2)}}\frac{\operatorname{sn}\tau}{\operatorname{cn}\tau}$$

$$\dot\phi = -\omega_1\left(\frac{J_2}{J_1-J_2}\operatorname{sn}^2\tau + \frac{J_3}{J_1-J_3}\operatorname{cn}^2\tau\right)^{-1}$$

$$\dot\psi = \frac{L}{J_1}\frac{(J_1/(J_1-J_2))\operatorname{sn}^2\tau + (J_1/(J_1-J_3))\operatorname{cn}^2\tau}{(J_2/(J_1-J_2))\operatorname{sn}^2\tau + (J_3/(J_1-J_3))\operatorname{cn}^2\tau}.$$

The only difference compared with the case $D<J_2$ is that now $\dot\phi$ and ω_1 have opposite signs. In every other respect the results are qualitatively the same. In particular, it is found that the orientation of the axis of permanent rotation $\boldsymbol{e}_1$ in inertial space is stable.

The investigation of the special case $D=J_2$ is left to the reader (see Problem 4.2). With the exception of this item the results presented in this and in the previous subsections represent the complete solution to the problem of motion of an unsymmetric torque-free rigid body.

Problem

4.2 Show that for permanent rotations about $\boldsymbol{e}_2$ the orientation of this axis in inertial space is unstable. Interpret the motion of the principal axis $\boldsymbol{e}_2$ in inertial space in the case $D=J_2$ in which the polhode is a separatrix. H i n t: Define θ as angle between $\boldsymbol{L}$ and $\boldsymbol{e}_2$.

4.2 The symmetric torque-free rigid body

The solutions developed in the previous section become particularly simple if the body under consideration has two equal principal moments of inertia as is often the case in technical applications. It is left to the reader to adapt the general solutions to this special case. Here, it is preferred to start again from the equations of motion and to develop from them the special solutions directly. It is assumed that the principal axis $\boldsymbol{e}_3$ is the symmetry axis of the body so that $J_1 = J_2 \neq J_3$. The principal moment of inertia J_3 can be either smaller or larger than J_1 (the trivial case $J_1 = J_2 = J_3$ will not be considered). With these assumptions Euler's equations reduce to

$$\begin{aligned} &J_1\dot{\omega}_1 - (J_1 - J_3)\omega_2\omega_3 = 0 \\ &J_1\dot{\omega}_2 - (J_3 - J_1)\omega_3\omega_1 = 0 \\ &J_3\dot{\omega}_3 \qquad\qquad\qquad\;\; = 0. \end{aligned} \tag{4.19}$$

They yield at once

$$\omega_3 \equiv \omega_{30} = \text{const}$$

and by substituting this into the first two equations

$$\dot{\omega}_1 - \nu\omega_2 = 0, \qquad \dot{\omega}_2 + \nu\omega_1 = 0$$

where ν is the constant

$$\nu = \frac{\omega_{30}(J_1 - J_3)}{J_1}. \tag{4.20}$$

The differential equations have as first integral $\omega_1^2 + \omega_2^2 = \Omega^2 = \text{const}$. The general solution for initial values ω_{10} and ω_{20} has the form

$$\begin{aligned} \omega_1 &= \omega_{10}\cos\nu(t - t_0) + \omega_{20}\sin\nu(t - t_0) = \Omega\sin\nu(t - t_0') \\ \omega_2 &= \omega_{20}\cos\nu(t - t_0) - \omega_{10}\sin\nu(t - t_0) = \Omega\cos\nu(t - t_0'). \end{aligned}$$

Next, the kinematic equations are considered. Again, Euler angles are used with θ being the angle between the symmetry axis and the inertially fixed angular momentum vector $\boldsymbol{L} = J_1\boldsymbol{\Omega} + J_3\boldsymbol{\omega}_{30}$. Eqs. (4.15) to (4.16) are valid, again, so that

$$\begin{aligned} &\cos\theta = \frac{\omega_{30}J_3}{L}, \qquad \tan\phi = \frac{\omega_1}{\omega_2} = \tan\nu(t - t_0'), \qquad \dot{\phi} = \nu \\ &\dot{\psi} = \frac{\omega_{30} - \dot{\phi}}{\cos\theta} = \frac{\omega_{30}J_3}{J_1\cos\theta} = \frac{L}{J_1}. \end{aligned} \tag{4.21}$$

Thus, θ, $\dot{\phi}$ and $\dot{\psi}$ turn out to be constant. For a better understanding of these results Poinsot's interpretation of the motion is again considered. The energy ellipsoid in Fig. 4.2 is now an ellipsoid of revolution with the symmetry axis $\boldsymbol{e}_3$. The polhodes are, therefore, circles (and so are the herpolhodes). The axis $\boldsymbol{e}_3$ is moving around a circular cone whose axis is the angular momentum vector $\boldsymbol{L}$. The angular velocity of this motion around the cone, called nutation angular velocity, is $\dot{\psi}$. The angular velocity vector $\boldsymbol{\omega}$ of the body always lies in the plane spanned by the symmetry axis and the angular momentum $\boldsymbol{L}$. Relative to the body the vector $\boldsymbol{\omega}$ is moving around a circular cone which is defined by the polhode. Relative to inertial space the vector $\boldsymbol{\omega}$ is also moving around a circular cone. The axis of this cone is the angular momentum $\boldsymbol{L}$. Following Chap. 2.2 (Fig. 2.7) the motion of the body can be visualized as a rolling motion without slipping of the body-fixed cone on the inertially fixed cone. In Fig. 4.3a, b the cone swept out by $\boldsymbol{e}_3$ and the two cones swept out by $\boldsymbol{\omega}$ are illustrated. Only

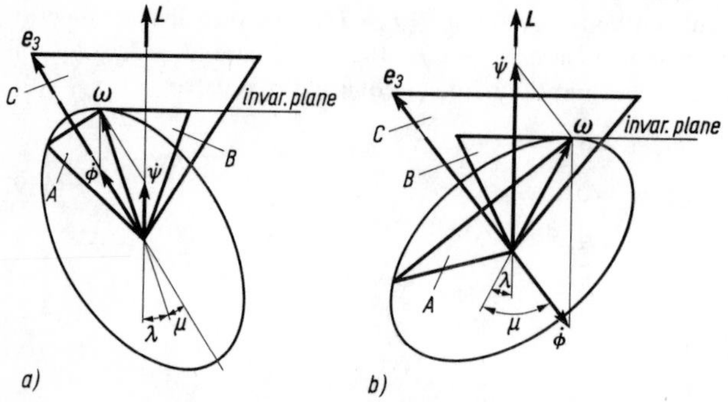

Fig. 4.3 The body-fixed $\boldsymbol{\omega}$-cone A is rolling on the space-fixed $\boldsymbol{\omega}$-cone B while the symmetry axis $\boldsymbol{e}_3$ generates the space-fixed cone C. The energy ellipsoid intersects cone A in a polhode and rolls on the invariable plane
a) belongs to a rod-shaped body and b) to a disc-shaped body

their projections onto the plane spanned by $\boldsymbol{e}_3$, $\boldsymbol{\omega}$ and $\boldsymbol{L}$ are shown. The ellipses represent contours of the energy ellipsoid. Two figures are necessary since rod-shaped bodies with $J_1 > J_3$ show another behavior than disc-shaped bodies with $J_1 < J_3$. For the former $\dot{\phi}$ is positive and for the latter negative (cf. Eq. (4.20)). Note that an observer of the motion can see the cone described by the symmetry axis $\boldsymbol{e}_3$ but that the other two cones are invisible (Magnus [8] describes an experimental trick which renders the motion of $\boldsymbol{\omega}$ around the body-fixed cone visible). It is left to the reader to verify that the angles λ and μ are related by the equation

$$\frac{\sin\lambda}{\sin\mu} = \frac{|J_1 - J_3|}{J_3}\cos\theta .$$

In technical applications motions of symmetric bodies usually differ very little from permanent rotations about the symmetry axis. For such motions ω_{30} is practically identical with $\omega = |\boldsymbol{\omega}|$, and θ is very small. The nutation angular velocity is then approximately

$$\dot{\psi} \approx \frac{\omega_{30} J_3}{J_1} \quad \text{for } \theta \ll 1. \tag{4.22}$$

4.3 The self-excited, symmetric rigid body

The subject of this section is a symmetric rigid body with principal moments of inertia J_1, $J_2=J_1$ and $J_3 \neq J_1$ which is under the action of a torque whose coordinates in the principal axes frame are given functions of time. Euler's equations of motion have the form

$$\begin{aligned} J_1\dot{\omega}_1-(J_1-J_3)\omega_2\omega_3&=M_1(t)\\ J_1\dot{\omega}_2-(J_3-J_1)\omega_3\omega_1&=M_2(t)\\ J_3\dot{\omega}_3\qquad\qquad\qquad\quad&=M_3(t)\,. \end{aligned} \tag{4.23}$$

We begin with the simple case in which $M_3(t)$ is identically zero. Then, $\omega_3=\text{const}$, and the first two equations reduce to

$$\begin{aligned} \dot{\omega}_1-\nu\omega_2&=m_1(t)\\ \dot{\omega}_2+\nu\omega_1&=m_2(t) \end{aligned} \tag{4.24}$$

with the constant $\nu=\omega_3(J_1-J_3)/J_1$ and with functions $m_1(t)=M_1(t)/J_1$ and $m_2(t)=M_2(t)/J_1$. By introducing the complex quantities $\omega^*=\omega_1+\mathrm{i}\omega_2$ and $m^*=m_1+\mathrm{i}m_2$ these equations can be combined in the single complex equation

$$\dot{\omega}^*+\mathrm{i}\nu\omega^*=m^*(t)\,.$$

It has the general solution

$$\omega^*(t)=\mathrm{e}^{-i\nu t}\left[\omega_0^*+\int_0^t m^*(\tau)\mathrm{e}^{\mathrm{i}\nu\tau}\mathrm{d}\tau\right].$$

When this is split, again, into real and imaginary parts the solutions for $\omega_1(t)$ and $\omega_2(t)$ are obtained:

$$\begin{aligned} \omega_1(t)&=\omega_{10}\cos\nu t+\omega_{20}\sin\nu t+\int_0^t[m_1(\tau)\cos\nu(\tau-t)+m_2(\tau)\sin\nu(\tau-t)]\,\mathrm{d}\tau,\\ \omega_2(t)&=\omega_{20}\cos\nu t-\omega_{10}\sin\nu t+\int_0^t[m_1(\tau)\sin\nu(\tau-t)+m_2(\tau)\cos\nu(\tau-t)]\,\mathrm{d}\tau\,. \end{aligned} \tag{4.25}$$

Next, the general case with $M_3(t)\neq 0$ is investigated. From the third of Eqs. (4.23) the solution for $\omega_3(t)$ is obtained

$$\omega_3(t)=\omega_{30}+\frac{1}{J_3}\int_0^t M_3(\tau)\mathrm{d}\tau\,.$$

An auxiliary variable $\alpha(t)$ is now introduced by the equation

$$\alpha(t)=\int_0^t\omega_3(\tau)\mathrm{d}\tau\,.$$

This variable is a known function of time. For the inverse function $t(\alpha)$ a closed-form expression may not exist. However, it is available at least numerically. In the first two of

Euler's equations $\dot{\omega}_1$ and $\dot{\omega}_2$ can be expressed in the form

$$\dot{\omega}_i = \frac{d\omega_i}{d\alpha}\dot{\alpha} = \omega_i'\omega_3 \qquad i=1,2$$

where the prime denotes differentiation with respect to α. With these expressions the two equations of motion take the form

$$\omega_1' - \nu\omega_2 = m_1(\alpha) \qquad \omega_2' + \nu\omega_1 = m_2(\alpha)$$

with the constant $\nu = (J_1 - J_3)/J_1$ and with functions $m_1(\alpha) = M_1(t(\alpha))/[J_1\omega_3(t(\alpha))]$ and $m_2(\alpha) = M_2(t(\alpha))/[J_1\omega_3(t(\alpha))]$. These equations are identical with Eq. (4.24) except that α is the independent variable instead of t. The solution has, therefore, the form of Eq. (4.25) provided t is replaced everywhere by $\alpha(t)$ and m_1 and m_2 are the functions of α defined above.

4.4 The symmetric heavy top

A rigid body is considered which is supported in inertial space at a single point which is not the body center of mass. The body is subject to gravity only. In the literature this system is known as heavy top. The general solution for its equations of motion is not known. It is known only for the special case in which the body is inertia-symmetric and in which, furthermore, the support point is located on the symmetry axis. In Fig. 4.4 such a symmetric heavy top is shown in a position in which the support point is at a lower level than the center of mass. This system the solutions of which were found by Lagrange is the subject of the following considerations.

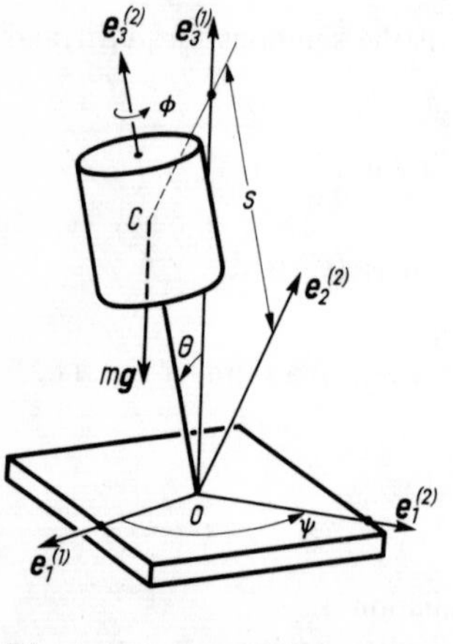

Fig. 4.4
A symmetric heavy top and its coordinates ψ, θ, ϕ

The torque caused by gravity is a function of the orientation of the body in inertial space. This functional relationship has the consequence that Euler's equations of motion are coupled with the kinematic differential equations which relate $\boldsymbol{\omega}$ to generalized coordinates. For this reason Euler's equations will not be used. Instead, it is preferred to establish second-order differential equations for appropriately chosen generalized coordinates and to solve these equations. On the body, as well as in inertial space, there exists one physically significant direction each, namely the symmetry axis on the body and the vertical line of action of gravity in inertial space. This suggests the use of Euler angles as generalized coordinates with θ being the angle between these two directions. The Euler angles ψ, θ and ϕ are defined as in Fig. 2.1. They relate a body-fixed base to a base fixed in inertial space. In Fig. 4.4 the body-fixed base is not shown. Shown are the base $\underline{e}^{(1)}$ which is fixed in inertial space and a base $\underline{e}^{(2)}$ which is fixed neither on the body nor in inertial space. Its base vector $\boldsymbol{e}_3^{(2)}$ lies in the symmetry axis, and $\boldsymbol{e}_1^{(2)}$ is always perpendicular to the vertical $\boldsymbol{e}_3^{(1)}$. This base is identical with the base $\underline{e}^{(2)'}$ of Fig. 2.1. Its absolute angular velocity differs from the absolute angular velocity $\boldsymbol{\omega}$ of the body by a component along the symmetry axis which is equal to $\dot{\phi}\boldsymbol{e}_3^{(2)}$. For reasons of symmetry the body moments of inertia are constant in $\underline{e}^{(2)}$ in spite of the motion of this base relative to the body.

Equations of motion are established from the law of moment of momentum in the general form of Eq. (3.15). As reference point in inertial space the support point O of the body is chosen. The absolute time derivative $\dot{\boldsymbol{L}}^O$ is expressed in terms of the time derivative ${}^{(2)}\mathrm{d}\boldsymbol{L}^O/\mathrm{d}t$ in the base $\underline{e}^{(2)}$. According to Eq. (2.23) the two derivatives are related by the equation

$$\dot{\boldsymbol{L}}^O = \frac{{}^{(2)}\mathrm{d}}{\mathrm{d}t}\boldsymbol{L}^O + \boldsymbol{\Omega} \times \boldsymbol{L}^O \tag{4.26}$$

where $\boldsymbol{\Omega}$ denotes the absolute angular velocity of the base $\underline{e}^{(2)}$. Fig. 4.4 yields

$$\boldsymbol{L}^O = J_1\dot{\theta}\boldsymbol{e}_1^{(2)} + J_1\dot{\psi}\sin\theta\,\boldsymbol{e}_2^{(2)} + J_3(\dot{\phi}+\dot{\psi}\cos\theta)\boldsymbol{e}_3^{(2)} \tag{4.27}$$

$$\boldsymbol{\Omega} = \dot{\theta}\boldsymbol{e}_1^{(2)} + \dot{\psi}\sin\theta\,\boldsymbol{e}_2^{(2)} + \dot{\psi}\cos\theta\,\boldsymbol{e}_3^{(2)} \tag{4.28}$$

$$\boldsymbol{M}^O = mgs\sin\theta\,\boldsymbol{e}_1^{(2)}.$$

Substitution into Eqs. (4.26) and (3.15) results in the desired scalar differential equations of motion

$$J_1\ddot{\theta} + [J_3(\dot{\phi}+\dot{\psi}\cos\theta) - J_1\dot{\psi}\cos\theta]\dot{\psi}\sin\theta - mgs\sin\theta = 0 \tag{4.29}$$

$$J_1\ddot{\psi}\sin\theta + 2J_1\dot{\psi}\dot{\theta}\cos\theta - J_3\dot{\theta}(\dot{\phi}+\dot{\psi}\cos\theta) = 0 \tag{4.30}$$

$$\ddot{\phi} + \ddot{\psi}\cos\theta - \dot{\psi}\dot{\theta}\sin\theta = 0.$$

The left hand side expression in the last equation is equal to

$$\frac{\mathrm{d}}{\mathrm{d}t}(\dot{\phi}+\dot{\psi}\cos\theta) = \dot{\omega}_3$$

from which follows the first integral

$$\dot{\phi} + \dot{\psi}\cos\theta \equiv \omega_3 = \text{const}. \tag{4.31}$$

It can also be concluded more directly from the third Euler equation which for $J_1 = J_2$ and $M_3 = 0$ reduces to $\dot{\omega}_3 = 0$. With this integral Eqs. (4.29) and (4.30) become

$$J_1\ddot{\theta} + (J_3\omega_3 - J_1\dot{\psi}\cos\theta)\dot{\psi}\sin\theta - mgs\sin\theta = 0 \tag{4.32}$$

$$J_1\ddot{\psi}\sin\theta + 2J_1\dot{\psi}\dot{\theta}\cos\theta - J_3\omega_3\dot{\theta} = 0. \tag{4.33}$$

These equations furnish two more algebraic integrals. When the second equation is multiplied by $\sin\theta$ it can be written in the form

$$\frac{\mathrm{d}}{\mathrm{d}t}(J_1\dot{\psi}\sin^2\theta+J_3\omega_3\cos\theta)=0. \tag{4.34}$$

If, on the other hand, Eq. (4.32) is multiplied by $\dot{\theta}$ and Eq. (4.33) by $\dot{\psi}\sin\theta$ and both expressions are summed it is found that

$$\frac{\mathrm{d}}{\mathrm{d}t}\left[\frac{J_1(\dot{\psi}^2\sin^2\theta+\dot{\theta}^2)}{2}+mgs\cos\theta\right]=0. \tag{4.35}$$

These two integrals can also be found directly without knowing the equations of motion. Since gravity does not produce a torque about the vertical $\boldsymbol{e}_3^{(1)}$ the angular momentum component in this direction must be constant. The magnitude L of this component is, in view of Eq. (4.27) and of the geometric relationships shown in Fig. 4.4,

$$J_1\dot{\psi}\sin^2\theta+J_3\omega_3\cos\theta=L. \tag{4.36}$$

This equation is equivalent to Eq. (4.34). The system is conservative so that its total energy E is constant. This yields

$$J_1(\omega_1^2+\omega_2^2)+J_3\omega_3^2+2mgs\cos\theta=2E$$

or with $\omega_1=\Omega_1$ and $\omega_2=\Omega_2$ from Eq. (4.28)

$$J_1(\dot{\psi}^2\sin^2\theta+\dot{\theta}^2)+2mgs\cos\theta=2E-J_3\omega_3^2=\text{const}. \tag{4.37}$$

This is equivalent to Eq. (4.35).

Before the general solution of the problem is developed two special types of motion will be considered which can be realized by a proper choice of initial conditions. One is the common plane pendulum motion with $\omega_3\equiv0$ in which θ is the only time dependent variable. In this case the equations of motion reduce to $J_1\ddot{\theta}-mgs\sin\theta=0$, and of the three integrals of motion only the energy equation $J_1\dot{\theta}^2+2mgs\cos\theta=2E$ is not trivial. These two equations represent, indeed, the differential equation and the energy integral, respectively, of a plane physical pendulum (note that, normally, $\alpha=\pi-\theta$ is used as variable). The second special type of motion is characterized by a time independent angle $\theta\equiv\theta_0$. With this condition Eq. (4.33) yields $\dot{\psi}=\text{const}$ and, furthermore, Eq. (4.31) leads to $\dot{\phi}=\text{const}$. This geometrically simple form of motion is referred to as regular precession. The symmetry axis of the body is moving with the constant precession angular velocity $\dot{\psi}$ around a circular cone the axis of which is the vertical $\boldsymbol{e}_3^{(1)}$. Eq. (4.32) with $\ddot{\theta}=0$ represents a quadratic equation for $\dot{\psi}$ which has the solutions

$$\dot{\psi}_{1,2}=\begin{cases}\dfrac{J_3\omega_3}{2J_1\cos\theta_0}\left(1\pm\sqrt{1-\dfrac{4J_1mgs\cos\theta_0}{J_3^2\omega_3^2}}\right) & \text{for } \cos\theta_0\neq0\\[2ex] \dfrac{mgs}{J_3\omega_3} & \text{for } \cos\theta_0=0.\end{cases} \tag{4.38}$$

At this point it should be noted that all results obtained thus far are valid also for the special case where s equals zero, i.e. for a symmetric body which is supported at its center of mass and which is, therefore, torque-free. From Sec. 4.2 it is known that under these conditions only two types of motion can occur, namely a permanent rotation about the symmetry axis and nutations with a nutation angular velocity $\dot{\psi}$ given by Eq. (4.21). Eq. (4.38) yields for $s=0$ the two solutions

$$\dot{\psi}_1 = \frac{J_3 \omega_3}{J_1 \cos\theta_0}, \qquad \dot{\psi}_2 = 0\,.$$

These are, indeed, the nutation angular velocity and a result for $\dot{\psi}_2$ which in the case $\omega_3 \neq 0$ can be interpreted as a permanent rotation about the symmetry axis. We now return to the general case of Eq. (4.38) with $s>0$. If the body is hanging ($\cos\theta_0<0$) both roots $\dot{\psi}_1$ and $\dot{\psi}_2$ are real for any value of θ_0. In positions with $\cos\theta_0>0$ for which Fig. 4.4 shows an example regular precessions are possible only if the body angular velocity ω_3 is sufficiently large to render the expression under the square root positive.

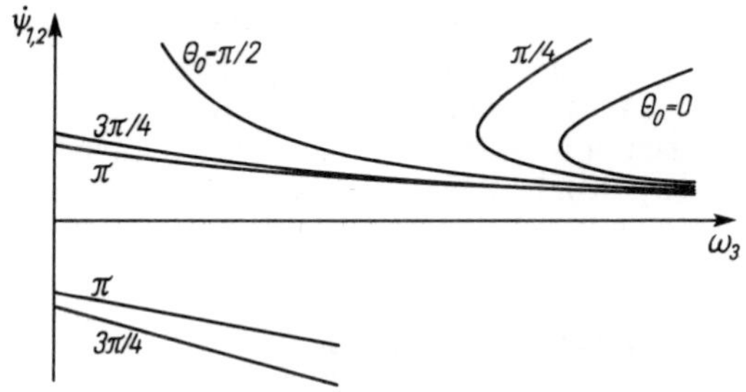

Fig. 4.5
The angular velocities $\dot{\psi}_1$ and $\dot{\psi}_2$ of regular precessions as functions of ω_3 and θ_0

In Fig. 4.5 the relationship between $\dot{\psi}_1$, $\dot{\psi}_2$ and ω_3 is schematically illustrated for various values of the parameter θ_0. For rapidly spinning bodies the roots tend toward

$$\lim_{\omega_3\to\infty} \dot{\psi}_1 = \frac{J_3 \omega_3}{J_1 \cos\theta_0}, \qquad \lim_{\omega_3\to\infty} \dot{\psi}_2 = \frac{mgs}{J_3 \omega_3}. \tag{4.39}$$

One of the asymptotic solutions is proportional to ω_3 and represents a *fast regular precession* while the other is proportional to $1/\omega_3$ and represents a *slow regular precession*. The fast regular precession angular velocity is identical with the nutation angular velocity of a torque-free symmetric rigid body and also with $\dot{\psi}_1$ for $s=0$ while the slow one is identical with the exact solution for $\dot{\psi}$ in the case $\cos\theta_0=0$.

We now turn to the general solution of the problem. We start out from the integrals of motion. Eq. (4.36) yields

$$\dot{\psi} = \frac{L - J_3 \omega_3 \cos\theta}{J_1 \sin^2\theta}. \tag{4.40}$$

Substitution into Eq. (4.37) results in the differential equation for θ

$$J_1 \dot{\theta}^2 = 2E - J_3 \omega_3^2 - 2mgs\cos\theta - \frac{(L - J_3 \omega_3 \cos\theta)^2}{J_1 \sin^2\theta}. \tag{4.41}$$

As soon as its solution $\theta(t)$ is known $\psi(t)$ and $\phi(t)$ can be found by simple integration from Eqs. (4.40) and (4.31). With the new variable

$$u = \cos\theta, \qquad \dot{u} = -\dot{\theta}\sin\theta \tag{4.42}$$

Eq. (4.41) takes after simple manipulations the form

$$\dot{u}^2 = \frac{(2E - J_3\omega_3^2 - 2mgsu)(1-u^2)}{J_1} - \frac{(L - J_3\omega_3 u)^2}{J_1^2}. \tag{4.43}$$

The expression on the right hand side is a cubic polynomial in u. About the location of its roots the following statements can be made. For $u=+1$ and also for $u=-1$ the polynomial has negative values. In the limit $u \to +\infty$ it tends toward plus infinity. It has, therefore, at least one real root $u_3 > 1$. Because of Eq. (4.42) only the interval $|u| \leqslant 1$ is of interest. Since for real solutions $\dot{u}^2(u)$ must be non-negative somewhere in this interval the polynomial must have either two real roots or one real double root in the interval. For parameter combinations corresponding to real solutions the diagram of the function $\dot{u}^2(u)$ has, therefore, a form which is shown schematically in Fig. 4.6. The roots u_1 and u_2 have either equal or opposite signs. That both cases are physically possible is demonstrated by the two special types of motion studied earlier. For pendulum motions the amplitude of θ can be chosen such that the sign of $u(t) = \cos\theta(t)$ is either always negative or alternating. For a regular precession u is constant. It is assumed that the roots u_1, u_2 and u_3 are ordered as shown in Fig. 4.6 ($u_1 \leqslant u_2 < u_3$). In terms of these roots Eq. (4.43) reads

$$\dot{u}^2 = \frac{2mgs}{J_1}(u-u_1)(u-u_2)(u-u_3).$$

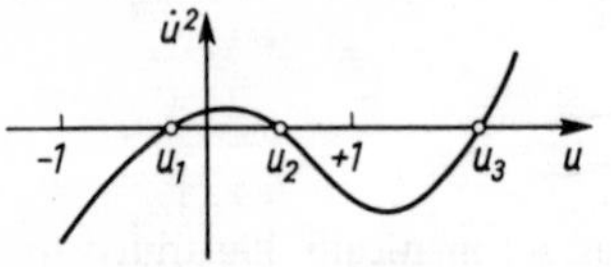

Fig. 4.6 Schematic view of the function $\dot{u}^2(u)$

When u is replaced by the new variable v defined by

$$u = u_1 + (u_2 - u_1)v^2 \tag{4.44}$$

this equation becomes after simple manipulations

$$\dot{v}^2 = \frac{mgs}{2J_1}(u_3 - u_1)(1-v^2)(1-k^2v^2)$$

$$\text{with} \quad 0 \leqslant k^2 = \frac{u_2 - u_1}{u_3 - u_1} \leqslant 1. \tag{4.45}$$

Separation of the variables leads to the elliptic integral of the first kind with modulus k

$$\int_{v_*}^{v} \frac{\mathrm{d}\bar{v}}{\sqrt{(1-\bar{v}^2)(1-k^2\bar{v}^2)}} = (t-t_0)\sqrt{\frac{(u_3-u_1)mgs}{2J_1}} = \tau.$$

It has the solution $v = \operatorname{sn}\tau$. The solution for $\theta(t)$ is, therefore, in view of Eqs. (4.44) and (4.42),

$$\cos\theta = \cos\theta_1 + (\cos\theta_2 - \cos\theta_1)\operatorname{sn}^2\tau. \tag{4.46}$$

The constants θ_1 and θ_2 are determined through $\cos\theta_1 = u_1$ and $\cos\theta_2 = u_2$, respectively. They respresent the minimum and maximum values of $\theta(t)$. From Eqs. (4.40) and (4.31) also $\dot{\psi}$ and $\dot{\phi}$ are obtained as functions of t:

$$\dot{\psi} = \frac{L - J_3\omega_3\cos\theta}{J_1(1-\cos^2\theta)}, \qquad \dot{\phi} = \omega_3 - \dot{\psi}\cos\theta. \tag{4.47}$$

All three quantities θ, $\dot{\psi}$ and $\dot{\phi}$ are, thus, shown to be elliptic functions of time. The period of these functions is half the period of $\operatorname{sn}\tau$, i.e. $2K(k)$ in the variable τ and

$$t_p = K\sqrt{\frac{8J_1}{mgs(u_3-u_1)}}$$

in the variable t, K being the complete elliptic integral

$$K(k) = \int_0^1 \frac{dv}{\sqrt{(1-v^2)(1-k^2v^2)}}.$$

The superposition of periodic changes in $\theta(t)$ onto a precession about the vertical with a (periodically changing) angular velocity $\dot{\psi}(t)$ is best visualized as follows. Imagine a sphere with the center at the support point of the top. The intersection point of the symmetry axis of the body with this sphere generates paths which render the periodic changes of $\theta(t)$ as well as of $\dot{\psi}(t)$ clearly visible. In Fig. 4.7 characteristic features of all physically realizable types of paths are illustrated schematically. The two lower paths belong to the special types of motion described as pendulum motion and regular precession. The upper three paths represent general cases of motion in which $\dot{\psi}(t)$ is changing either between negative and positive values (path a) or between zero and a positive maximum (b) or between two positive extreme values (c). The periodic nodding motion of the top with $\theta(t)$ which is superimposed on the precessional motion is called nutation. For a more detailed discussion of the solutions the reader is referred to Arnold/Maunder [9] and to Magnus [6].

Many technical gyroscopic instruments represent, in principle, a symmetric heavy top (see Magnus [6]). Such instruments are operated under special conditions. For one thing, they are in rapid rotation. By this is meant that the part $J_3\omega_3^2/2$ of the kinetic energy of the body is very large compared with the potential energy mgs. Second, such instruments are set into motion in such a way that initially the precession angular velocity $\dot{\psi}$ is very small compared with ω_3 (usually, θ and ψ are held fixed until the body has reached its full spin. Only then the constraints on θ and ψ are lifted keeping the initial values of $\dot{\theta}$ and $\dot{\psi}$ as small as possible). Under these conditions motions are observed which can hardly be distinguished from regular precessions. In reality, these

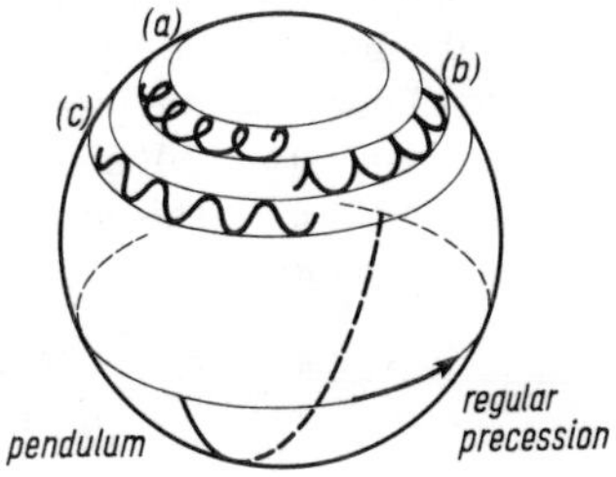

Fig. 4.7
Paths generated by the symmetry axis on a sphere surrounding the support point of the heavy top

motions are governed by Eqs. (4.46) and (4.47). The nutation amplitude $(\theta_2-\theta_1)/2$ is extremely small, however, and θ is oscillating rapidly. The precession angular velocity $\dot{\psi}(t)$ is very small, and it appears to be constant although it is also undergoing rapid oscillations. Such motions are called pseudo-regular precessions. Their characteristic properties can be developed from the general solution by means of approximation formulas. For this purpose it is assumed that the top is started with the initial conditions $\theta(0)=\theta_1$, $\dot{\theta}(0)=0$, $\dot{\psi}(0)=\dot{\psi}_1$ and ω_3 (constant throughout the motion). The angular momentum and energy integrals are (Eqs. (4.36) and (4.37))

$$\begin{aligned} L&=J_1\dot{\psi}_1(1-u_1^2)+J_3\omega_3u_1 \\ 2E&=J_1\dot{\psi}_1^2(1-u_1^2)+2mgsu_1+J_3\omega_3^2. \end{aligned} \tag{4.48}$$

These expressions have to be substituted into the cubic in Eq. (4.43). Because of the initial condition $\dot{\theta}(0)=0$ the angle θ_1 is one of the two extreme values of $\theta(t)$. This means that $u_1=\cos\theta_1$ is a root of the cubic. Division by $(u-u_1)$ leads after some algebraic manipulations to

$$\dot{u}^2=\frac{2mgs}{J_1}(u-u_1)\{u^2-2au-1+2a[u_1+b(1-u_1^2)(2-bu_1-b)]\}$$

where a and b are the dimensionless quantities

$$a=\frac{J_3^2\omega_3^2}{4mgsJ_1}, \qquad b=\frac{J_1\dot{\psi}_1}{J_3\omega_3}. \tag{4.49}$$

The quadratic function of u in curled brackets has the roots

$$u_{2,3}=a\mp\sqrt{a^2+1-2a[u_1+b(1-u_1^2)(2-bu_1-b)]}\,.$$

Under the assumed conditions the quantity a is much larger than one, and the absolute value of b is much smaller than one. This allows the approximations

$$u_{2,3}\approx a\mp\sqrt{a^2+1-2a[u_1+2b(1-u_1^2)]}\,.$$

A Taylor series expansion (up to second order terms) yields

$$u_2\approx u_1-(1-u_1^2)\left(\frac{1}{2a}-2b\right), \qquad u_3\approx 2a. \tag{4.50}$$

With this result for u_2 and with the Taylor formula $\cos\theta_2\approx\cos\theta_1-(\theta_2-\theta_1)\sin\theta_1$ the nutation amplitude $(\theta_2-\theta_1)/2$ becomes approximately

$$\theta_2-\theta_1\approx\left(\frac{1}{2a}-2b\right)\sin\theta_1=\frac{2J_1}{J_3\omega_3}\left(\frac{mgs}{J_3\omega_3}-\dot{\psi}_1\right)\sin\theta_1\,.$$

This is, indeed, a very small quantity. Note that it becomes zero if the initial value $\dot{\psi}_1$ equals the angular velocity of a slow regular precession (cf. Eq. (4.39)). The modulus k of the elliptic functions given by Eq. (4.45) is very small compared with unity as can be seen from Eqs. (4.50) and (4.49). The complete elliptic integral $K(k)$ is, therefore, approximately $\pi/2$. This yields for the period length of the functions $\theta(t)$ and $\dot{\psi}(t)$ the approximation

$$t_p\approx\frac{\pi}{2}\sqrt{\frac{8J_1}{mgsu_3}}=\frac{2\pi J_1}{J_3\omega_3}.$$

The corresponding circular frequency $2\pi/t_p \approx \omega_3 J_3/J_1$ is very large. It equals the nutation angular velocity of a torque-free symmetric body in the case of small nutation amplitudes (cf. Eq. (4.22)). Finally, an approximation formula for $\dot\psi$ can be developed from Eqs. (4.47) and (4.48):

$$\dot\psi \approx \frac{J_1\dot\psi_1(1-u_1^2)+J_3\omega_3(u_1-u)}{J_1(1-u_1^2)} = \dot\psi_1 + \frac{J_3\omega_3}{J_1(1-u_1^2)}(u_1-u).$$

For the difference $u_1-u=\cos\theta_1-\cos\theta$ Eq. (4.46) yields (with the approximation $\operatorname{sn}\tau\approx\sin\tau$ (valid for $k\approx 0$)) $u_1-u\approx(u_1-u_2)\sin^2\tau$. For (u_1-u_2) Eq. (4.50) is used. This leads to

$$u_1-u\approx(1-u_1^2)\left(\frac{1}{2a}-2b\right)\sin^2\tau$$

and, furthermore, to

$$\dot\psi\approx\dot\psi_1+2\left(\frac{mgs}{J_3\omega_3}-\dot\psi_1\right)\sin^2\tau=\frac{mgs}{J_3\omega_3}-\left(\frac{mgs}{J_3\omega_3}-\dot\psi_1\right)\cos 2\tau.$$

This result indicates that $\dot\psi$ is oscillating about the mean value $mg\,s/(J_3\omega_3)$ which is the angular velocity of a slow regular precession. The amplitude of the oscillation is zero if the initial value $\dot\psi_1$ equals this mean value. The approximation formulas just established confirm the initial statement that motions of a rapidly rotating symmetric heavy top differ only slightly from regular precessions. The motion can be interpreted as superposition of a fast nutation with very small amplitude $(\theta_2-\theta_1)/2$ onto a slow regular precession.

4.5 The symmetric heavy body in a cardan suspension

Fig. 4.8 depicts a symmetric rigid rotor in a two-gimbal suspension. All three rotation axes intersect in one point O. This point is the center of mass of the rotor as well as of the inner gimbal. When all three rotation axes are perpendicular to one another these axes represent principal axes of inertia for the rotor and for the inner gimbal. The axis of the outer gimbal is mounted in a vertical position on the Earth (which is assumed to represent inertial space). To the symmetry axis of the rotor, at a distance s from O,

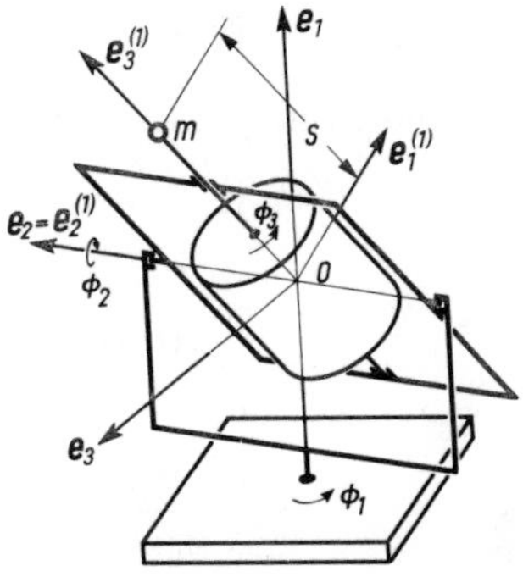

Fig. 4.8
A symmetric rotor in a cardan suspension with vertical outer gimbal axis and with offset point mass m

a point mass m is attached. Such a system has many features in common with a symmetric heavy top (see Fig. 4.4). The only differences are the presence of gimbals with inertia properties and the constraint torque normal to the outer gimbal axis which is transmitted to the system by the bearings on this axis. In the following investigation Bryant angles ϕ_1, ϕ_2 and ϕ_3 are used as generalized coordinates. They are, in this order, the rotation angle of the outer gimbal relative to inertial space, of the inner gimbal relative to the outer gimbal and of the rotor relative to the inner gimbal. The angle ϕ_2 is zero when the plane of the inner gimbal is in a horizontal position. This definition of Bryant angles is the same as in Sec. 2.1.2. The angles are related to the Euler angles ψ, θ and ϕ which were used in the previous section for the symmetric heavy top through the equations

$$\dot{\phi}_1=\dot{\psi}, \qquad \phi_2=\frac{\pi}{2}-\theta, \qquad \dot{\phi}_3=\dot{\phi}. \tag{4.51}$$

These equations will be used in order to render the similarities and differences between the two systems more distinct. As reference point for angular momenta and torques the point O is chosen. If there is no friction in the bearings which is supposed to be the case then the resultant torque on the rotor has no component along the rotor axis. The coordinate ω_3 in this direction of the absolute rotor angular velocity is, therefore, constant. The resultant torque on the entire system, composed of the constraint torque in the outer gimbal axis and of the torque caused by the weight of the point mass, has no vertical component. Hence, the coordinate L in this direction of the absolute angular momentum of the entire system is a constant. In addition to these two integrals of motion a third integral exists which states that the total energy E of the system is constant. These are the same integrals on which the analysis of the symmetric heavy top was based. Note that the angular momentum integral exists only if the axis of the outer gimbal is mounted vertically! To the angular momentum coordinate L and to the energy E contributions are made by the two gimbals. In order to formulate these quantities the absolute angular velocity of each of the three bodies is resolved in the respective principal axes frame. For the outer gimbal this axes frame is the base called $\underline{e}$ in Fig. 4.8, and for the other two bodies the base called $\underline{e}^{(1)}$ can be used. The two bases are related by the equation

$$\underline{e}^{(1)}=\begin{bmatrix}\cos\phi_2 & 0 & -\sin\phi_2\\ 0 & 1 & 0\\ \sin\phi_2 & 0 & \cos\phi_2\end{bmatrix}\underline{e}=\underline{A}\,\underline{e}\,.$$

The coordinate matrices of the absolute angular velocities are

$$\begin{aligned}
\underline{\omega}^{(1)}&=[\dot{\phi}_1\ \ 0\ \ 0]^{\mathrm{T}} && \text{outer gimbal in } \underline{e}\\
\underline{\omega}^{(2)}&=\underline{A}\,\underline{\omega}^{(1)}+[0\ \ \dot{\phi}_2\ \ 0]^{\mathrm{T}}\\
&=[\dot{\phi}_1\cos\phi_2\ \ \dot{\phi}_2\ \ \dot{\phi}_1\sin\phi_2]^{\mathrm{T}} && \text{inner gimbal in } \underline{e}^{(1)}\\
\underline{\omega}^{(r)}&=\underline{\omega}^{(2)}+[0\ \ 0\ \ \dot{\phi}_3]^{\mathrm{T}}\\
&=[\dot{\phi}_1\cos\phi_2\ \ \dot{\phi}_2\ \ \dot{\phi}_1\sin\phi_2+\dot{\phi}_3]^{\mathrm{T}} && \text{rotor in } \underline{e}^{(1)}.
\end{aligned}$$

For the rotor the third coordinate $\omega_3^{(r)}=\dot{\phi}_1\sin\phi_2+\dot{\phi}_3$ is a constant as was stated earlier. Using Eq. (4.51) this can be written in the form $\dot\psi\cos\theta+\dot\phi=\omega_3=\text{const}$ which is identical with Eq. (4.31). The angular momentum integral can be formulated as follows. Let J_3^1 be the moment of inertia of the outer gimbal about its vertical axis. The principal moments of inertia in the base $\underline{e}^{(1)}$ are denoted J_1^2, J_2^2, J_3^2 for the inner gimbal and J_1, $J_2=J_1$, J_3 for the rotor plus point mass. The absolute angular momenta of the individual bodies then have the coordinate matrices

$$\underline{L}^{(1)}=[J_1^1\dot\phi_1 \quad 0 \quad 0]^{\mathrm{T}} \qquad \text{outer gimbal in } \underline{e}$$
$$\underline{L}^{(2)}=[J_1^2\dot\phi_1\cos\phi_2 \quad J_2^2\dot\phi_2 \quad J_3^2\dot\phi_1\sin\phi_2]^{\mathrm{T}} \qquad \text{inner gimbal in } \underline{e}^{(1)}$$
$$\underline{L}^{(r)}=[J_1\dot\phi_1\cos\phi_2 \quad J_1\dot\phi_2 \quad J_3\omega_3]^{\mathrm{T}} \qquad \text{rotor in } \underline{e}^{(1)}.$$

The coordinate L of the total angular momentum in the vertical direction is the first element of the column matrix $\underline{L}^{(1)}+\underline{A}^{\mathrm{T}}(\underline{L}^{(2)}+\underline{L}^{(r)})$. A simple rearrangement of terms yields the expression

$$(J_1^1+J_1^2+J_1)\dot\phi_1\cos^2\phi_2+J_3\omega_3\sin\phi_2+(J_1^1+J_3^2)\dot\phi_1\sin^2\phi_2=L$$

or, in view of Eq. (4.51),

$$(J_1^1+J_1^2+J_1)\dot\psi\sin^2\theta+J_3\omega_3\cos\theta+(J_1^1+J_3^2)\dot\psi\cos^2\theta=L. \tag{4.52}$$

The total kinetic energy of the system is calculated from $\underline{\omega}^{(1)}, \underline{\omega}^{(2)}$ and $\underline{\omega}^{(r)}$. When the potential energy of the point mass is added the energy equation becomes

$$(J_1^1+J_1^2+J_1)\dot\phi_1^2\cos^2\phi_2+(J_2^2+J_1)\dot\phi_2^2+2mgs\sin\phi_2+(J_1^1+J_3^2)\dot\phi_1^2\sin^2\phi_2$$
$$=2E-J_3\omega_3^2$$

or, in view of Eq. (4.51)

$$(J_1^1+J_1^2+J_1)\dot\psi^2\sin^2\theta+(J_2^2+J_1)\dot\theta^2+2mgs\cos\theta+(J_1^1+J_3^2)\dot\psi^2\cos^2\theta$$
$$=2E-J_3\omega_3^2. \tag{4.53}$$

Eqs. (4.52) and (4.53) correspond to Eqs. (4.36) and (4.37), respectively, for the symmetric heavy top. They become identical with these equations if all gimbal moments of inertia are set equal to zero. The general solution of the problem is attempted by using the same approach as for the symmetric heavy top. Eq. (4.52) is solved for $\dot\psi$

$$\dot\psi=\frac{L-J_3\omega_3\cos\theta}{(J_1^1+J_1^2+J_1)\sin^2\theta+(J_1^1+J_3^2)\cos^2\theta}.$$

Substitution into Eq. (4.53) yields for θ the differential equation

$$(J_2^2+J_1)\dot\theta^2=2E-J_3\omega_3^2-2mgs\cos\theta-\frac{(L-J_3\omega_3\cos\theta)^2}{(J_1^1+J_1^2+J_1)\sin^2\theta+(J_1^1+J_3^2)\cos^2\theta}$$

This corresponds to Eq. (4.41). As before, θ is substituted by the new variable $u=\cos\theta$. This results in the differential equation for u

$$\dot u^2=\frac{(2E-J_3\omega_3^2-2mgsu)(1-u^2)}{J_2^2+J_1}-\frac{(L-J_3\omega_3u)^2(1-u^2)}{(J_2^2+J_1)[(J_1^1+J_1^2+J_1)(1-u^2)+(J_1^1+J_3^2)u^2]}.$$

When all gimbal moments of inertia are zero this is identical with the differential equation for u which was obtained for the symmetric heavy top. The gimbal inertia has the effect that the right hand side expression is a rational algebraic fraction of u instead of a cubic polynomial. The solution has, therefore, a completely different character. It cannot be expressed in terms of known special functions. It is interesting to note that the equation is not even soluable in the special case where no point mass and, thus, no torque caused by gravity is present. For more details about the dynamic effects of suspension systems the reader is referred to Magnus [6], Arnold/Maunder [9] and Saidov [10].

4.6 The gyrostat. General considerations

A gyrostat is a mechanical system which is composed of more than one body and yet has the rigid body property that its inertia components are time independent constants. In its simplest form a gyrostat consists of two bodies as shown in Fig. 4.9. A symmetric rigid rotor is supported in rigid bearings on another rigid body called the carrier.

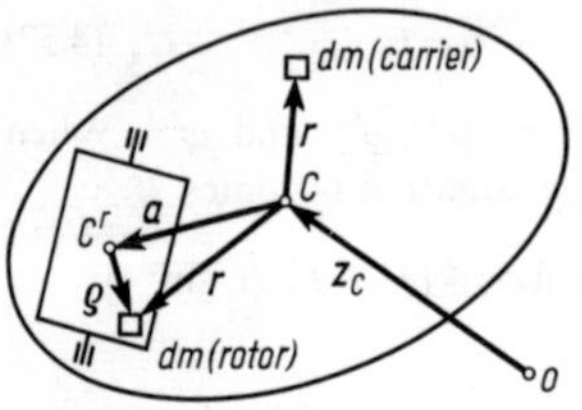

Fig. 4.9
Radius vectors of mass particles on the carrier and rotor of a gyrostat

We shall, first, establish equations of motion for this particular system. Later it will be simple to formulate equations for gyrostats which consist of more than two bodies. About the central principal moments of inertia of the composite system and about the orientation of the rotor axis relative to the principal axes of inertia no special assumptions are made. Let $\boldsymbol{\omega}$ denote the absolute angular velocity of the carrier and $\boldsymbol{\omega}_{\text{rel}}$ the angular velocity of the rotor relative to the carrier. The absolute angular momentum of the total system with respect to a point O fixed in inertial space is (cf. Eqs. (3.5) and (3.6) and Fig. 3.1)

$$\boldsymbol{L}^O = \int_{m_c} (\boldsymbol{z}_C+\boldsymbol{r})\times(\boldsymbol{v}_C+\boldsymbol{\omega}\times\boldsymbol{r})\mathrm{d}m + \int_{m_r} (\boldsymbol{z}_C+\boldsymbol{r})\times(\boldsymbol{v}_C+\boldsymbol{\omega}\times\boldsymbol{r}+\boldsymbol{\omega}_{\text{rel}}\times\boldsymbol{\varrho})\mathrm{d}m.$$

The first integral extends over the mass of the carrier and the second over the mass of the rotor. The vectors $\boldsymbol{z}_C$, $\boldsymbol{r}$ and $\boldsymbol{\varrho}$ are explained in Fig. 4.9. The points C and C^r designate the composite system center of mass and the rotor center of mass, respectively. The symbol $\boldsymbol{v}_C$ denotes the absolute velocity of C. Under the second integral, $\boldsymbol{r}$ is expressed as $\boldsymbol{a}+\boldsymbol{\varrho}$. Then,

$$\begin{aligned}\boldsymbol{L}^O &= \int_{m_c+m_r} (\boldsymbol{z}_C+\boldsymbol{r})\times(\boldsymbol{v}_C+\boldsymbol{\omega}\times\boldsymbol{r})\mathrm{d}m + \int_{m_r} (\boldsymbol{z}_C+\boldsymbol{a}+\boldsymbol{\varrho})\times(\boldsymbol{\omega}_{\text{rel}}\times\boldsymbol{\varrho})\mathrm{d}m\\ &= \boldsymbol{z}_C\times\boldsymbol{v}_C m + \int_{m_c+m_r} \boldsymbol{r}\times(\boldsymbol{\omega}\times\boldsymbol{r})\mathrm{d}m + \int_{m_r} \boldsymbol{\varrho}\times(\boldsymbol{\omega}_{\text{rel}}\times\boldsymbol{\varrho})\mathrm{d}m\\ &= \boldsymbol{z}_C\times\boldsymbol{v}_C m+\boldsymbol{J}\cdot\boldsymbol{\omega}+\boldsymbol{J}^r\cdot\boldsymbol{\omega}_{\text{rel}}\,.\end{aligned}$$

In this expression m is the total system mass, $\boldsymbol{J}$ the inertia tensor of the total system with respect to C and $\boldsymbol{J}^r$ the rotor inertia tensor with respect to C^r. Both inertia tensors have constant coordinates in a vector base fixed on the carrier. The vector $\boldsymbol{J}^r \cdot \boldsymbol{\omega}_{\text{rel}}$ is the rotor angular momentum relative to the carrier. It is abbreviated $\boldsymbol{h}$. Since the rotor axis is a principal axis of inertia of the rotor $\boldsymbol{h}$ has the direction of this axis. With $\boldsymbol{L}^O$ in the form

$$\boldsymbol{L}^O = \boldsymbol{z}_C \times \boldsymbol{v}_C m + \boldsymbol{J} \cdot \boldsymbol{\omega} + \boldsymbol{h}$$

the law of moment of momentum (Eq. (3.15)) yields the equation of motion

$$\boldsymbol{J} \cdot \dot{\boldsymbol{\omega}} + \mathring{\boldsymbol{h}} + \boldsymbol{\omega} \times (\boldsymbol{J} \cdot \boldsymbol{\omega} + \boldsymbol{h}) = \boldsymbol{M} \tag{4.54}$$

which represents a generalization of Euler's equation for the rigid body. The symbol $\mathring{\boldsymbol{h}}$ is an abbreviation for the time derivative of $\boldsymbol{h}$ in a base fixed on the carrier, and $\boldsymbol{M}$ is the resultant external torque with respect to C.

Of considerable technical importance is the special case where the angular velocity of the rotor relative to the carrier is a prescribed function of time. The coordinates of $\boldsymbol{h}(t)$ and of $\mathring{\boldsymbol{h}}(t)$ in a carrier-fixed base are then known functions of time. The rotor has no degree of freedom of motion of its own, and Eq. (4.54) together with kinematic differential equations for the carrier fully describes the system. The term $-\mathring{\boldsymbol{h}}(t)$ can be treated as if it were an external torque:

$$\boldsymbol{J} \cdot \dot{\boldsymbol{\omega}} + \boldsymbol{\omega} \times (\boldsymbol{J} \cdot \boldsymbol{\omega} + \boldsymbol{h}(t)) = \boldsymbol{M} - \mathring{\boldsymbol{h}}(t)\,. \tag{4.55}$$

In the simplest case of this kind the relative angular velocity of the rotor is kept constant. The equation then reads

$$\boldsymbol{J} \cdot \dot{\boldsymbol{\omega}} + \boldsymbol{\omega} \times (\boldsymbol{J} \cdot \boldsymbol{\omega} + \boldsymbol{h}) = \boldsymbol{M}\,.$$

In another case of technical importance the component along the rotor axis of the torque which is exerted on the rotor by the carrier is a prescribed function of time. In this case the rotor has one degree of freedom of motion of its own so that one additional scalar equation of motion is needed. This equation is established as follows. The law of moment of momentum for the rotor alone reads

$$\frac{\mathrm{d}}{\mathrm{d}t}[\boldsymbol{J}^r \cdot (\boldsymbol{\omega} + \boldsymbol{\omega}_{\text{rel}})] = \boldsymbol{M}^r$$

with $\boldsymbol{M}^r$ being the resultant torque on the rotor with respect to the center of mass C^r. Carrying out the differentiation in a carrier fixed base we get

$$\boldsymbol{J}^r \cdot \dot{\boldsymbol{\omega}} + \mathring{\boldsymbol{h}} + \boldsymbol{\omega} \times (\boldsymbol{J}^r \cdot \boldsymbol{\omega} + \boldsymbol{h}) = \boldsymbol{M}^r\,. \tag{4.56}$$

It is reasonable to assume that the external torque $\boldsymbol{M}$ which acts on the gyrostat as a whole does not contribute to the component of $\boldsymbol{M}^r$ along the rotor axis. This component is then caused by interaction from the carrier alone and by assumption it is a given function of time. Let it be called $M^r(t)$. If $\boldsymbol{u}$ is a unit vector along the rotor axis the desired scalar equation of motion is obtained by scalar multiplication of Eq. (4.56) by $\boldsymbol{u}$:

$$\boldsymbol{u} \cdot (\boldsymbol{J}^r \cdot \dot{\boldsymbol{\omega}} + \mathring{\boldsymbol{h}}) = \boldsymbol{u} \cdot \boldsymbol{M}^r = M^r(t)\,. \tag{4.57}$$

Note that the third term on the left in Eq. (4.56) does not give a contribution since $\boldsymbol{u}$ and $\boldsymbol{h}$ are parallel and since for a symmetric body the vectors $\boldsymbol{u}$, $\boldsymbol{\omega}$ and $\boldsymbol{J}^r\cdot\boldsymbol{\omega}$ are coplanar. The equation can be integrated once. For this purpose it is given the form

$$\frac{\mathrm{d}}{\mathrm{d}t}\left[\boldsymbol{u}\cdot(\boldsymbol{J}^r\cdot\boldsymbol{\omega}+\boldsymbol{h})\right]=M^r(t)$$

which is identical since the product $(\boldsymbol{\omega}\times\boldsymbol{u})\cdot(\boldsymbol{J}^r\cdot\boldsymbol{\omega}+\boldsymbol{h})$ is zero. Integration then yields

$$\boldsymbol{u}\cdot(\boldsymbol{J}^r\cdot\boldsymbol{\omega}+\boldsymbol{h}) = \int M^r(t)\mathrm{d}t = L^r(t) \tag{4.58}$$

where the known function $L^r(t)$ represents the coordinate along the rotor axis of the absolute rotor angular momentum. In order to describe the motion completely it is necessary to supplement Eqs. (4.54) and (4.58) by kinematic differential equations which relate $\boldsymbol{\omega}$ to generalized coordinates for the angular orientation of the carrier. Having described the simplest possible type of gyrostat it is now a straightforward procedure to formulate equations of motion for gyrostats in which more than one rotor is mounted on the carrier. Let there be $m+n$ rotors altogether and let them be identified by an index i which is attached to all rotor quantities. It is assumed that for the rotors labeled $i=1\dots m$ the axial torque component $M_i^r(t)$ is a known function of time whereas for the remaining rotors $i=m+1\dots m+n$ the coordinates of $\boldsymbol{h}_i(t)$ in a carrier-fixed base are given functions of time. The equations of motion consist of one vector equation of the kind of Eq. (4.54) and of a set of m scalar equations of the form of Eq. (4.58), one for each of the rotors number $1\dots m$:

$$\boldsymbol{J}\cdot\dot{\boldsymbol{\omega}}+\sum_{i=1}^{m}\mathring{\boldsymbol{h}}_i+\boldsymbol{\omega}\times\left[\boldsymbol{J}\cdot\boldsymbol{\omega}+\sum_{i=1}^{m}\boldsymbol{h}_i+\sum_{i=m+1}^{m+n}\boldsymbol{h}_i(t)\right]=\boldsymbol{M}-\sum_{i=m+1}^{m+n}\mathring{\boldsymbol{h}}_i(t) \tag{4.59}$$

$$\boldsymbol{u}_i\cdot(\boldsymbol{J}_i^r\cdot\boldsymbol{\omega}+\boldsymbol{h}_i) = \int M_i^r(t)\mathrm{d}t=L_i^r(t) \qquad i=1\dots m. \tag{4.60}$$

The scalar equations are equivalent to

$$\boldsymbol{u}_i\cdot(\boldsymbol{J}_i^r\cdot\dot{\boldsymbol{\omega}}+\mathring{\boldsymbol{h}}_i)=M_i^r(t) \qquad i=1\dots m. \tag{4.61}$$

This set of equations can be simplified substantially. To begin with, Eqs. (4.60) and (4.61) are reformulated. Since $\boldsymbol{u}_i$ is an eigenvector of $\boldsymbol{J}_i^r$ the product $\boldsymbol{u}_i\cdot\boldsymbol{J}_i^r\cdot\boldsymbol{\omega}$ can be rewritten in the form $J_i^r\boldsymbol{u}_i\cdot\boldsymbol{\omega}$ where J_i^r is the principal moment of inertia of the i-th rotor about the rotor axis. Furthermore, $\boldsymbol{u}_i\cdot\mathring{\boldsymbol{h}}_i=|\mathring{\boldsymbol{h}}_i|$. Multiplication of Eq. (4.61) by $\boldsymbol{u}_i$ yields, therefore, the vector equation

$$J_i^r\boldsymbol{u}_i\boldsymbol{u}_i\cdot\dot{\boldsymbol{\omega}}+\mathring{\boldsymbol{h}}_i=\boldsymbol{u}_iM_i^r(t) \qquad i=1\dots m \tag{4.62}$$

in which $J_i^r\boldsymbol{u}_i\boldsymbol{u}_i$ is an inertia tensor. In a similar manner multiplication of Eq. (4.60) by $\boldsymbol{\omega}\times\boldsymbol{u}_i$ produces the vector equation

$$\boldsymbol{\omega}\times(J_i^r\boldsymbol{u}_i\boldsymbol{u}_i)\cdot\boldsymbol{\omega}+\boldsymbol{\omega}\times\boldsymbol{h}_i=\boldsymbol{\omega}\times\boldsymbol{u}_iL_i^r(t) \qquad i=1\dots m. \tag{4.63}$$

Eqs. (4.62) and (4.63) are summed over $i=1\dots m$, and both sums are subtracted from Eq. (4.59):

$$\left(\boldsymbol{J}-\sum_{i=1}^{m} \boldsymbol{J}_i^r \boldsymbol{u}_i \boldsymbol{u}_i\right)\cdot\dot{\boldsymbol{\omega}}+\boldsymbol{\omega}\times\left[\left(\boldsymbol{J}-\sum_{i=1}^{m} J_i^r \boldsymbol{u}_i \boldsymbol{u}_i\right)\cdot\boldsymbol{\omega}+\sum_{i=1}^{m} \boldsymbol{u}_i L_i^r(t)+\sum_{i=m+1}^{m+n} \boldsymbol{h}_i(t)\right]$$
$$=\boldsymbol{M}-\sum_{i=1}^{m} \boldsymbol{u}_i M_i^r(t)-\sum_{i=m+1}^{m+n} \mathring{\boldsymbol{h}}_i(t). \tag{4.64}$$

Next, the new quantities

$$\boldsymbol{J}^* = \boldsymbol{J}-\sum_{i=1}^{m} J_i^r \boldsymbol{u}_i \boldsymbol{u}_i, \qquad \boldsymbol{h}^*(t)=\sum_{i=1}^{m} \boldsymbol{u}_i L_i^r(t)+\sum_{i=m+1}^{m+n} \boldsymbol{h}_i(t)$$

are introduced. Because of the identity $M_i^r(t)=\mathrm{d}L_i^r(t)/\mathrm{d}t$ one has

$$\mathring{\boldsymbol{h}}^*(t)=\sum_{i=1}^{m} \boldsymbol{u}_i M_i^r(t)+\sum_{i=m+1}^{m+n} \mathring{\boldsymbol{h}}_i(t).$$

In a carrier-fixed base the tensor $\boldsymbol{J}^*$ has constant coordinates, and $\boldsymbol{h}^*(t)$ as well as $\mathring{\boldsymbol{h}}^*(t)$ have coordinates which are known functions of time. In terms of these quantities Eq. (4.64) becomes

$$\boldsymbol{J}^*\cdot\dot{\boldsymbol{\omega}}+\boldsymbol{\omega}\times[\boldsymbol{J}^*\cdot\boldsymbol{\omega}+\boldsymbol{h}^*(t)]=\boldsymbol{M}-\mathring{\boldsymbol{h}}^*(t). \tag{4.65}$$

This equation, together with kinematic differential equations for the carrier, fully describes the motion of the carrier. The equation has the same form as Eq. (4.55) for a gyrostat with a single rotor and with given functions $\boldsymbol{h}(t)$ and $\mathring{\boldsymbol{h}}(t)$. Note, however, that in contrast to $\boldsymbol{h}(t)$ in Eq. (4.55) the vector $\boldsymbol{h}^*(t)$ in Eq. (4.65) does not have, in general, a fixed direction in the carrier. The two equations are, therefore, not equivalent. They are equivalent only in special cases such as the following:

(i) For arbitrary m and n all rotor axes are aligned parallel.

(ii) $m=1$ and $n=0$. Thus, a gyrostat with a single rotor with given torque $M^r(t)$ and a gyrostat with a single rotor with given relative angular momentum $\boldsymbol{h}(t)$ are equivalent.

(iii) $m=0$, n arbitrary and $|\boldsymbol{h}_i(t)|=\lambda_i h(t)$ for $i=1\ldots n$ with constants $\lambda_1\ldots\lambda_n$ and with an arbitrary function $h(t)$. These conditions are fulfilled if all rotors are connected by gear wheels.

(iv) m and n arbitrary, $M_i^r(t)\equiv 0$ for $i=1\ldots m$ and $|\boldsymbol{h}_i(t)|=\text{const}$ for $i=m+1\ldots m+n$.

Carrier bodies with built-in rotors are not the only multi-body systems with time independent inertia components. This property is conserved if the carrier, in addition to rotors, contains cavities which are completely filled with homogeneous fluids. Various technical instruments and vehicles with rotating engines, with fuel tanks and with hydraulic systems represent gyrostats of this nature. Their dynamic behavior has been studied by Moiseev/Rumjancev [11].

Problem

4.3 In the absence of external torques Eq. (4.65) represents the equation of a self-excited rigid body with the driving torque $-\boldsymbol{\omega}\times\boldsymbol{h}^*(t)-\mathring{\boldsymbol{h}}^*(t)$. Integrate the equation in closed form in the following case. A gyrostat with one rotor has principal moments of inertia $J_1=J_2\neq J_3$. The axis of the rotor whose moment of inertia about the symmetry axis is J^r is mounted parallel to the principal axis about which J_3 is measured. The carrier transmits a torque $M^r(t)$ to the rotor about the rotor axis. Solve the equation also if the rotor torque has the form $M^r(t)-ah$ ($a>0$, const) where $M^r(t)$ is, again, a given function of time and the term $-ah$ represents a viscous damping torque.

4.7 The torque-free gyrostat

In this section Eq. (4.65) is investigated for the special case where the external torque $\boldsymbol{M}$ on the gyrostat is identically zero and where, furthermore, the relative angular momentum $\boldsymbol{h}^*$ has constant coordinates in a carrier-fixed base. Omitting the asterisk we have the equation

$$\boldsymbol{J}\cdot\dot{\boldsymbol{\omega}}+\boldsymbol{\omega}\times(\boldsymbol{J}\cdot\boldsymbol{\omega}+\boldsymbol{h})=\boldsymbol{0}. \tag{4.66}$$

It governs torque-free gyrostats whose rotors are operated under the condition $M_i^r(t)\equiv 0$ for $i=1\ldots m$ (arbitrary) and $|\boldsymbol{h}_i(t)|=\text{const}$ for $i=m+1\ldots m+n$ (n arbitrary). For the sake of simplicity the equation will always be interpreted as that of a gyrostat with a *single rotor of constant relative angular momentum* $\boldsymbol{h}$.

Eq. (4.66) possesses two algebraic integrals which are found by multiplication with $\boldsymbol{\omega}$ and with $\boldsymbol{J}\cdot\boldsymbol{\omega}+\boldsymbol{h}$, respectively:

$$\boldsymbol{\omega}\cdot\boldsymbol{J}\cdot\boldsymbol{\omega}=2T=\text{const} \qquad (\boldsymbol{J}\cdot\boldsymbol{\omega}+\boldsymbol{h})^2=L^2=\text{const}. \tag{4.67}$$

They are an energy and an angular momentum integral. Note that T is not the total kinetic energy but only that portion of it which remains when the carrier is rotating with its actual angular velocity $\boldsymbol{\omega}$ while the rotor is "frozen" in the carrier. The quantity L is the magnitude of the absolute angular momentum of the composite system. Resolved in the principal axes frame of the composite system Eqs. (4.66) and (4.67) read

$$\begin{aligned}
&J_1\dot{\omega}_1-(J_2-J_3)\omega_2\omega_3+\omega_2h_3-\omega_3h_2=0\\
&J_2\dot{\omega}_2-(J_3-J_1)\omega_3\omega_1+\omega_3h_1-\omega_1h_3=0\\
&J_3\dot{\omega}_3-(J_1-J_2)\omega_1\omega_2+\omega_1h_2-\omega_2h_1=0
\end{aligned} \tag{4.68}$$

$$\sum_{\alpha=1}^{3} J_\alpha\omega_\alpha^2 = 2T \tag{4.69}$$

$$\sum_{\alpha=1}^{3} (J_\alpha\omega_\alpha+h_\alpha)^2 = L^2=2DT. \tag{4.70}$$

The parameter D with the physical dimension of moment of inertia is introduced for convenience and also in order to show the similarity with Eq. (4.4) for the torque-free rigid body.

The following investigation will be concerned only with the most general case in which all three principal moments of inertia are different from one another and, furthermore, all three angular momentum coordinates h_1, h_2 and h_3 are different from zero[1].

---

[1] For the special cases not treated here and also for additional details about the general case see Wittenburg [12].

Without loss of generality it is assumed that $J_3 < J_2 < J_1$. The same assumption was made in Chap. 4.1 on the dynamics of the torque-free rigid body.

4.7.1 Polhodes and permanent rotations

Eqs. (4.69) and (4.70) are analogous to the integrals of motion for a torque-free rigid body (Eqs. (4.3) and (4.4)) in that they define two ellipsoids which are fixed on the carrier and which, both, represent a geometric locus for the vector $\boldsymbol{\omega}$. The angular velocity is, therefore, confined to the lines of intersection of the ellipsoids. As for the rigid body these lines are called polhodes. Their investigation is considerably more complicated than in the case of a rigid body since the centers of the ellipsoids are not coincident. This has the consequence, for instance, that projections of the polhodes along principal axes are not ellipses or hyperbolas but fourth-order curves. As in Sec. 4.1.1 it is imagined that the energy ellipsoid is given and that the angular momentum ellipsoid is "blown up" by increasing D. In this process the family of all physically realizable polhodes is produced on the energy ellipsoid. Of particular interest are, again, those degenerate polhodes which have singular points. These points mark permanent rotations with constant angular velocities $\boldsymbol{\omega} \equiv \boldsymbol{\omega}^* = \text{const}$. To each singular point belongs a particular value D^* of D which determines the corresponding angular momentum ellipsoid. The location of the singular points on the energy ellipsoid as well as relationships between $\boldsymbol{\omega}^*$, D^* and the system parameters J_α, h_α $(\alpha = 1,2,3)$ und $2T$ can be found from the differential equations and from the integrals of motion. According to Eq. (4.66) solutions $\boldsymbol{\omega} \equiv \boldsymbol{\omega}^* = \text{const}$ require that either $\boldsymbol{\omega}^* = \mathbf{0}$ or $\boldsymbol{J} \cdot \boldsymbol{\omega}^* + \boldsymbol{h} = \mathbf{0}$ or $\boldsymbol{J} \cdot \boldsymbol{\omega}^* + \boldsymbol{h} = \lambda^* \boldsymbol{\omega}^*$ with an as yet unknown scalar λ^*. The first two conditions represent trivial cases. In the first case the energy ellipsoid degenerates into a single point. Only the rotor is moving, not the carrier. In the second case the angular momentum ellipsoid degenerates into a single point. The third condition implies that the ellipsoids are not degenerate and that they have a common tangential plane in the singular point which is marked by $\boldsymbol{\omega}^*$. The condition does not yield an eigenvalue problem as in the corresponding case for the rigid body (cf. Eq. (4.7)). Let u_1, u_2 and u_3 be the coordinates of the unit vector $\boldsymbol{u}$ along the rotor axis so that $h_\alpha = h u_\alpha$ $(\alpha = 1,2,3)$. The coordinate form of the third condition then yields

$$\omega_\alpha^* = \frac{h u_\alpha}{\lambda^* - J_\alpha} \qquad \alpha = 1,2,3\,. \tag{4.71}$$

The unknown λ^* is determined from the equation

$$\sum_{\alpha=1}^{3} \frac{u_\alpha^2 J_\alpha}{(\lambda^* - J_\alpha)^2} = \frac{2T}{h^2} = \frac{1}{J_0} \tag{4.72}$$

which is found by substituting ω_α^* into Eq. (4.69). Through the equation for λ^* the parameter J_0 is introduced. It has the physical dimension of moment of inertia. Eq. (4.72) represents a sixth-order equation for λ^*. To every real solution there belongs one permanent rotation. The value D^* of D which defines the corresponding angular momentum integral follows from Eq. (4.70):

$$D^* = J_0 \lambda^{*2} \sum_{\alpha=1}^{3} \frac{u_\alpha^2}{(\lambda^* - J_\alpha)^2}. \tag{4.73}$$

In order to be able to make statements about the number of axes of permanent rotation the function

$$F(\lambda) = \sum_{\alpha=1}^{3} \frac{u_\alpha^2 J_\alpha}{(\lambda - J_\alpha)^2}$$

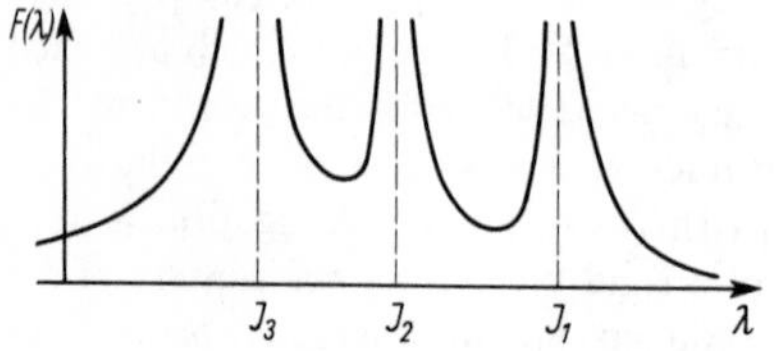

Fig. 4.10 The function $F(\lambda)$

is considered (see Fig. 4.10). It has second-order poles at $\lambda = J_\alpha\,(\alpha = 1, 2, 3)$. Furthermore, it is positive everywhere and it has exactly one minimum in each of the intervals $J_3 < \lambda < J_2$ and $J_2 < \lambda < J_1$ since $\mathrm{d}^2 F(\lambda)/\mathrm{d}\lambda > 0$. The location of these minima is found from the condition $\mathrm{d}F/\mathrm{d}\lambda = 0$, i.e. from

$$\sum_{\alpha=1}^{3} \frac{u_\alpha^2 J_\alpha}{(\lambda - J_\alpha)^3} = 0. \tag{4.74}$$

This represents a sixth-order equation for λ which has exactly two real solutions. From these properties of the function $F(\lambda)$ the following conclusions can be drawn. Eq. (4.72) has either six or four or two real solutions λ^* depending on the choice of system parameters. Double roots which are then also roots of Eq. (4.74) can occur. To each real solution corresponds a permanent angular velocity with coordinates calculated from Eq. (4.71). None of these coordinates can be zero so that axes of permanent rotation are neither parallel to principal axes of inertia nor parallel to planes spanned by two principal axes of inertia. Any other axis can—for given moments of inertia J_1, J_2 and J_3—be made an axis of permanent rotation by a proper choice of h_1, h_2 and h_3. For a given set of parameters J_α, u_α $(\alpha = 1, 2, 3)$ and J_0 no two angular velocity vectors $\boldsymbol{\omega}^*$ have opposite directions. It is interesting to look at two degenerate cases. For $h = 0$ the gyrostat is a rigid body. It should then have six axes of permanent rotation which coincide with the principal axes of inertia (three if permanent rotations of opposite directions are not counted separately). In the limit $h \to 0$, i.e. for $J_0 \to 0$ the roots of Eq. (4.72) tend toward double roots $\lambda^* = J_\alpha\,(\alpha = 1, 2, 3)$. For each root two out of three coordinates of $\boldsymbol{\omega}^*$ are zero according to Eq. (4.71). This means that all six permanent angular velocity vectors have, indeed, directions of principal axes of inertia. In the second degenerate case h is infinitely large while $2T$ remains finite. This is the case of a gyrostat with a slowly rotating carrier and with an infinitely rapidly spinning rotor. The rotor then dominates the dynamic behavior of the gyrostat. It should, therefore, be expected that only two axes of permanent rotation exist and that these axes coincide with the symmetry axis of the rotor. This is, indeed, the case. In the limit $J_0 \to \infty$ Eq. (4.72) has only two real solutions, namely $\lambda^* \to \pm\infty$. For the corresponding permanent angular velocity coordinates the relationship $\omega_1^* : \omega_2^* : \omega_3^* = u_1 : u_2 : u_3$ is found from Eq. (4.71) which means that $\boldsymbol{\omega}^*$ is parallel to the rotor axis.

4.7.2 The solution of the dynamic equations of motion

In this subsection closed-form solutions will be obtained for the differential equations (4.68) by a method which was suggested by Wangerin [13] and further developed by Wittenburg [12]. The approach leads to real functions of time $\omega_1(t)$, $\omega_2(t)$ and $\omega_3(t)$. Another method developed by Volterra [14] (see also [12]) results in complex functions of time.

We start out from the integrals of motion. Multiply Eq. (4.69) by an as yet undetermined scalar λ with the dimension of moment of inertia and subtract Eq. (4.70). The result can be written in the form

$$\sum_{\alpha=1}^{3} J_\alpha(\lambda - J_\alpha)\left(\omega_\alpha - \frac{hu_\alpha}{\lambda - J_\alpha}\right)^2 = 2T[f(\lambda) - D] \tag{4.75}$$

with $$f(\lambda) = \lambda\left[1 + J_0 \sum_{\alpha=1}^{3} \frac{u_\alpha^2}{\lambda - J_\alpha}\right]. \tag{4.76}$$

The function $f(\lambda)$ has first-order poles at $\lambda = J_\alpha$ $(\alpha = 1,2,3)$. From this follows that in each of the intervals $J_3 < \lambda < J_2$ and $J_2 < \lambda < J_1$ all values from $-\infty$ to $+\infty$ are assumed at least once. We choose for λ a value λ_0 for which

$$f(\lambda_0) = D \quad \text{and} \quad J_3 < \lambda_0 < J_2. \tag{4.77}$$

If several values exist which satisfy these conditions one of them is chosen arbitrarily. The equation $f(\lambda_0) = D$ represents a fourth-order equation. When new variables w_α defined as

$$w_\alpha = \omega_\alpha - \frac{hu_\alpha}{\lambda_0 - J_\alpha} \qquad \alpha = 1,2,3 \tag{4.78}$$

are introduced Eq. (4.75) becomes

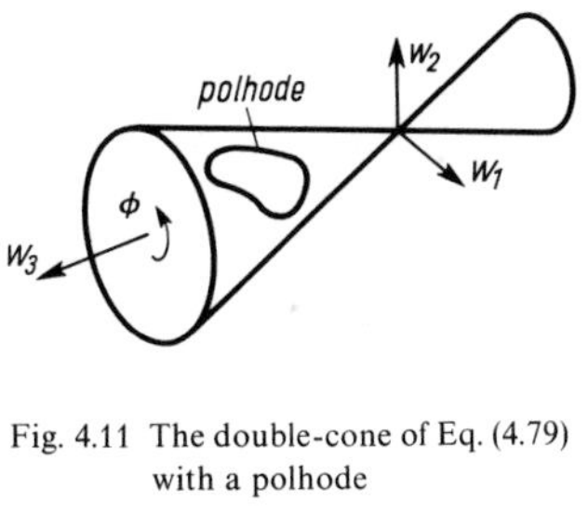

Fig. 4.11 The double-cone of Eq. (4.79) with a polhode

$$\left(\frac{w_1}{k_1 w_3}\right)^2 + \left(\frac{w_2}{k_2 w_3}\right)^2 = 1 \tag{4.79}$$

with real coefficients

$$k_1 = \sqrt{\frac{J_3(\lambda_0 - J_3)}{J_1(J_1 - \lambda_0)}}, \quad k_2 = \sqrt{\frac{J_3(\lambda_0 - J_3)}{J_2(J_2 - \lambda_0)}}. \tag{4.80}$$

This equation defines a double-cone of elliptic cross section whose axis is the w_3-axis (see Fig. 4.11). This cone, too, is a geometric locus of the polhodes since its equation is a linear combination of the equations which define the ellipsoids. Eq. (4.79) can be replaced by the parameter presentation for the cone

$$w_1 = k_1 w_3 \sin\phi, \qquad w_2 = k_2 w_3 \cos\phi. \tag{4.81}$$

Through these equations in combination with Eq. (4.78) also ω_1 and ω_2 are functions of w_3 and ϕ. Substituting the expressions for ω_1 and ω_2 into the energy equation

(4.69) one finds for w_3 the quadratic equation

$$a_1(\phi)w_3^2 - 2a_2(\phi)w_3 + a_3 = 0$$

with the coefficients

$$a_1(\phi) = J_1 k_1^2 \sin^2\phi + J_2 k_2^2 \cos^2\phi + J_3 > 0 \tag{4.82}$$

$$a_2(\phi) = \frac{h_1 J_1 k_1}{J_1 - \lambda_0} \sin\phi + \frac{h_2 J_2 k_2}{J_2 - \lambda_0} \cos\phi + \frac{h_3 J_3}{J_3 - \lambda_0} \tag{4.83}$$

$$a_3 \quad = 2TJ_0 \left[\sum_{\alpha=1}^{3} \frac{u_\alpha^2 J_\alpha}{(\lambda_0 - J_\alpha)^2} - \frac{1}{J_0} \right]. \tag{4.84}$$

The solution for w_3 has the form

$$w_3(\phi) = \frac{a_2(\phi) \pm \sqrt{a_2^2(\phi) - a_1(\phi)a_3}}{a_1(\phi)}. \tag{4.85}$$

This equation in combination with Eqs. (4.81) and (4.78) establishes all three angular velocity coordinates as functions of the single variable ϕ. For completing the solution it remains to show how ϕ is related to time. This can be done as follows. The time derivative of Eq. (4.69) reads

$$J_1\omega_1\dot\omega_1 + J_2\omega_2\dot\omega_2 + J_3\omega_3\dot\omega_3 = 0.$$

In it $\dot\omega_1$ and $\dot\omega_2$ are substituted by the following expressions which result from Eqs. (4.78) and (4.81)

$$\dot\omega_1 = \dot w_1 = k_1(\quad w_3\dot\phi\cos\phi + \dot w_3 \sin\phi) = k_1\left(\frac{w_2\dot\phi}{k_2} + \dot\omega_3 \sin\phi\right)$$

$$\dot\omega_2 = \dot w_2 = k_2(-w_3\dot\phi\sin\phi + \dot w_3 \cos\phi) = k_2\left(\frac{-w_1\dot\phi}{k_1} + \dot\omega_3 \cos\phi\right).$$

This yields

$$\left(\frac{k_1}{k_2} J_1\omega_1 w_2 - \frac{k_2}{k_1} J_2\omega_2 w_1\right)\dot\phi + (J_1 k_1\omega_1 \sin\phi + J_2 k_2\omega_2\cos\phi + J_3\omega_3)\dot\omega_3 = 0. \tag{4.86}$$

In the expression in front of $\dot\omega_3$ for ω_1, ω_2 and ω_3 the respective functions of w_3 and ϕ are substituted which are obtained from Eqs. (4.78) and (4.81):

$$\begin{aligned} &J_1 k_1\omega_1 \sin\phi + J_2 k_2\omega_2 \cos\phi + J_3\omega_3 \\ &\quad = w_3(J_1 k_1^2 \sin^2\phi + J_2 k_2^2 \cos^2\phi + J_3) + \\ &\quad\quad + \frac{h_1 J_1 k_1}{\lambda_0 - J_1} \sin\phi + \frac{h_2 J_2 k_2}{\lambda_0 - J_2} \cos\phi + \frac{h_3 J_3}{\lambda_0 - J_3} \\ &\quad = w_3 a_1(\phi) - a_2(\phi) = \pm\sqrt{a_2^2(\phi) - a_1(\phi)a_3}. \end{aligned} \tag{4.87}$$

In the second and third steps of this reformulation Eqs. (4.82), (4.83) and (4.85) have been used. Next, the expression in front of $\dot\phi$ in Eq. (4.86) is rewritten. For w_1 and w_2 Eq. (4.78) is substituted. With k_1 and k_2 from Eq. (4.80) this yields

$$\frac{k_1}{k_2}J_1\omega_1 w_2 - \frac{k_2}{k_1}J_2\omega_2 w_1$$
$$= \sqrt{\frac{J_1 J_2}{(J_1-\lambda_0)(J_2-\lambda_0)}}\left[-(J_1-J_2)\omega_1\omega_2+\omega_1 h_2-\omega_2 h_1\right].$$

Up to this point only the two integrals of motion have been used. Now, the third of the three differential equations of motion (4.68) is brought into play. When it is solved for $J_3\dot{\omega}_3$ the negative of the expression in square brackets in the last equation above is obtained. Hence,

$$\frac{k_1}{k_2}J_1\omega_1 w_2 - \frac{k_2}{k_1}J_2\omega_2 w_1 = -J_3\dot{\omega}_3\sqrt{\frac{J_1 J_2}{(J_1-\lambda_0)(J_2-\lambda_0)}}.$$

Substituting this together with Eq. (4.87) into Eq. (4.86) one gets, after separation of the variables, for ϕ the equation

$$\pm(t-t_0)\frac{1}{J_3}\sqrt{\frac{(J_1-\lambda_0)(J_2-\lambda_0)}{J_1 J_2}} = \int_{\phi_0}^{\phi}\frac{\mathrm{d}\bar{\phi}}{\sqrt{a_2^2(\bar{\phi})-a_1(\bar{\phi})a_3}}. \tag{4.88}$$

From Eqs. (4.82) to (4.84) it is seen that the expression under the square root has the form

$$a_2^2-a_1a_3=c_1\sin^2\bar{\phi}+c_2\cos^2\bar{\phi}+c_3+c_4\sin\bar{\phi}+c_5\cos\bar{\phi}+c_6\sin\bar{\phi}\cos\bar{\phi}$$

where $c_1\ldots c_6$ are abbreviations for constant coefficients. The integral is, therefore, an elliptic integral of the first kind. Its reduction to a normal form requires a sequence of substitutions of new variables in the course of which another fourth-order equation must be solved (the first one led to the root λ_0 of Eq. (4.77)). Additional difficulties arise from the necessity to distinguish between four different combinations of signs of certain constants in order to prevent the solution $\phi(t)$ from being a complex function of time. For details the reader is referred to Wittenburg [12]. At this place we content ourselves with the statement that the angular velocity coordinates ω_1, ω_2 and ω_3 are real elliptic functions of time.

A number of mathematical expressions in the analysis just presented are similar to other expressions which play a role in connection with permanent rotations. Examples are Eqs. (4.78) and (4.71) and Eqs. (4.84) and (4.72). This similarity suggests an investigation of the question whether the root λ_0 of Eq. (4.77) can be identical with a root λ^* of Eq. (4.72). The function $f(\lambda)$ defined in Eq. (4.76) has the derivative with respect to λ

$$f'(\lambda)=\frac{\mathrm{d}f}{\mathrm{d}\lambda}=-J_0\left[\sum_{\alpha=1}^{3}\frac{u_\alpha^2 J_\alpha}{(\lambda-J_\alpha)^2}-\frac{1}{J_0}\right].$$

Comparison with Eq. (4.72) reveals the identity

$$f'(\lambda^*)=-J_0\left[F(\lambda^*)-\frac{1}{J_0}\right]=0.$$

This relationship between $f(\lambda)$ and $F(\lambda)$ is illustrated in Fig. 4.12. The function $f(\lambda)$

is stationary for those values λ^* of λ which are roots of Eq. (4.72) and which determine the coordinates of permanent angular velocities. Furthermore, according to Eq. (4.77) $f(\lambda=\lambda^*)$ equals the parameter D^* which is associated with λ^* by Eq. (4.73). Fig. 4.12 is based on a set of parameters for which six axes of permanent rotation exist. The roots λ^* are labeled in ascending order. Related quantities are identified by the respective indices. For parameter combinations for which only four axes of permanent rotation exist the function $f(\lambda)$ has stationary values in only one of the intervals $J_3<\lambda<J_2$ and $J_2<\lambda<J_1$. It has no stationary values in any one of these intervals if only two axes of permanent rotation exist. In conclusion the following can be said. If a gyrostat is in a state of permanent rotation and if, in addition, the associated root λ^* of Eq. (4.72) lies in the interval $J_3<\lambda^*<J_2$ then this λ^* is also a root of Eq. (4.77) and it can be used as λ_0 in all subsequent equations. Under no other conditions is λ_0 a root of Eqs. (4.77) and (4.72) simultaneously.

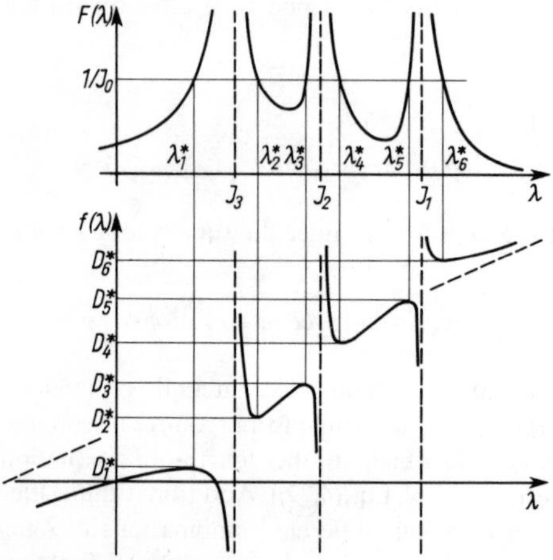

Fig. 4.12 The relationship between the functions $F(\lambda)$ and $f(\lambda)$

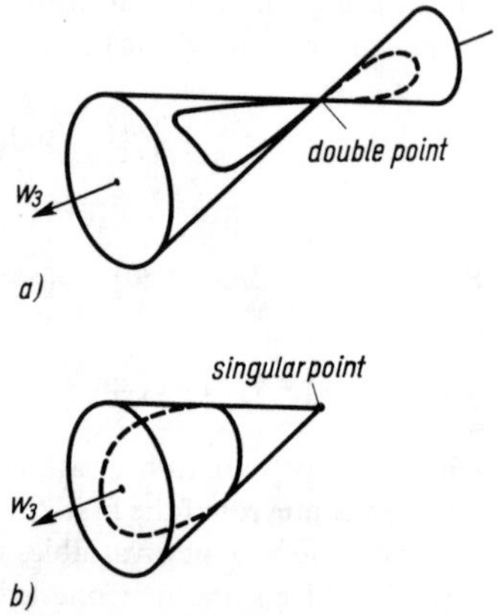

Fig. 4.13
Polhodes on the double-cone associated with a) unstable and b) stable permanent rotations

Using the identity $\lambda_0=\lambda^*$ we can develop stability criteria for permanent rotations with $J_3<\lambda^*<J_2$ (λ_2^* and λ_3^* in Fig. 4.12) on the basis of the exact solutions for the equations of motion. From Eqs. (4.84) and (4.72) follows $a_3=0$ so that the solutions for $w_3(\phi)$ given by Eq. (4.85) are $w_3(\phi)\equiv 0$ and

$$w_3(\phi) = \frac{2a_2(\phi)}{a_1(\phi)}. \tag{4.89}$$

The first solution yields, because of Eq. (4.81), $w_1=w_2=0$ and, hence, with Eq. (4.78) $\omega_\alpha=hu_\alpha/(\lambda^*-J_\alpha)$ $(\alpha=1,2,3)$. This represents the permanent rotation associated with λ^* (cf. Eq. (4.71)). From $w_1=w_2=w_3=0$ follows that the apex of the cone in Fig. 4.11 lies on the energy ellipsoid. In addition to this singular point the cone and the energy ellipsoid intersect each other in a polhode which is described by Eq. (4.89). The form of this polhode decides whether the permanent rotation is stable or unstable. If the apex is a double point of the polhode (Fig. 4.13a) then this polhode represents a separatrix, and the permanent rotation is unstable. If, on the contrary, the polhode is a closed curve which is isolated from the singular point at the apex (Fig. 4.13b) then the permanent rotation is stable. The nature of the polhode is determined as follows. With $a_3=0$ Eq. (4.88) becomes

$$c(t-t_0) = \int_{\phi_0}^{\phi} \frac{\mathrm{d}\bar{\phi}}{s_1 \sin\bar{\phi} + s_2 \cos\bar{\phi} + s_3}$$

with

$$c = \sqrt{\frac{(J_1-\lambda^*)(J_2-\lambda^*)(\lambda^*-J_3)}{J_1 J_2 J_3}} \qquad s_1 = \frac{h_1}{J_1-\lambda^*}\sqrt{\frac{J_1}{J_1-\lambda^*}}$$

$$s_2 = \frac{h_2}{J_2-\lambda^*}\sqrt{\frac{J_2}{J_2-\lambda^*}} \qquad s_3 = \frac{h_3}{J_3-\lambda^*}\sqrt{\frac{J_3}{\lambda^*-J_3}}. \tag{4.90}$$

This equation has the solution

$$\tan\frac{\phi}{2} = \begin{cases} \dfrac{s_1 - \sqrt{p}\tan\tau}{s_2-s_3} & \text{for } p>0 \\[2ex] \dfrac{s_1 - \sqrt{-p}\tanh\tau}{s_2-s_3} & \text{for } p<0 \\[2ex] \dfrac{s_1 + 2/\tau}{s_2-s_3} & \text{for } p=0 \end{cases} \tag{4.91}$$

where τ is a linear function of time and

$$p = s_3^2 - (s_1^2 + s_2^2).$$

For $p>0$ the solution is periodic. The polhode is then of the type shown in Fig. 4.13b, and the permanent rotation represented by the apex of the cone is stable. For $p<0$ and also for $p=0$ the solution is aperiodic. Using the expressions for $\tan\phi/2$ it is straightforward to show that in either case the function $a_2(\phi) = \text{const} \times (s_1 \sin\phi + s_2 \cos\phi + s_3)$ tends toward zero for $\tau \to \infty$. Because of Eq. (4.89), then also $w_3(\phi)$ tends toward zero. This means that the motion along the polhode approaches asymptotically the apex of the cone. The polhode is, therefore, of the type shown in Fig. 4.13a, and the permanent rotation is unstable. For the quantity p Eq. (4.90) yields

$$p = -\sum_{\alpha=1}^{3} \frac{h_\alpha^2 J_\alpha}{(J_\alpha - \lambda^*)^3} = -2h^2 \left.\frac{\mathrm{d}F}{\mathrm{d}\lambda}\right|_{\lambda=\lambda^*}.$$

Thus, the sign of the rate of change of the curve in the upper diagram of Fig. 4.12 determines the stability behavior. The permanent rotation belonging to the smaller root λ_2^* is stable and the one belonging to λ_3^* is unstable. Unstable is also the case of a double root $\lambda_2^* = \lambda_3^*$ in which p equals zero.

In an analogous way stability criteria can be developed for permanent rotations which are associated with roots λ^* in the interval $J_2 < \lambda^* < J_1$. It was shown that also these roots are roots of $f(\lambda) = D^*(\lambda^*)$. The details are left to the reader. It is, first, necessary to develop new forms for Eqs. (4.79) to (4.88) which are based on a root λ_0

of the equation $f(\lambda_0)=D$ in the interval $J_2<\lambda_0<J_1$. This can be done by a simple permutation of indices. Starting from the new equations it can be shown that permanent rotations for $J_2<\lambda^*<J_1$ are stable if $F'(\lambda^*)$ is positive and unstable otherwise. Thus, the permanent rotation associated with λ_5^* in Fig. 4.12 is stable and the one associated with λ_4^* is unstable.

The stability of permanent rotations associated with the roots $\lambda_1^*<J_3$ and $\lambda_6^*>J_1$ cannot be investigated in a similar way. There is another and even simpler method, however. It is, first, noted that the values D^* belonging to these permanent rotations (D_1^* and D_6^* in Fig. 4.12) are the smallest and the largest, respectively, of all values D^*. This can be proved as follows. For each value D^* belonging to one of the permanent rotations with $J_3<\lambda^*<J_1$ a polhode is obtained which consists of a singular point and, in addition, of a closed curve (see Fig. 4.13). As long as the angular momentum ellipsoid intersects the energy ellipsoid in a closed curve it is possible to reduce (or to enlarge) the angular momentum ellipsoid by decreasing (increasing) the parameter D and still to get an intersection curve. The intersection curve degenerates into a singular point when D reaches a certain minimal (maximum) value. These extreme values are the values D^* which belong to $\lambda^*<J_3$ and $\lambda^*>J_1$. For parameters D which are not in the interval $D_1^*\leqslant D\leqslant D_6^*$ no real polhodes exist. After these preparatory remarks the stability of the permanent rotations under consideration can be proved as follows. For $\lambda=\lambda^*$ Eq. (4.75) becomes in view of Eq. (4.71)

$$\sum_{\alpha=1}^{3} J_\alpha(\lambda^*-J_\alpha)(\omega_\alpha-\omega_\alpha^*)^2=2T(D^*-D).$$

On the left hand side $\Omega_\alpha=\omega_\alpha-\omega_\alpha^*$ ($\alpha=1,2,3$) represents the deviation of ω_α from the angular velocity of permanent rotation belonging to D^*. The function

$$V=\sum_{\alpha=1}^{3} J_\alpha(\lambda^*-J_\alpha)\Omega_\alpha^2$$

is negative definite for $\lambda_1^*<J_3$ and positive definite for $\lambda_6^*>J_1$. Furthermore, the total time derivative of V is zero since V is a linear combination of two integrals of motion. With these properties V can serve as a Ljapunov function proving the stability of the two permanent rotations.

The stability analysis has provided us with information about the pattern of polhodes on the energy ellipsoid. In Fig. 4.14a the pattern is schematically illustrated. The separatrices belonging to the two unstable permanent rotations are the "figures eight" shown as heavy lines. They subdivide the ellipsoid into five regions, namely two "eyes" for each figure eight and the region between the two figures eight. The larger eye of one figure eight covers the entire reverse as well as the outer part of the front side of the ellipsoid. The intersection points of the axes of permanent rotation

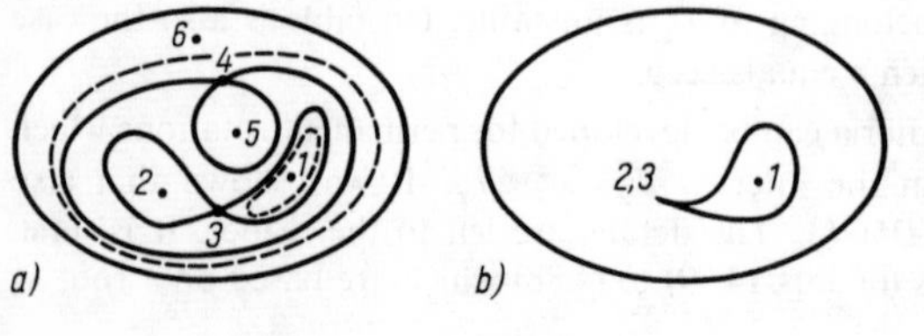

Fig. 4.14
The pattern of polhodes on the energy ellipsoid (schematically)
a) The general case of six different real roots of Eq. (4.72)
b) A separatrix belonging to a real double root of Eq. (4.72)

carry the same indices as related quantities in Fig. 4.12. The polhodes shown as dashed lines are associated with the permanent rotations number 2 and 5. The one which circles around the point number 1 belongs to the solution in Eq. (4.91) for $p>0$ while the solution for $p<0$ belongs to the separatrix which passes through the point number 3. When the system parameters are chosen such that Eq. (4.72) has a double root $\lambda_2^* = \lambda_3^*$ the points number 2 and 3 on the energy ellipsoid coincide, and also the dashed polhode around point 1 and the separatrix passing through point 3 become identical. The shape of such a particular polhode is drawn schematically in Fig. 4.14b. To it belongs the solution in Eq. (4.91) for $p=0$. In Fig. 4.15 parallel projections are shown of polhode families which were numerically calculated from the exact solutions of the equations of motion. The parameters for the two gyrostats in the figure differ only by the direction of the relative angular momentum $\boldsymbol{h}$ in the carrier. The moments of inertia J_1, J_2 and J_3, the magnitude of $\boldsymbol{h}$ and the energy constant $2T$ are the same in both cases. The gyrostat on the left has six and the other has two axes of permanent rotation.

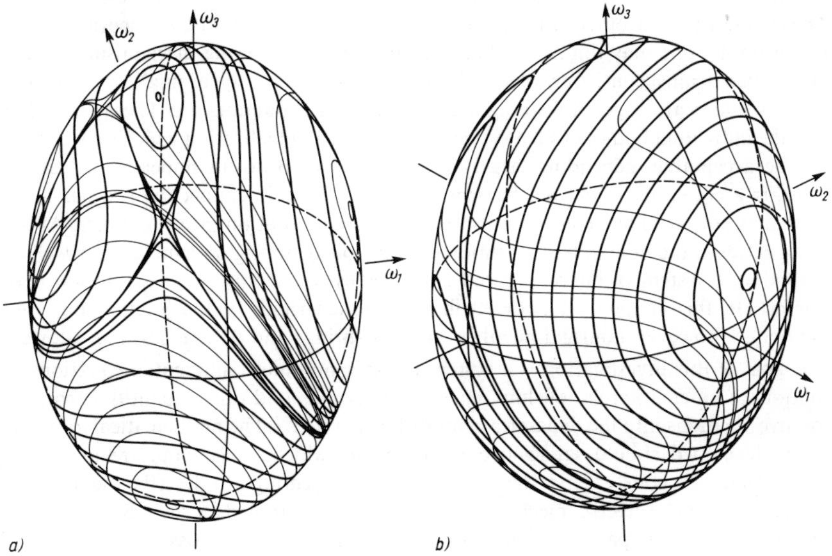

Fig. 4.15 Polhodes on the energy ellipsoid in the case of six (a)) and of two axes of permanent rotation (b)). Both computer graphics are based on the parameters $J_1 = 7\,\mathrm{kgm}^2$, $J_2 = 5\,\mathrm{kgm}^2$, $J_3 = 3\,\mathrm{kgm}^2$, $J_0 = 0.48\,\mathrm{kgm}^2$ and $2T = 75\,\mathrm{kgm}^2\mathrm{s}^{-2}$. Different are the directions of $\boldsymbol{u}$. Its coordinates are 0.4, 0.1, $+\sqrt{0.83}$ in a) and 0.6, 0.4, $+\sqrt{0.48}$ in b)

5 General Multi-Body Systems

5.1 Introductory remarks

In the preceding chapter mechanical systems were investigated which consist of either one single rigid body or several rigid bodies in some particularly simple geometric configuration. The important role they play in classical mechanics is due to the fact that their equations of motion can be integrated in closed form. This is not possible, in general, if a system is constructed of many rigid bodies in some arbitrary configuration. The engineer is confronted with an endless variety of such systems. To mention only a few examples one may think of linkages in machines, of steering mechanisms in cars, of railroad trains consisting of elastically connected cars, of a single railroad car with its undercarriadge, of walking machines and manipulators etc. The assumption that the individual bodies of such systems are rigid is an idealization which may or may not be acceptable, depending largely on the kind of problem under investigation. Thus, in a crank-and-slider mechanism the seemingly rigid connecting rod has to be treated as an elastic member when its forced bending vibrations are of concern. At the other extreme, the human body composed of obviously nonrigid members may well be treated as a system of interconnected rigid bodies when only its gross motion is of interest. In this chapter all bodies will be assumed rigid. In the joints connecting the bodies, however, nonrigid members such as springs and dampers will be allowed. The goal of this investigation is a system of exact nonlinear differential equations of motion, of kinematic relationships, energy expressions and other quantities required for investigations into the dynamics of multi-body systems. The mathematical formulas to be developed should satisfy two requirements which, in general, are not easily fulfilled simultaneously. First, they should be general enough to describe the dynamic behavior of such diverse mechanical systems as mentioned earlier. Second, their application to any particular mechanical system should be possible with only a minimum amount of preparatory work. Lagrange's equations of the second kind, for example,

$$\frac{\mathrm{d}}{\mathrm{d}t}\frac{\partial L}{\partial \dot{q}_k} - \frac{\partial L}{\partial q_k} = Q_k \qquad k=1\ldots n$$

satisfy only the first requirement, since when applying them to any particular mechanical system a substantial amount of labor is required to formulate the Lagrangian L and its derivatives. The equations of motion to be developed in this chapter are considerably more explicit. They will be required to have the standard form

$$\ddot{q}_k = \ddot{q}_k(q_1 \ldots q_n, \dot{q}_1 \ldots \dot{q}_n, t) \qquad k=1\ldots n$$

with explicitly formulated functions on the right hand side. At the same time, however, the user of the equations must have complete freedom, as with Lagrange's equations, to specify the nature of the variables $q_1 \dots q_n$ on the basis of the particular mechanical system under consideration. For equations of motion formulated in such an explicit form a general-purpose computer program for numerical integrations can be written. It is one of the goals of this chapter to enable the reader to do this. The formalism should, however, be useful for nonnumerical investigations as well. This will be achieved by applying suitable mathematical notations which lead to formulas which can easily be interpreted in physical terms. Examples of nonnumerical investigations will be given in Sec. 5.2.5 and in Chap. 6.

For the complete description of a multi-body system a large number of parameters is required. They must specify the geometry and mass distribution, as well as the nature of forces acting from outside the system and internally in the joints between bodies. Those describing geometry and mass distribution can be subdivided into the following groups:

(1) the number of bodies
(2) parameters specifying the interconnection structure of the system
(3) parameters specifying kinematic constraints
(4) parameters specifying the location of hinges on the bodies
(5) masses and inertia components of the bodies.

Before going into any detail, some definitions and introductory comments are required. Fig. 5.1 illustrates a four-body system. Between certain pairs of bodies there is a direct interaction by internal forces. Thus, for example, between the bodies labeled number 2 and 3 there exists direct force interaction caused by the kinematic constraints in the joint connecting these two bodies. Between bodies number 3 and 4 there is a direct interaction by magnetic forces. Bodies number 2 and 4, on the other hand, do not act directly upon each other. Their interaction is only indirect via another body. Two bodies are said to be contiguous if and only if they exert force on each other directly. The coupling between two contiguous bodies is called a hinge. This definition gives the word hinge a broader meaning than it has in ordinary English. Here, it is used for any kind of coupling allowing relative rotational and/or translational motion between contiguous bodies. A hinge might not even be a material interconnection (see, for instance, the hinge between bodies 3 and 4 in Fig. 5.1). In the hinge between two contiguous bodies all interaction forces between these bodies are combined so that there exists exactly one hinge for each pair of contiguous bodies. In Fig. 5.1, for example, the hinge between bodies 1 and 2 comprises the ball-and-socket joint as well as the spring. Furthermore, for each hinge there exists only one pair of contiguous bodies. This means that if, for example, three bodies at first sight

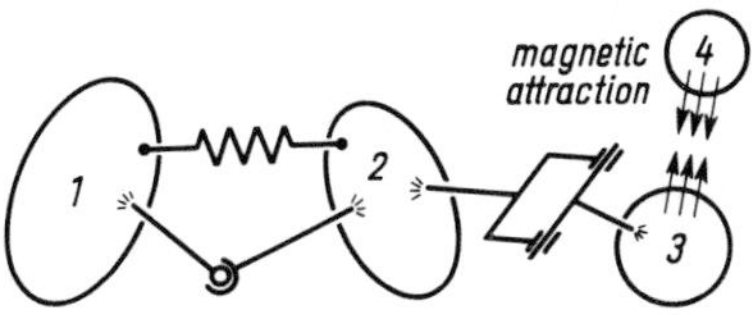

Fig. 5.1
A four-body system

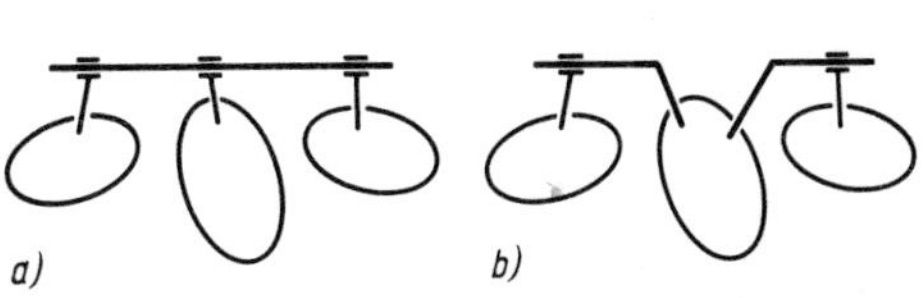

Fig. 5.2 The bodies in a) are coupled by two hinges as shown in b)

appear to be interconnected by a single hinge then this hinge will be counted as two separate hinges, each connecting two of the bodies. The system of Fig. 5.2a with three bodies mounted on a single massless shaft illustrates this. The shaft actually provides two hinges, as in shown more clearly in Fig. 5.2b. The description of the interconnection structure of the system—item (2) in the list of parameters—gives complete information about which bodies of the system are interconnected by hinges. The physical properties of the hinges are not included in this description. Their kinematic properties are, however, the subject of item (3) in the list. Kinematic constraints in the hinges may be of any kind, i.e. scleronomic, rheonomic, holonomic or non-holonomic. All constraints must be ideal, in that constraint forces do no work during virtual displacements. In practice this means, among other things, that no dry friction is allowed in the hinges.

Kinematic constraints are introduced not only by the individual hinges but also through the interconnection structure of the system. This is illustrated by the plane crank-and-slider mechanism in Fig. 5.3 whose bodies are interconnected by three

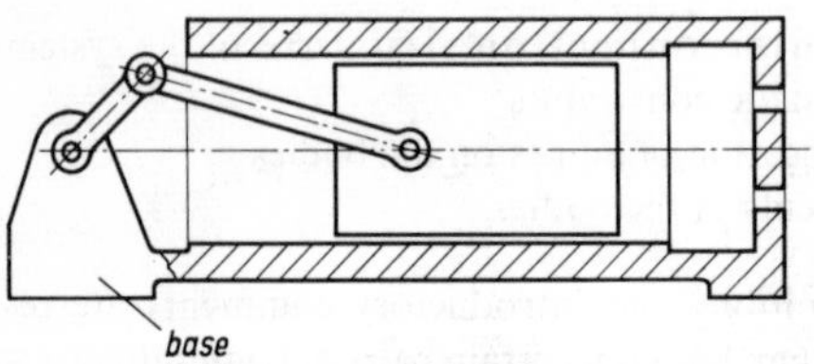

Fig. 5.3
Plane crank-and-slider mechanism, a closed kinematic chain

pin joints and one sliding joint. The body labeled base is assumed to be fixed in inertial space. The total number of degrees of freedom is one. It remains unchanged if one pin joint is replaced by a ball-and-socket joint. On the other hand, it becomes zero if the axes of the three pin joints are not mounted parallel to each other. The crank-and-slider mechanism is a simple example of a broad and important class of multi-body systems categorized as systems with closed kinematic chains. In such systems the number of degrees of freedom depends on more than just the kinematic properties of the individual hinges. In order to define a closed kinematic chain it is necessary to introduce first the notion of path between two bodies. Consider any two bodies in a multi-body system, for instance bodies i and j in Fig. 5.4a. Proceed from one body to the other along a sequence of bodies and hinges in such a way that no hinge is passed more than once. The set of hinges thus defined is called the path between bodies i and j. If for all pairs of bodies the path between them is uniquely

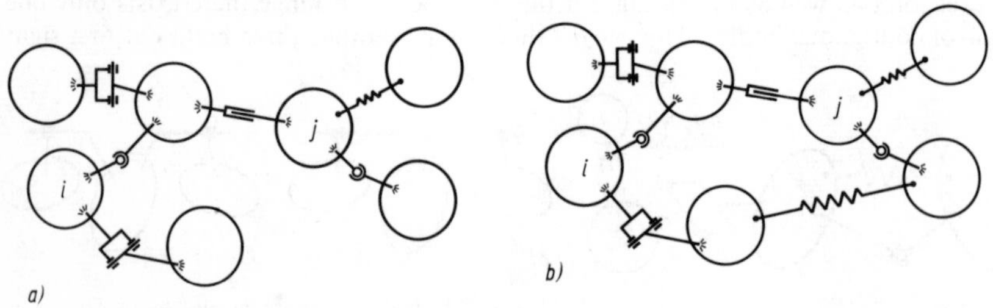

Fig. 5.4 a) A system with tree structure and b) a system with a closed non-kinematic chain

defined as is the case for the system of Fig. 5.4a then the system is said to have *tree structure*. If, on the other hand, between two bodies two different paths exist then these two paths form a closed chain. The system of Fig. 5.4b contains a closed chain. If, in particular, every hinge in a closed chain contains at least one kinematic constraint then the closed chain is called a *closed kinematic chain*. The closed chain in Fig. 5.4b is not a closed kinematic chain because in one of its hinges there are no kinematic constraints. In contrast, the crank-and-slider mechanism in Fig. 5.3 is a closed kinematic chain.

Multi-body systems are found in practice under two basically different conditions of operation. In most systems one or more bodies are connected by hinges to an *external body* whose position in inertial space is a prescribed function of time. Typical examples are a double-pendulum with a moving suspension point (Fig. 5.5a), the human body with one or both feet resting on an escalator (Fig. 5.5b) and most linkages in machines where the frame of the machine is the external body. It is obvious that the dimensions and inertia properties of the external body are irrelevant since its motion is prescribed. For this reason the external body will not be counted as another body of the system, but will be represented by a moving base rigidly attached to it. In Figs. 5.5a, b this base is called $\underline{e}^{(0)}$. The prescribed motion of the base as well as the properties of the hinges between the base and the system will enter the equations of motion to be developed.

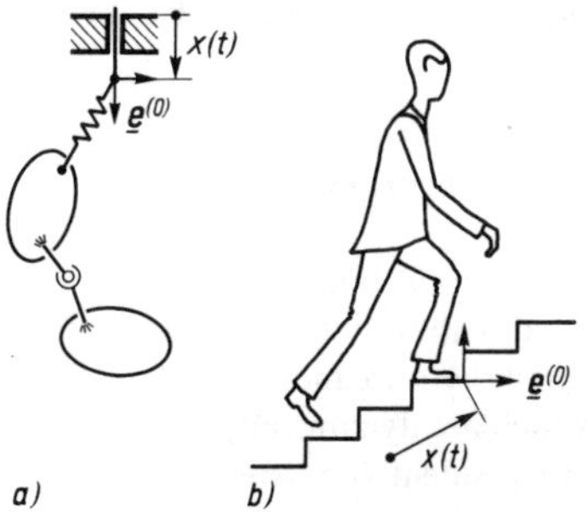

Fig. 5.5 Two systems with tree structure which are coupled to an external body whose motion is prescribed

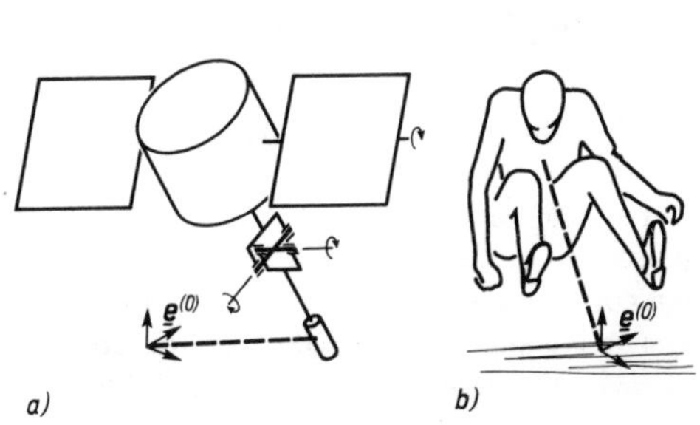

Fig. 5.6 Two systems with tree structure without coupling to an external body whose motion is prescribed

Comparatively rare is the mode of operation in which none of the bodies of a system is connected to an external body whose motion is prescribed. Typical examples for such systems are an orbiting multi-body satellite (Fig. 5.6a) and the human body in a phase of motion without contact to the ground (Fig. 5.6b). For the formulation of scalar differential equations of motion for such systems some common frame of reference is needed in which vectors and tensors can be decomposed. Depending on the particular problem under consideration this reference frame will be moving relative to inertial space according to some appropriately chosen function of time. In Figs. 5.6a, b the moving base is called $\underline{e}^{(0)}$. The position of the multi-body system in inertial space is uniquely specified if the position of contiguous bodies relative to each other is known for all hinges and if, in addition, the position in $\underline{e}^{(0)}$ is known for one arbitrarily chosen body of the system. This suggests the introduction of a *fictitious* hinge between the moving base and the one arbitrarily chosen body (indicated by a dashed line in Figs. 5.6a, b). With this hinge in which, of course, there act no internal forces, and with the moving base $\underline{e}^{(0)}$ the situation is now the same as for the systems in Figs. 5.5a, b. The mathematical description of the interconnection structure of a system will, therefore, be the same for both modes of operation.

5.2 Equations of motion for systems with tree structure

Multi-body systems with tree structure are less frequent in practice than systems with closed chains. There are two reasons, however, to treat this class of systems first. One reason is the greater simplicity of the mathematical description of the interconnection structure and of the system kinematics. The second reason is that any system with closed chains can be transformed into a system with tree structure by cutting suitably selected hinges. Thus, in order to obtain the equations of motion for a system with closed chains, all that is necessary is to add internal forces and kinematic constraints for the cut hinges to the equations of motion for a system with tree structure. This procedure will be described in Chap. 5.3. The present section on systems with tree structure will be developed in great detail. The problem will not be treated at once in full generality but rather in steps of increasing complexity. This allows, among other things, a demonstration of the application of, both, the Newton-Euler approach and d'Alembert's principle.

5.2.1 The mathematical description of the interconnection structure

If a system with tree structure is connected with an external body whose motion is prescribed as a function of time then it can be assumed without loss of generality that it is connected with this body by exactly one hinge. For if there are more hinges then the system is actually composed of several subsystems which are dynamically independent of each other (see, for example, Fig. 5.7 with two independent subsystems). If, on the

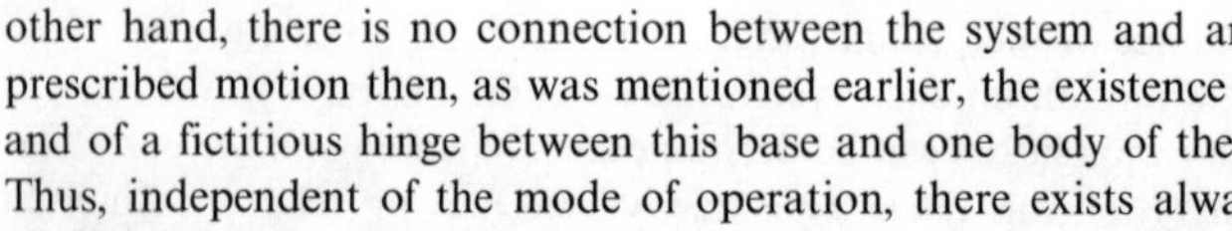

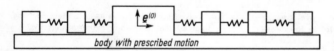

Fig. 5.7 A system which is subdivided by body 0 into two dynamically independent subsystems

other hand, there is no connection between the system and an external body with prescribed motion then, as was mentioned earlier, the existence of a moving base $\underline{e}^{(0)}$ and of a fictitious hinge between this base and one body of the system are assumed. Thus, independent of the mode of operation, there exists always a base $\underline{e}^{(0)}$ whose motion relative to inertial space is prescribed as a function of time, and this base is connected by a hinge with one body of the system. In what follows the body represented by the base $\underline{e}^{(0)}$ will be referred to as body number zero.

Let n be the number of bodies in the system (not counting body 0). Then there will always be n hinges if the hinge between body 0 and the system is counted. Bodies and hinges are labeled separately from 1 to n each. The order in which the numbers are assigned is arbitrary except that the body contiguous to body 0 and the hinge between these two are labeled body 1 and hinge 1, respectively. Fig. 5.8a shows as example a system with $n=7$. Remember that the description of the interconnection structure must provide full information about which bodies are coupled by hinges and that in this description the physical properties of the hinges are irrelevant. The interconnection structure is best displayed by the system graph. This graph consists of points called vertices and of lines connecting the vertices called edges. The vertices represent the bodies of the system and also the moving base $\underline{e}^{(0)}$, and the edges represent the hinges. Obviously, the system graph will also have tree structure. Fig. 5.8b shows the graph corresponding to the system of Fig. 5.8a. The vertex representing the base $\underline{e}^{(0)}$ is called s_0. The other vertices and the edges are called $s_1 \dots s_n$ and $u_1 \dots u_n$, respectively, with both sequences of indices being determined by the labeling of bodies and hinges in the mechanical system. In order to be able to distinguish between an index of a vertex and an index of an edge letters i, j and k will be used for the former and a, b and c for the latter.

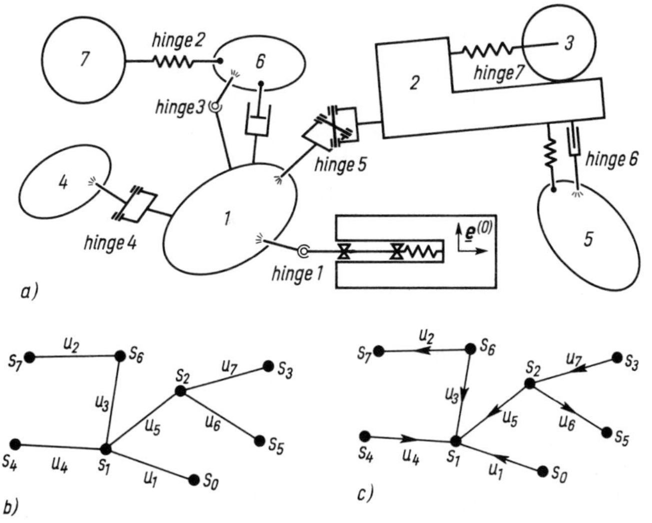

Fig. 5.8
a) A system with tree structure, b) its system graph and c) its directed system graph

To every edge of the graph a sense of direction is now arbitrarily assigned. This produces a directed graph. Its directed edges are called arcs. Fig. 5.8c shows a directed graph for the system of Fig. 5.8a. The sense of direction of an arc is indicated by an arrow. The purpose of defining a sense of direction is to render distinguishable the two vertices connected by an arc and, thus, the two bodies connected by a hinge. This is necessary for the following reasons. When describing the kinematics of motion of two contiguous bodies relative to one another it must be specified unambiguously which motion relative to which body is meant. Internal forces in the hinges act with opposite signs on contiguous bodies. When describing the system dynamics it must be specified unambiguously on which body a force is acting with a positive sign and on which with a negative.

The expressions tree, graph, vertex, edge and arc are adopted from mathematical graph theory[1]. Two other useful expressions are "incidence" and "path between two vertices". The relationship between an arc (or edge since the sense of direction is irrelevant in this context) and the two vertices connected by it is expressed by the phrase "the arc (edge) is incident with the two vertices". In Fig. 5.8c, for instance, u_5 is incident with s_1 and s_2. The path between two vertices s_i and s_j is defined as follows (compare with the definition of path between two bodies given earlier). Proceed from s_i to s_j along a sequence of vertices and arcs (regardless of sense of direction) in such a way that no arc is passed more than once. Then, the (unordered) set of arcs thus defined is called the path between s_i and s_j. In a graph with tree structure a path between s_i and s_j is uniquely defined for each combination of i and j. In Fig. 5.8c, for instance, the path between s_6 and s_3 is the set of arcs u_3, u_5 and u_7. A vertex s_k is said to be on the path between s_i and s_j if at least one arc belonging to this path is incident with s_k. According to this definition then, the vertices s_i and s_j themselves are on the path between s_i and s_j.

With the help of these notions the following weak ordering relationships for vertices are defined. The symbol $s_i \leqslant s_j$ means that s_i is on the path between s_0 and s_j. The relationship $s_i < s_j$ means that s_i is on the path between s_0 and s_j, but that it is not identical with s_j. Finally, $s_i \nleqslant s_j$ is the negation of $s_i \leqslant s_j$. Note that for two vertices s_i and s_j, both relationships, $s_i \nleqslant s_j$ and $s_j \nleqslant s_i$ can be satisfied simultaneously. Consider, for instance, s_6 and s_2 in Fig. 5.8c.

The interconnection structure of the directed graph uniquely defines two integer functions $i^+(a)$ and $i^-(a)$ which establish relationships between arc indices and vertex indices. For $a=1 \ldots n$ $i^+(a)$ is the index of the vertex from which the arc u_a is pointing away, and $i^-(a)$ is the index of the vertex toward which the arc u_a is pointing. For the directed graph of Fig. 5.8c the two functions read as follows

a	1	2	3	4	5	6	7
$i^+(a)$	0	6	6	4	2	2	3
$i^-(a)$	1	7	1	1	1	5	2

If for a directed graph with tree structure the functions $i^+(a)$ and $i^-(a)$ $(a=1 \ldots n)$ are given then it is possible to reconstruct the graph. To do this $n+1$ vertices $s_0 \ldots s_n$ are marked on a sheet of paper. Then, for $a=1 \ldots n$ an arc is drawn pointing from $s_{i^+(a)}$ to $s_{i^-(a)}$. The result of this procedure is the original graph. Consequently, there exists a one-to-one relationship between a directed graph and its pair of functions $i^+(a)$ and $i^-(a)$. This does not mean, however, that for any arbitrarily chosen pair of integer functions $i^+(a)$ and $i^-(a)$ $(a=1 \ldots n)$ a directed graph with tree structure exists! As an example, consider the functions

a	1	2	3	4	5	6	7
$i^+(a)$	1	3	6	5	2	2	2
$i^-(a)$	0	1	4	5	3	1	0

[1] As application-oriented textbook on graph theory see Busacker and Saaty [15].

They define something that is also called a graph in mathematics, but it is one without tree structure. It is, therefore, necessary to distinguish between admissible and inadmissible pairs of functions. The admissibility conditions need not be considered here, however, since all function pairs under consideration will always be derived from given directed graphs with tree structure.

The same information as in the pair of functions $i^+(a)$ and $i^-(a)$ is contained in the incidence matrix of the directed graph. This matrix has $n+1$ rows and n columns which correspond to the vertices and arcs, respectively. Its elements are called S_{ia} (remember that the letters i, j and k denote vertex indices and a, b and c arc indices). They are defined as follows

$$S_{ia} = \begin{cases} +1 & \text{if arc } u_a \text{ is pointing away from vertex } s_i \\ -1 & \text{if arc } u_a \text{ is pointing toward vertex } s_i \\ 0 & \text{otherwise} \end{cases} \qquad i=0\ldots n;\, a=1\ldots n.$$

This can also be expressed in the form

$$S_{ia} = \begin{cases} +1 & \text{if } i=i^+(a) \\ -1 & \text{if } i=i^-(a) \quad i=0\ldots n;\, a=1\ldots n. \\ 0 & \text{otherwise} \end{cases} \tag{5.1}$$

Each column of the incidence matrix contains exactly one element $+1$ and one element -1. The matrix is partitioned into two submatrices $\underline{S}_0$ and $\underline{S}$ with $\underline{S}_0$ being the row matrix

$$\underline{S}_0 = [S_{01} \ldots S_{0n}] \tag{5.2}$$

and $\underline{S}$ being the square matrix

$$\underline{S} = \begin{bmatrix} S_{11} \ldots S_{1n} \\ \vdots \\ S_{n1} \ldots S_{nn} \end{bmatrix}. \tag{5.3}$$

In $\underline{S}_0$ only the first element S_{01} is nonzero. For the directed graph of Fig. 5.8c the two matrices are

$$\underline{S}_0 = [+1 \quad 0 \quad 0 \quad 0 \quad 0 \quad 0 \quad 0]$$

$$\underline{S} = \begin{bmatrix} -1 & 0 & -1 & -1 & -1 & 0 & 0 \\ 0 & 0 & 0 & 0 & +1 & +1 & -1 \\ 0 & 0 & 0 & 0 & 0 & 0 & +1 \\ 0 & 0 & 0 & +1 & 0 & 0 & 0 \\ 0 & 0 & 0 & 0 & 0 & -1 & 0 \\ 0 & +1 & +1 & 0 & 0 & 0 & 0 \\ 0 & -1 & 0 & 0 & 0 & 0 & 0 \end{bmatrix}.$$

There is another matrix called $\underline{T}$ with elements $+1$, -1 and zero that can be constructed from the directed graph. Like $\underline{S}$ it is an $(n \times n)$ matrix. This time, however, the rows correspond to the arcs $u_1 \ldots u_n$ and the columns to the vertices $s_1 \ldots s_n$

so that its elements are called T_{ai}. Their definition is

$$T_{ai}=\begin{cases} +1 & \text{if } u_a \text{ belongs to the path between } s_0 \text{ and } s_i \\ & \text{and is directed toward } s_0 \\ -1 & \text{if } u_a \text{ belongs to the path between } s_0 \text{ and } s_i \\ & \text{and is directed away from } s_0 \\ 0 & \text{if } u_a \text{ does not belong to the path between} \\ & s_0 \text{ and } s_i \end{cases} \quad a,i=1\ldots n. \tag{5.4}$$

For the directed graph of Fig. 5.8c this matrix reads

$$\underline{T}=\begin{bmatrix} -1 & -1 & -1 & -1 & -1 & -1 & -1 \\ 0 & 0 & 0 & 0 & 0 & 0 & -1 \\ 0 & 0 & 0 & 0 & 0 & +1 & +1 \\ 0 & 0 & 0 & +1 & 0 & 0 & 0 \\ 0 & +1 & +1 & 0 & +1 & 0 & 0 \\ 0 & 0 & 0 & 0 & -1 & 0 & 0 \\ 0 & 0 & +1 & 0 & 0 & 0 & 0 \end{bmatrix}.$$

Between the matrices $\underline{T}$, $\underline{S}_0$ and $\underline{S}$ there exist the following important relationships

$$\underline{T}^{\mathrm{T}}\underline{S}_0^{\mathrm{T}}=-\underline{1}_n \tag{5.5}$$

$$\underline{T}\,\underline{S}=\underline{S}\,\underline{T}=\underline{E} \tag{5.6}$$

with $\underline{E}$, the $(n\times n)$ unit matrix and $\underline{1}_n$, a column matrix of n elements, each of which is one.

Proof. The proof for Eq. (5.5) follows from the fact that in $\underline{S}_0$ only the first element S_{01} is nonzero and that by definition of T_{ai} all elements in the first row of $\underline{T}$ are equal to $-S_{01}$. As for Eq. (5.6) it is sufficient to prove that the product $\underline{T}\,\underline{S}$ is the unit matrix. This product is an $(n\times n)$ matrix with elements $(\underline{T}\,\underline{S})_{ab}=\sum_{i=1}^{n} T_{ai}S_{ib}$ $(a,b=1\ldots n)$. Because of Eq. (5.1) S_{ib} is equal to $+1$ for $i=i^+(b)$, to -1 for $i=i^-(b)$ and to zero otherwise. Therefore, $(\underline{T}\underline{S})_{ab}=T_{ai^+(b)}-T_{ai^-(b)}$. Let us consider, first, the case $a=b$. The arc $u_a=u_b$ is either pointing towards s_0 or away from s_0. If the former is true then $T_{ai^+(b)}=1$, $T_{ai^-(b)}=0$ and if the latter then $T_{ai^+(b)}=0$, $T_{ai^-(b)}=-1$. Hence, in either case, $(\underline{T}\,\underline{S})_{aa}=1$. Next, let a and b be different. Consider the two paths between s_0 and $s_{i^+(b)}$ and between s_0 and $s_{i^-(b)}$, respectively. The arc u_a belongs either to both paths or to none of them. In either case, $T_{ai^+(b)}=T_{ai^-(b)}$ and, therefore, $(\underline{T}\,\underline{S})_{ab}=0$. ■

From the definition of T_{ai} follows that in the j-th column of $\underline{T}$ the set of row indices of all nonzero elements is also the set of indices of all arcs which belong to the path between s_0 and s_j. As an example column 7 of the matrix $\underline{T}$ for the graph of Fig. 5.8c yields the set of arcs u_1, u_2 and u_3. As this example shows, the order in which the arcs are arranged along the path from s_0 to s_j cannot be determined from the j-th column of $\underline{T}$ alone. It can be determined, however, from the entire matrix $\underline{T}$. This follows from the fact that $\underline{S}$ is determined by $\underline{T}$, that with $\underline{S}$ the functions $i^+(a)$ and $i^-(a)$ are known and that from these functions the directed graph can be constructed. There is a simple way employing both matrices, $\underline{T}$ and $\underline{S}$, to determine the order of arcs along the path between s_0 and s_j. In order to describe it the terms inboard arc of a vertex and inboard vertex of a vertex must be introduced. The inboard arc of a vertex s_k $(k \neq 0)$ is the arc which belongs to the path between s_0 and s_k and which, furthermore, is incident with s_k. The inboard vertex of the vertex s_k $(k \neq 0)$ is the vertex which is connected with s_k by the inboard arc of s_k. To give an example, u_3 and s_1 are the inboard arc and the inboard vertex, respectively, of s_6 in Fig. 5.8c. In these terms the method can be described as follows. It is recursive. At each step the inboard arc u_a and the inboard vertex s_i are determined for a certain vertex s_k. At the beginning of the first step the vertex s_k is the vertex s_j. At each successive step s_k is the previously determined inboard vertex s_i. The procedure stops when s_i is equal to s_0. The ordered sequence of inboard arcs thus determined is the sequence in which the arcs are arranged along the path from s_j to s_0. It remains to show how the inboard arc u_a and the inboard vertex s_i of s_k $(k \neq 0)$ are found from the matrices $\underline{T}$ and $\underline{S}$. The arc u_a is the only arc for which both S_{ka} and T_{ak} are nonzero. It follows that a is the intersection of two sets of indices, namely the set of all column indices b for which $S_{kb} \neq 0$ and the set of all row indices c for which $T_{ck} \neq 0$. The inboard vertex s_i is one of the two vertices $s_{i^+(a)}$ and $s_{i^-(a)}$, namely the one not identical with s_k. Thus, s_i is found from the a-th column of $\underline{S}$.

In any tree graph the vertices and arcs can be labeled in such a way that the following conditions are fulfilled. For all vertices s_k $(k \neq 0)$ the number of the inboard arc of s_k is k and the number of the inboard vertex of s_k is smaller than k. In general, there is more than one way in which numbers can be assigned fulfilling these conditions. Any such labeling is called regular. For a given graph with given vertex s_0 a regular labeling can be achieved as follows. The graph contains at least one peripheral vertex. Peripheral vertices are all vertices except s_0 with which only one arc is incident. To these peripheral vertices the highest numbers n, $n-1$, $n-2$ etc. are assigned. The same numbers are given to the corresponding inboard arcs. Then all vertices and arcs already labeled (except s_0) are cut off from the graph. This results in a smaller graph with new peripheral vertices to which, in turn, the highest numbers still available are assigned. This recursive procedure is continued until all vertices and arcs have been labeled. Proceeding in this manner the only vertex which is contiguous to s_0 and the arc connecting these two vertices are called s_1 and u_1, respectively, as before. The matrices $\underline{S}$ and $\underline{T}$ of a directed tree graph with regular labeling have some important properties. Of the numbers $i^+(a)$ and $i^-(a)$ which are assigned to the two vertices connected by the arc u_a $(a=1 \dots n)$ one is identical with a and the other is smaller than a. As a consequence, all diagonal elements of $\underline{S}$ are nonzero, and all other nonzero elements of $\underline{S}$ are above the main diagonal. Furthermore, for $a=1 \dots n$ the arc u_a belongs to the path between s_0 and s_a. It follows that all main diagonal elements of $\underline{T}$ also are nonzero. Finally, the arc u_a $(a=1 \dots n)$ can only belong to paths between s_0 and such vertices s_k for which k is greater than or equal to a. From this it follows that in the matrix $\underline{T}$ as in $\underline{S}$ no nonzero elements occur below the main diagonal. Above the main diagonal of $\underline{T}$ nonzero elements occur only in the first $n-n'$ rows where n' is the number of peripheral vertices in the graph. An example illustrates these properties. Fig. 5.9 shows the directed graph of Fig. 5.8c with a new and regular labeling. The sense of direction of the arcs has not been changed. The matrices $\underline{S}$ and $\underline{T}$ are now

$$\underline{S} = \begin{bmatrix} -1 & -1 & -1 & 0 & 0 & 0 & -1 \\ 0 & +1 & 0 & 0 & 0 & +1 & 0 \\ 0 & 0 & +1 & -1 & +1 & 0 & 0 \\ 0 & 0 & 0 & +1 & 0 & 0 & 0 \\ 0 & 0 & 0 & 0 & -1 & 0 & 0 \\ 0 & 0 & 0 & 0 & 0 & -1 & 0 \\ 0 & 0 & 0 & 0 & 0 & 0 & +1 \end{bmatrix}$$

$$\underline{T} = \begin{bmatrix} -1 & -1 & -1 & -1 & -1 & -1 & -1 \\ 0 & +1 & 0 & 0 & 0 & +1 & 0 \\ 0 & 0 & +1 & +1 & +1 & 0 & 0 \\ 0 & 0 & 0 & +1 & 0 & 0 & 0 \\ 0 & 0 & 0 & 0 & -1 & 0 & 0 \\ 0 & 0 & 0 & 0 & 0 & -1 & 0 \\ 0 & 0 & 0 & 0 & 0 & 0 & +1 \end{bmatrix}.$$

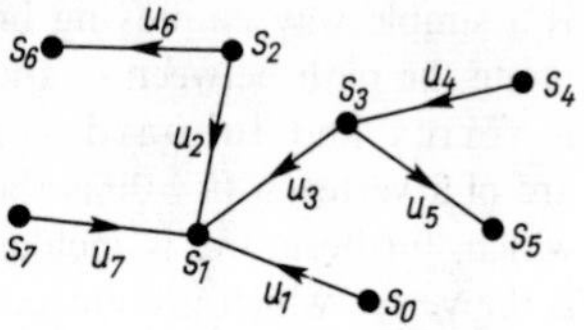

Fig. 5.9
A directed system graph with regular labeling for the system of Fig. 5.8a

In this particular case n' equals 3. If in a graph with regular labeling all arcs point towards s_0 then all nonzero elements of $\underline{T}$ and all elements on the main diagonal of $\underline{S}$ are $+1$. If, on the other hand, all arcs point away from s_0 then all matrix elements just mentioned are -1. Consider, again, the problem of determining the order in which the arcs are arranged along the path from s_0 to s_j ($j=1 \dots n$). A general method based on the matrices $\underline{S}$ and $\underline{T}$ was described earlier. In a graph with regular labeling the indices of the arcs are monotonically increasing along this path. The order can, therefore, be obtained directly from the j-th column of $\underline{T}$.

Problems

5.1 For the system graph of Fig. 5.8b specify the sets of vertices s_i which satisfy the following conditions (one at a time) for $s_k = s_2$ and for $s_k = s_3$:

1. $s_i < s_k$, 3. $s_k < s_i$, 5. $s_k \nleq s_i$,
2. $s_i \leq s_k$, 4. $s_k \leq s_i$, 6. $s_k \nleq s_i$ and $s_i \nleq s_k$.

5.2 Construct a labeled system graph for which $s_i \leq s_j$ implies $i \leq j$ for all combinations of i and j.

5.3 Give a direct proof for the statement $\underline{S}\,\underline{T} = \underline{E}$ in Eq. (5.6).

5.2.2 Systems with ball-and-socket joints. One body is coupled to an external body whose motion is prescribed

Fig. 5.10 shows an example of a system with tree structure consisting of seven bodies labeled 1 to 7. The system is coupled to a body number zero whose position in inertial space is prescribed as a function of time. All hinges—labeled 1 to 7—are ball-and-socket joints. This simplifies the development of equations of motion considerably.

The motion of any two contiguous bodies relative to each other is a pure rotation with three degrees of freedom about the geometric center of the connecting ball-and-socket joint. This point will be called the hinge point. The interconnection structure of the system and the labeling of bodies and hinges in Fig. 5.10 are the same as in Fig. 5.8a. Therefore, the directed system graph of Fig. 5.8c and its functions $i^+(a)$ and

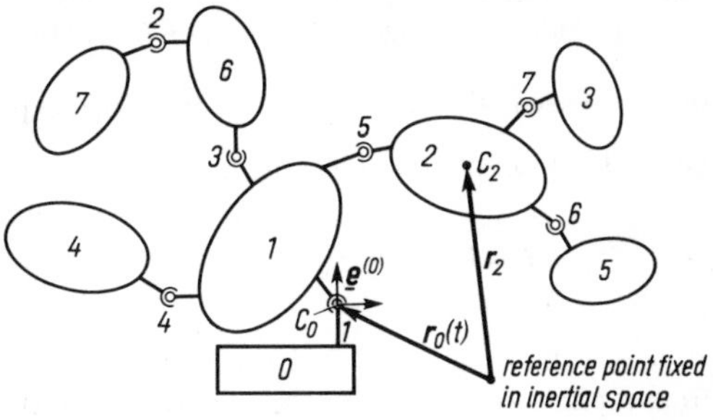

Fig. 5.10
A system with ball-and-socket joints. The motion of the base $\underline{e}^{(0)}$ is prescribed

$i^-(a)$ as well as its matrices $\underline{S}_0$, $\underline{S}$ and $\underline{T}$ can be used again for the new system. It is convenient to attach the base $\underline{e}^{(0)}$ to body zero such that its origin coincides with the hinge point 1 on this body (see Fig. 5.10). This hinge point is called C_0. The radius vector $\boldsymbol{r}_0$ of C_0 from a reference point fixed in inertial space is a prescribed function of time.

Equations of motion will be derived from Newtons law and from the law of moment of momentum for a single rigid body. As a preparatory step cuts are applied to all hinges of the system. This results in $n+1$ single rigid bodies. Only the bodies number 1 to n need be considered since for body number zero no equations of motion are needed. Each of the bodies 1 to n is subject to unspecified external forces and torques as well as to internal forces and torques which act in the cut hinges located on the body. The internal forces are kinematic constraint forces. They assure that each hinge point is fixed on two contiguous bodies simultaneously. Internal torques could be caused, for example, by viscous damping in the joints. On body i ($i=1 \dots n$) all external forces and torques are combined in a resultant external force $\boldsymbol{F}_i$ with its line of action passing through the body center of mass C_i and a resultant external torque $\boldsymbol{M}_i$ (Fig. 5.11a). The internal forces and torques acting in hinge a ($a=1 \dots n$) are combined in a resultant internal hinge force $\boldsymbol{X}_a^c$ with its line of action passing through the hinge point a and a resultant internal hinge torque $\boldsymbol{Y}_a$. The superscript c in $\boldsymbol{X}_a^c$ points to the fact that this force is a constraint force. The following sign convention will be adopted. On body $i^+(a)$ act the force $+\boldsymbol{X}_a^c$ and the torque $+\boldsymbol{Y}_a$ and, consequently, on body $i^-(a)$ the force $-\boldsymbol{X}_a^c$ and the torque $-\boldsymbol{Y}_a$. This is illustrated by Fig. 5.11a.

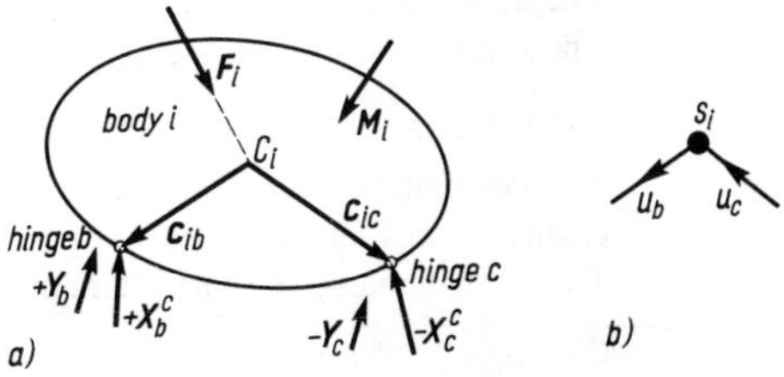

Fig. 5.11
a) Free-body diagram for body i, b) the pertinent section of the directed graph

On the body number i shown in this figure there are two hinges labeled b and c. In the directed graph whose relevant section is shown in Fig. 5.11 b the arc u_b is pointing away from the vertex s_i while the arc u_c is pointing toward s_i. It follows that i is equal to $i^+(b)$ and to $i^-(c)$. This explains why $+\boldsymbol{X}_b^c$ and $-\boldsymbol{X}_c^c$ are shown to act on body i. One can also say that the forces $S_{ib}\boldsymbol{X}_b^c$ and $S_{ic}\boldsymbol{X}_c^c$ are acting since by definition in Eq. (5.1) S_{ib} and S_{ic} are $+1$ and -1, respectively. Furthermore, since S_{ia} is equal to zero if a is the index of a hinge not located on body i the resultant of all internal forces acting on this body is the sum $\sum_{\alpha=1}^{n} S_{ia}\boldsymbol{X}_a^c$. With this expression Newton's law for the translational motion of body i has the form

$$m_i\ddot{\boldsymbol{r}}_i=\boldsymbol{F}_i+\sum_{a=1}^{n} S_{ia}\boldsymbol{X}_a^c \qquad i=1\ldots n. \tag{5.7}$$

In it m_i is the body mass and $\boldsymbol{r}_i$ the radius vector of its center of mass C_i measured from a point fixed in inertial space (see Fig. 5.10). For the formulation of the law of moment of momentum with respect to C_i the levers of the internal hinge forces must be specified. In Fig. 5.11 a the levers are called $\boldsymbol{c}_{ib}$ and $\boldsymbol{c}_{ic}$. The first index identifies the body on which the vector is fixed and the second index the hinge point to which the vector is pointing. The general definition is as follows. The vector $\boldsymbol{c}_{ia}$ $(i,a=1\ldots n)$ is zero if i equals neither $i^+(a)$ nor $i^-(a)$, i.e. if hinge a is not located on body i. Otherwise, $\boldsymbol{c}_{ia}$ is the vector from the body i center of mass to the hinge point a on this body. By this definition all nonzero vectors $\boldsymbol{c}_{ia}$ are body-fixed vectors. The resultant torque about C_i caused by the internal hinge forces and torques—in Fig. 5.11 a it is $(\boldsymbol{c}_{ib}\times\boldsymbol{X}_b^c+\boldsymbol{Y}_b)-(\boldsymbol{c}_{ic}\times\boldsymbol{X}_c^c+\boldsymbol{Y}_c)$—can now be expressed in the form $\sum_{a=1}^{n} S_{ia}(\boldsymbol{c}_{ia}\times\boldsymbol{X}_a^c+\boldsymbol{Y}_a)$. With this the law of moment of momentum takes the form

$$\dot{\boldsymbol{L}}_i=\boldsymbol{M}_i+\sum_{a=1}^{n} S_{ia}(\boldsymbol{c}_{ia}\times\boldsymbol{X}_a^c+\boldsymbol{Y}_a) \qquad i=1\ldots n \tag{5.8}$$

where $\boldsymbol{L}_i$ denotes the absolute moment of momentum of body i with respect to C_i. Both sets of equations, (5.7) and (5.8), are most compactly expressed in matrix form as

$$\underline{m}\,\ddot{\underline{\boldsymbol{r}}}=\underline{\boldsymbol{F}}+\underline{S}\,\underline{\boldsymbol{X}}^c \tag{5.9}$$

$$\dot{\underline{\boldsymbol{L}}}=\underline{\boldsymbol{M}}+\underline{\boldsymbol{C}}\times\underline{\boldsymbol{X}}^c+\underline{S}\,\underline{\boldsymbol{Y}}. \tag{5.10}$$

The various matrices are defined as follows.

Scalar matrices

The diagonal mass matrix $\underline{m}$ with elements $m_{ij}=\delta_{ij}m_i$ $(i,j=1\ldots n)$.
The submatrix $\underline{S}$ of the incidence matrix (see Eq. (5.3)).

Vectorial matrices

The column matrices $\ddot{\underline{\boldsymbol{r}}}$, $\underline{\boldsymbol{F}}$, $\underline{\boldsymbol{X}}^c$, $\underline{\boldsymbol{Y}}$, $\dot{\underline{\boldsymbol{L}}}$ and $\underline{\boldsymbol{M}}$ with n elements each. The elements are vectors. Example: $\ddot{\underline{\boldsymbol{r}}}=[\ddot{\boldsymbol{r}}_1\ldots\ddot{\boldsymbol{r}}_n]^T$.
The $(n\times n)$ matrix $\underline{\boldsymbol{C}}$ with elements

$$\boldsymbol{C}_{ia}=S_{ia}\boldsymbol{c}_{ia} \qquad i,a=1\ldots n. \tag{5.11}$$

In the equations of motion (5.9) and (5.10) appear, among other quantities, the constraint forces $\boldsymbol{X}_1^c \ldots \boldsymbol{X}_n^c$. They are of no interest if only motions of the mechanical system are of concern. It is, therefore, desirable, to eliminate these forces. This is possible with the help of Eq. (5.6) which states that $\underline{T}$ is the inverse of $\underline{S}$. When Eq. (5.9) is premultiplied by $\underline{T}$ the constraint forces $\underline{\boldsymbol{X}}^c$ are obtained explicitly in the form

$$\underline{\boldsymbol{X}}^c = \underline{T}(\underline{m}\,\underline{\ddot{\boldsymbol{r}}} - \underline{\boldsymbol{F}})\,.$$

Once $\underline{\ddot{\boldsymbol{r}}}$ has been found as a function of time from equations of motion this expression can be used to determine stresses occuring in the hinges during motions. Substituting the expression into Eq. (5.10) we obtain

$$\underline{\dot{\boldsymbol{L}}} - \underline{\boldsymbol{C}}\,\underline{T} \times (\underline{m}\,\underline{\ddot{\boldsymbol{r}}} - \underline{\boldsymbol{F}}) = \underline{\boldsymbol{M}} + \underline{S}\,\underline{\boldsymbol{Y}}. \tag{5.12}$$

The symbol indicating vector-cross multiplication has been shifted to the right across the factor $\underline{T}$. This is allowed since $\underline{T}$ is a scalar quantity (see Problem 1.2). Eq. (5.12) constitutes a set of n vectorial equations of motion, i.e. of $3n$ scalar equations. This corresponds to the total number of degrees of freedom of the entire system—three in each of the n hinges. The equations are not yet in a form suitable for application, however, since so far nothing has been said about the kinematic relationship between the translational and angular accelerations appearing in the terms $\underline{\ddot{\boldsymbol{r}}}$ and $\underline{\dot{\boldsymbol{L}}}$, respectively. Before writing down any equation it is obvious from Fig. 5.10 what this relationship will look like, in principle. The radius vector $\boldsymbol{r}_2$, for instance, is the sum of $\boldsymbol{r}_0$ which is a known function of time and of two body-fixed vectors. The first vector connects C_0 with the hinge point of hinge number 5 and the second this hinge point with C_2. Generally speaking, the radius vector $\boldsymbol{r}_i$ of body i ($i=1 \ldots n$) can be represented as a sum of $\boldsymbol{r}_0$ and of a chain of vectors each of which is fixed on one of the bodies which form the path between the bodies 0 and i. Hence, $\ddot{\boldsymbol{r}}_i$ will be a sum of $\ddot{\boldsymbol{r}}_0(t)$ and of terms containing the said body-fixed vectors together with angular velocities and accelerations of bodies. Thus, $\underline{\ddot{\boldsymbol{r}}}$ can be expressed in terms of a known function of time and of quantities describing only rotational motions of the bodies. These quantities are the same which appear in $\underline{\dot{\boldsymbol{L}}}$. The relationships just described will now be formulated mathematically.

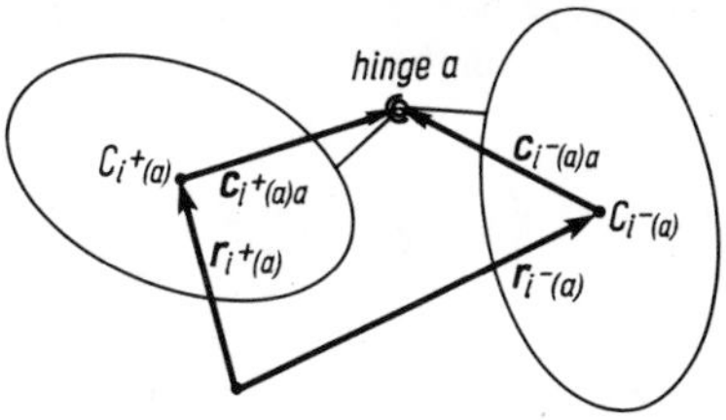

Fig. 5.12
Vectors used for describing the kinematics of two contiguous bodies

Fig. 5.12 shows a pair of contiguous bodies number $i^+(a)$ and $i^-(a)$ for some hinge a ($a=1 \ldots n$). In the case $a=1$ one of the two bodies is body 0. The point C_0 on this body was defined to be the hinge point 1 (see Fig. 5.10) so that the vector

$$\boldsymbol{c}_{01} = \boldsymbol{0} \tag{5.13}$$

must be defined in order to render the figure applicable to this case. For $a=2 \ldots n$ the vectors $\boldsymbol{c}_{0a}$ are also defined to be zero. After these preparatory remarks Fig. 5.12 yields

$$(\boldsymbol{r}_{i^+(a)}+\boldsymbol{c}_{i^+(a)a})-(\boldsymbol{r}_{i^-(a)}+\boldsymbol{c}_{i^-(a)a})=\boldsymbol{0} \qquad a=1\dots n\,.$$

With Eq. (5.1) this can be rewritten in the form

$$\sum_{i=0}^{n} S_{ia}(\boldsymbol{r}_i+\boldsymbol{c}_{ia})=\boldsymbol{0} \qquad a=1\dots n$$

or with Eqs. (5.13) and (5.11)

$$S_{0a}\boldsymbol{r}_0+\sum_{i=1}^{n}(S_{ia}\boldsymbol{r}_i+\boldsymbol{C}_{ia})=\boldsymbol{0} \qquad a=1\dots n\,.$$

These n vectorial equations are combined in the single matrix equation

$$\boldsymbol{r}_0\underline{S}_0^{\mathrm{T}}+\underline{S}^{\mathrm{T}}\underline{\boldsymbol{r}}+\underline{\boldsymbol{C}}^{\mathrm{T}}\underline{1}_n=\underline{\boldsymbol{0}}$$

in which $\underline{\boldsymbol{r}}$ denotes the column matrix $[\boldsymbol{r}_1\dots\boldsymbol{r}_n]^{\mathrm{T}}$, and $\underline{S}_0$ and $\underline{S}$ are the submatrices of the incidence matrix defined in Eqs. (5.2) and (5.3). Premultiplying this by $\underline{T}^{\mathrm{T}}$ and recognizing Eqs. (5.5) and (5.6) we get for $\underline{\boldsymbol{r}}$ the explicit expression

$$\underline{\boldsymbol{r}}=\boldsymbol{r}_0\underline{1}_n-(\underline{\boldsymbol{C}}\,\underline{T})^{\mathrm{T}}\underline{1}_n\,. \tag{5.14}$$

For the elements of the matrix product $\underline{\boldsymbol{C}}\,\underline{T}$ the abbreviation $\boldsymbol{d}_{ij}$ is introduced:

$$\boldsymbol{d}_{ij}=(\underline{\boldsymbol{C}}\,\underline{T})_{ij}=\sum_{a=1}^{n} T_{aj}S_{ia}\boldsymbol{c}_{ia} \qquad i,j=1\dots n\,. \tag{5.15}$$

Since for a given value of i all vectors $S_{ia}\boldsymbol{c}_{ia}$ are fixed on body i $\boldsymbol{d}_{ij}$ is also fixed on body i. With these vectors Eq. (5.14) yields for a single vector $\boldsymbol{r}_i$ the expression

$$\boldsymbol{r}_i=\boldsymbol{r}_0-\sum_{j=1}^{n}\boldsymbol{d}_{ji} \qquad i=1\dots n\,. \tag{5.16}$$

In other words: The radius vector $\boldsymbol{r}_i$ is the sum of $\boldsymbol{r}_0$ and of n vectors $-\boldsymbol{d}_{ji}$, each fixed on another body j. In order to interpret the physical meaning of these vectors Eq. (5.15) is rewritten in the form

$$\boldsymbol{d}_{ji}=\sum_{a=1}^{n} T_{ai}S_{ja}\boldsymbol{c}_{ja} \qquad i,j=1\dots n \tag{5.17}$$

with interchanged indices i and j. The products $T_{ai}S_{ja}$ are different from zero only for those arcs u_a which belong to the path between s_0 and s_i ($T_{ai}\neq 0$) and which, furthermore, are incident with s_j ($S_{ja}\neq 0$). It is, therefore, necessary to distinguish between the cases $s_j \not\leq s_i$, $s_j<s_i$ and $s_j=s_i$. In the first case none of the arcs is contributing to the sum so that $\boldsymbol{d}_{ji}$ is zero. In the second case only two arcs are contributing. Let their indices be b and c with u_b being the inboard arc of vertex s_j (upper part of Fig. 5.13a). Independent of the sense of direction of these arcs $T_{bi}S_{jb}=+1$ and $T_{ci}S_{jc}=-1$ and, hence, $\boldsymbol{d}_{ji}=\boldsymbol{c}_{jb}-\boldsymbol{c}_{jc}$ (see lower part of Fig. 5.13a). In the third case ($s_j=s_i$) only the inboard arc u_b of s_i is contributing to the sum. Thus, the arguments used before yield $\boldsymbol{d}_{ii}=\boldsymbol{c}_{ib}$ (Fig. 5.13b). In Fig. 5.14 all vectors $\boldsymbol{d}_{ji}$ ($j=1\dots n$) in the system of Fig. 5.10 are shown for the special case $i=2$. Only $\boldsymbol{d}_{12}$ and $\boldsymbol{d}_{22}$ are different from zero. These results are in

agreement with the verbal description given earlier for the relationship between $\boldsymbol{r}_i$, $\boldsymbol{r}_0$ and body-fixed vectors[1]).
The second time derivative of the expression for $\underline{\boldsymbol{r}}$ in Eq. (5.14) is now substituted into the equation of motion (5.12):

$$\dot{\underline{\boldsymbol{L}}}+(\underline{\boldsymbol{C}}\,\underline{T})\times\underline{m}(\ddot{\underline{\boldsymbol{C}}}\,\underline{T})^{\mathrm{T}}\underline{1}_n-(\underline{\boldsymbol{C}}\,\underline{T})\times(\ddot{\boldsymbol{r}}_0\underline{m}\,\underline{1}_n-\underline{\boldsymbol{F}})=\underline{\boldsymbol{M}}+\underline{S}\,\underline{\boldsymbol{Y}}\,. \tag{5.18}$$

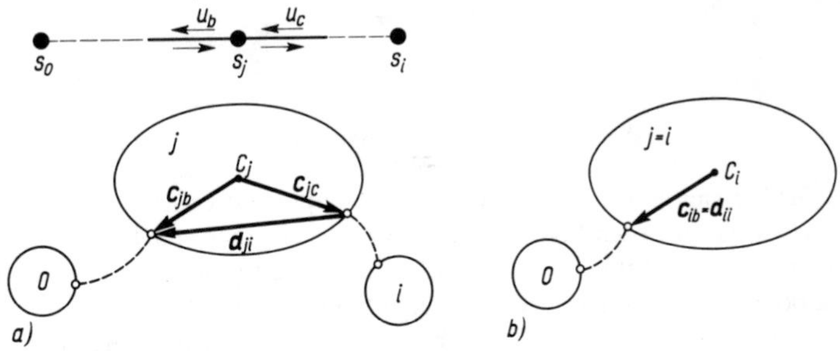

Fig. 5.13 The vector $\boldsymbol{d}_{ji}$ on body i and the pertinent section of the directed system graph; a) the case $s_j < s_i$ and b) the case $s_j = s_i$

In this equation the vectors $\boldsymbol{d}_{ij}$ play an essential role as is shown by the repeated occurence of the matrix product $\underline{\boldsymbol{C}}\,\underline{T}$. In what follows the $(n \times n)$ matrix $(\underline{\boldsymbol{C}}\,\underline{T})\times\underline{m}(\ddot{\underline{\boldsymbol{C}}}\,\underline{T})^{\mathrm{T}}$ will be examined. A single element, abbreviated $\boldsymbol{g}_{ij}$, is the vector

$$\boldsymbol{g}_{ij}=\sum_{k=1}^{n} m_k \boldsymbol{d}_{ik}\times\ddot{\boldsymbol{d}}_{jk} \qquad i,j=1\ldots n.$$

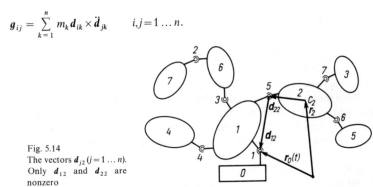

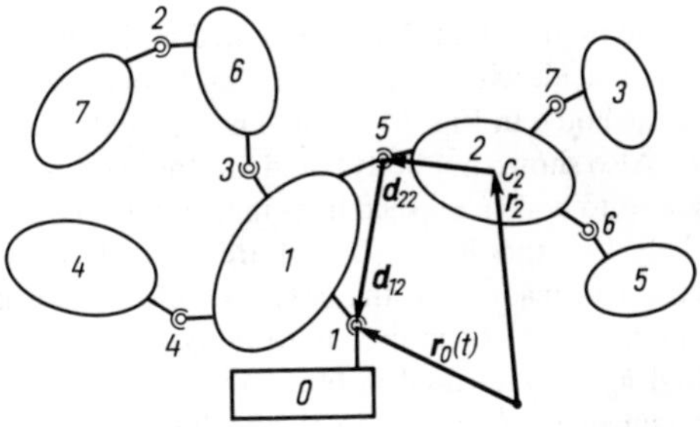

Fig. 5.14
The vectors $\boldsymbol{d}_{j2}$ $(j=1\ldots n)$. Only $\boldsymbol{d}_{12}$ and $\boldsymbol{d}_{22}$ are nonzero

Distinguish the following cases: (i) $s_i = s_j$, (ii) $s_i < s_j$, (iii) $s_j < s_i$ and (iv) otherwise. Because of the properties of the vectors $\boldsymbol{d}_{ij}$ in case (ii) only those vertices s_k contribute to $\boldsymbol{g}_{ij}$ for which $s_j \leqslant s_k$ (for all others $\boldsymbol{d}_{jk}$ equals zero). For them $\boldsymbol{d}_{ik}$ is, independently of k, identical with $\boldsymbol{d}_{ij}$. Likewise, in case (iii) only vertices s_k with $s_i \leqslant s_k$ contribute to

1) In Problem 5.3 a direct proof of the theorem $\underline{S}\,\underline{T}=\underline{E}$ was asked for. A single element of the matrix product is $(\underline{S}\,\underline{T})_{ji}=\sum_{a=1}^{n} T_{ai}S_{ja}$ $(j,i=1\ldots n)$. The right hand side has the same form as that of Eq. (5.17) if $\boldsymbol{c}_{ja}$ is replaced by one. The chain of arguments used for the interpretation of the vectors $\boldsymbol{d}_{ji}$ immediately yields the desired result $(\underline{S}\,\underline{T})_{ji}=\delta_{ji}$.

$\boldsymbol{g}_{ij}$, and for them $\boldsymbol{d}_{jk}$ is identical with $\boldsymbol{d}_{ji}$. Finally, in case (iv) at least one of the two vectors $\boldsymbol{d}_{ik}$ and $\boldsymbol{d}_{jk}$ is zero. Hence

$$\boldsymbol{g}_{ij} = \begin{cases} \sum_{k=1}^{n} m_k \boldsymbol{d}_{ik} \times \ddot{\boldsymbol{d}}_{ik} & s_i = s_j \\ \boldsymbol{d}_{ij} \times \sum_{k=1}^{n} m_k \ddot{\boldsymbol{d}}_{jk} & s_i < s_j \\ \sum_{k=1}^{n} m_k \boldsymbol{d}_{ik} \times \ddot{\boldsymbol{d}}_{ji} & s_j < s_i \\ \boldsymbol{0} & \text{otherwise.} \end{cases} \tag{5.19}$$

A further simplification is possible if the important concept of augmented bodies is introduced. For each body of the system an augmented body is constructed. For body i ($i=1 \dots n$) this is done as follows. To the tip of each vector $\boldsymbol{c}_{ia}$ fixed on body i a point mass is attached which is equal to the sum of the masses of all bodies except body 0 which are connected with body i either directly or indirectly via the respective hinge a. Two examples are given to illustrate this. For the system of Fig. 5.10 the augmented body 2 is obtained from the original body 2 by attaching the point masses $m_1+m_4+m_6+m_7$, m_3 and m_5 to the tips of the vectors $\boldsymbol{c}_{25}$, $\boldsymbol{c}_{27}$ and $\boldsymbol{c}_{26}$, respectively. The augmented body 7 is obtained by attaching to body 7 the point mass $m_1+m_2+m_3+m_4+m_5+m_6$ at the tip of $\boldsymbol{c}_{72}$. From the definition follows that the augmented bodies are rigid bodies and that they all have the same mass

$$M = \sum_{i=1}^{n} m_i$$

of the total system. The augmented body i has a center of mass which, in general, does not coincide with the center of mass C_i of the original body i. It is called the barycenter of the body. In Fig. 5.15 body i is depicted with its center of mass C_i and its barycenter B_i. Also shown are body 0 and another body k, both of which are, in general, not contiguous to body i. Dashed lines indicate the paths between the bodies. On the augmented body i vectors $\boldsymbol{b}_{ij}$ ($j=0 \dots n$) are defined. They point from the barycenter B_i to the center of mass C_i in the case $j=i$ and to the tip of the body-fixed vector $\boldsymbol{c}_{ia}$ leading either directly or indirectly to body j in the case $j \neq i$. As examples the vectors $\boldsymbol{b}_{ii}$, $\boldsymbol{b}_{i0}$ and $\boldsymbol{b}_{ik}$ are indicated in Fig. 5.15. The vectors $\boldsymbol{b}_{ij}$ with $j \neq i$ play the same role for the augmented bodies as do the body-fixed vectors $\boldsymbol{c}_{ia}$ for the original bodies. Notice, however, the difference in notation. The second index of $\boldsymbol{c}_{ia}$ is that of a hinge whereas

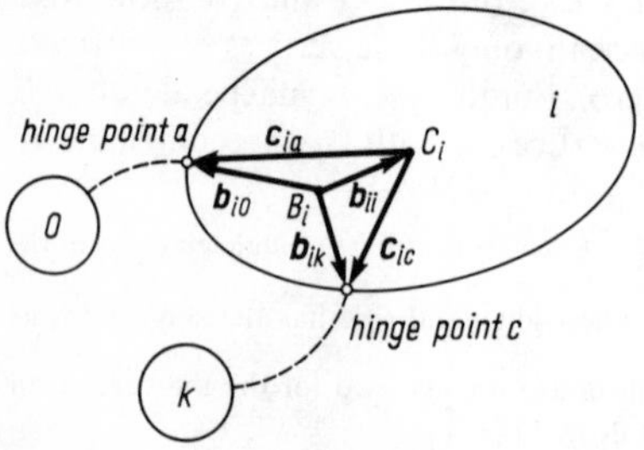

Fig. 5.15
Body i with hinge points 1 and 2, center of mass C_i, barycenter B_i and vectors $\boldsymbol{b}_{ii}$, $\boldsymbol{b}_{i0}$ and $\boldsymbol{b}_{ik}$

the second index of $\boldsymbol{b}_{ij}$ corresponds to a body. All vectors $\boldsymbol{b}_{ij}$ ($i=1\ldots n$; $j=0\ldots n$) are fixed on bodies. The number of different vectors $\boldsymbol{b}_{ij}$ is smaller than the number of different combinations of indices i,j. In the system of Fig. 5.10, for instance, the identities $\boldsymbol{b}_{12}=\boldsymbol{b}_{13}=\boldsymbol{b}_{15}$ hold. From the definition of the vectors follows that the equations

$$\sum_{j=1}^{n} \boldsymbol{b}_{ij} m_j = \boldsymbol{0} \qquad i=1\ldots n \tag{5.20}$$

are fulfilled.

Between the vectors $\boldsymbol{b}_{ij}$ and $\boldsymbol{d}_{ij}$ there exists the relationship

$$\boldsymbol{d}_{ij} = \boldsymbol{b}_{i0} - \boldsymbol{b}_{ij} \qquad i,j=1\ldots n \tag{5.21}$$

which is easily verified for all possible combinations of indices (an example: In the case $s_j < s_i$ it was shown that $\boldsymbol{d}_{ij}$ equals zero. In agreement with this, $\boldsymbol{b}_{ij}$, indeed, equals $\boldsymbol{b}_{i0}$). With the help of Eqs. (5.20) and (5.21) the expressions for $\boldsymbol{g}_{ij}$ in Eq. (5.19) can be further simplified. In the case $s_i < s_j$, for instance, substitution yields

$$\sum_{k=1}^{n} m_k \ddot{\boldsymbol{d}}_{jk} = \sum_{k=1}^{n} m_k (\ddot{\boldsymbol{b}}_{j0} - \ddot{\boldsymbol{b}}_{jk}) = M \ddot{\boldsymbol{b}}_{j0}$$

(remember that M is the total system mass). An analogous result is obtained for the sum in the case $s_j < s_i$. Together they yield

$$\boldsymbol{g}_{ij} = \begin{cases} \sum_{k=1}^{n} m_k \boldsymbol{d}_{ik} \times \ddot{\boldsymbol{d}}_{ik} & s_i = s_j \\ M \boldsymbol{d}_{ij} \times \ddot{\boldsymbol{b}}_{j0} & s_i < s_j \\ M \boldsymbol{b}_{i0} \times \ddot{\boldsymbol{d}}_{ji} & s_j < s_i \\ \boldsymbol{0} & \text{otherwise.} \end{cases} \tag{5.22}$$

These expressions are now substituted into Eq. (5.18). At this point the matrix formulation is abandoned replacing the equation again by n individual vectorial equations. They read

$$\dot{\boldsymbol{L}}_i + \sum_{j=1}^{n} \boldsymbol{g}_{ij} - \sum_{j=1}^{n} \boldsymbol{d}_{ij} \times (m_j \ddot{\boldsymbol{r}}_0 - \boldsymbol{F}_j) = \boldsymbol{M}_i + \sum_{a=1}^{n} S_{ia} \boldsymbol{Y}_a \qquad i=1\ldots n$$

or explicitly

$$\dot{\boldsymbol{L}}_i + \sum_{k=1}^{n} m_k \boldsymbol{d}_{ik} \times \ddot{\boldsymbol{d}}_{ik} + M \Bigl(\sum_{j:s_i<s_j} \boldsymbol{d}_{ij} \times \ddot{\boldsymbol{b}}_{j0} + \boldsymbol{b}_{i0} \times \sum_{j:s_j<s_i} \ddot{\boldsymbol{d}}_{ji} \Bigr) -$$
$$- \sum_{j=1}^{n} \boldsymbol{d}_{ij} \times (m_j \ddot{\boldsymbol{r}}_0 - \boldsymbol{F}_j) = \boldsymbol{M}_i + \sum_{a=1}^{n} S_{ia} \boldsymbol{Y}_a \qquad i=1\ldots n. \tag{5.23}$$

The symbol $\sum_{j:}$ denotes a sum over all values of j which satisfy the condition specified behind the colon. The two leading terms on the left hand side can be combined. The absolute angular momentum $\boldsymbol{L}_i$ of body i with respect to C_i is the integral $\int_m \boldsymbol{\varrho} \times \dot{\boldsymbol{\varrho}} \, \mathrm{d}m$

where ϱ is the radius vector of the mass particle dm measured from C_i (see Fig. 5.16). The time derivative is $\dot{\boldsymbol{L}}_i = \int_m \boldsymbol{\varrho} \times \ddot{\boldsymbol{\varrho}}\, dm$. Let ϱ' be the radius vector of dm shown in the figure. It starts at the hinge point of body i leading to body 0. This point will be referred

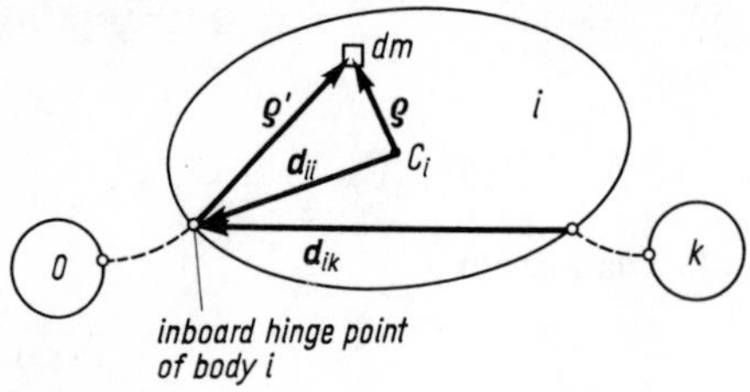

Fig. 5.16
Vectors locating a mass particle on body i

to as the inboard hinge point of body i. The time derivative of the body i angular momentum with respect to this point is the integral $\int_m \boldsymbol{\varrho}' \times \ddot{\boldsymbol{\varrho}}'\, \mathrm{d}m$. With $\boldsymbol{\varrho}' = \boldsymbol{\varrho} - \boldsymbol{d}_{ii}$ and $\int_m \boldsymbol{\varrho}\, \mathrm{d}m = \boldsymbol{0}$ it becomes

$$\begin{aligned}\int_{m_i} \boldsymbol{\varrho}' \times \ddot{\boldsymbol{\varrho}}'\, \mathrm{d}m &= \int_{m_i} (\boldsymbol{\varrho} - \boldsymbol{d}_{ii}) \times (\ddot{\boldsymbol{\varrho}} - \ddot{\boldsymbol{d}}_{ii})\mathrm{d}m \\ &= \int_{m_i} \boldsymbol{\varrho} \times \ddot{\boldsymbol{\varrho}}\,\mathrm{d}m + m_i \boldsymbol{d}_{ii} \times \ddot{\boldsymbol{d}}_{ii} = \dot{\boldsymbol{L}}_i + m_i \boldsymbol{d}_{ii} \times \ddot{\boldsymbol{d}}_{ii}\,.\end{aligned}$$

If to this expression the sum $\sum\limits_{\substack{k=1\\ \neq i}}^{n} m_k \boldsymbol{d}_{ik} \times \ddot{\boldsymbol{d}}_{ik}$ is added the two leading terms in Eq. (5.23) are obtained. This explains their physical significance. Together they represent the absolute time derivative of the absolute angular momentum of the augmented body i with respect to its inboard hinge point. The sum over $k \neq i$ is the contribution of the $n-1$ point masses which are attached to the original body i in order to produce the augmented body. Note that $\boldsymbol{d}_{ik}$ is the vector from the attachment point of point mass m_k to the inboard hinge point. Let $\boldsymbol{K}_i$ be the inertia tensor of the augmented body i with respect to its inboard hinge point. It is related to the central inertia tensor $\boldsymbol{J}_i$ of the original body i by the equation

$$\boldsymbol{K}_i = \boldsymbol{J}_i + \sum_{k=1}^{n} m_k (\boldsymbol{d}_{ik}^2 \boldsymbol{E} - \boldsymbol{d}_{ik} \boldsymbol{d}_{ik}) \qquad i = 1 \ldots n. \tag{5.24}$$

If, furthermore, $\boldsymbol{\omega}_i$ denotes the absolute angular velocity of body i then the two leading terms in Eq. (5.23) can be expressed in the form

$$\dot{\boldsymbol{L}}_i + \sum_{k=1}^{n} m_k \boldsymbol{d}_{ik} \times \ddot{\boldsymbol{d}}_{ik} = \boldsymbol{K}_i \cdot \dot{\boldsymbol{\omega}}_i + \boldsymbol{\omega}_i \times \boldsymbol{K}_i \cdot \boldsymbol{\omega}_i\,. \tag{5.25}$$

The term involving $\ddot{\boldsymbol{r}}_0$ in Eq. (5.23) is reduced with the help of Eqs. (5.20) and (5.21) to

$$\sum_{j=1}^{n} \boldsymbol{d}_{ij} m_j \times \ddot{\boldsymbol{r}}_0 = \sum_{j=1}^{n} (\boldsymbol{b}_{i0} - \boldsymbol{b}_{ij}) m_j \times \ddot{\boldsymbol{r}}_0 = \boldsymbol{b}_{i0} \times M \ddot{\boldsymbol{r}}_0\,. \tag{5.26}$$

In the expression involving the external forces $\boldsymbol{F}_j$ the factor $\boldsymbol{d}_{ij}$ is different from zero only for values of j which satisfy the relationship $s_i \leqslant s_j$. With this and with Eqs. (5.25) and (5.26) the equations of motion (5.23) take the form

$$\boldsymbol{K}_i\cdot\dot{\boldsymbol{\omega}}_i+\boldsymbol{\omega}_i\times\boldsymbol{K}_i\cdot\boldsymbol{\omega}_i+M\left[\sum_{j:s_i<s_j}\boldsymbol{d}_{ij}\times\ddot{\boldsymbol{b}}_{j0}+\boldsymbol{b}_{i0}\times\left(-\ddot{\boldsymbol{r}}_0+\sum_{j:s_j<s_i}\ddot{\boldsymbol{d}}_{ji}\right)\right]+\sum_{j:s_i\leqslant s_j}\boldsymbol{d}_{ij}\times\boldsymbol{F}_j$$

$$=\boldsymbol{M}_i+\sum_{a=1}^{n}S_{ia}\boldsymbol{Y}_a \qquad i=1\dots n\,. \tag{5.27}$$

For the second derivatives of $\boldsymbol{b}_{j0}$ and $\boldsymbol{d}_{ji}$ the expressions

$$\begin{aligned}\ddot{\boldsymbol{b}}_{j0}&=\dot{\boldsymbol{\omega}}_j\times\boldsymbol{b}_{j0}+\boldsymbol{\omega}_j\times(\boldsymbol{\omega}_j\times\boldsymbol{b}_{j0})\\ \ddot{\boldsymbol{d}}_{ji}&=\dot{\boldsymbol{\omega}}_j\times\boldsymbol{d}_{ji}+\boldsymbol{\omega}_j\times(\boldsymbol{\omega}_j\times\boldsymbol{d}_{ji})\end{aligned}\qquad i,j=1\dots n \tag{5.28}$$

will later be substituted. The equations allow a simple interpretation when written in the form

$$M(-\boldsymbol{b}_{i0})\times\left(\ddot{\boldsymbol{r}}_0-\sum_{j:s_j<s_i}\ddot{\boldsymbol{d}}_{ji}\right)+\boldsymbol{K}_i\cdot\dot{\boldsymbol{\omega}}_i+\boldsymbol{\omega}_i\times\boldsymbol{K}_i\cdot\boldsymbol{\omega}_i=\boldsymbol{M}_i^P \tag{5.29}$$

with $\quad \boldsymbol{M}_i^P=\sum_{a=1}^{n}S_{ia}\boldsymbol{Y}_a+\boldsymbol{M}_i-\boldsymbol{d}_{ii}\times\boldsymbol{F}_i-\sum_{j:s_i<s_j}\boldsymbol{d}_{ij}\times(M\ddot{\boldsymbol{b}}_{j0}+\boldsymbol{F}_j)\,.$

It will be shown now that this is the formulation of the law of moment of momentum for a single rigid body if as reference point for moment of momentum and for external torques a body-fixed point other than the center of mass is chosen. In Eq. (3.20) this was written in the form

$$m\boldsymbol{r}_C\times\ddot{\boldsymbol{z}}_P+\boldsymbol{J}^P\cdot\dot{\boldsymbol{\omega}}+\boldsymbol{\omega}\times\boldsymbol{J}^P\cdot\boldsymbol{\omega}=\boldsymbol{M}^P$$

where P is the body-fixed reference point, $\ddot{\boldsymbol{z}}_P$ its absolute acceleration, $\boldsymbol{J}^P$ and $\boldsymbol{M}^P$ the inertia tensor and the external torque with respect to P and $\boldsymbol{r}_C$ the vector from P to the center of mass. Eq. (5.29) has this form if the rigid body is understood to be the augmented body i and if, furthermore, the inboard hinge point of the body is chosen as reference point P. The mass of the body is then M and its center of mass is the barycenter. The vector from the inboard hinge point to the barycenter is $-\boldsymbol{b}_{i0}$ (see Fig. 5.15). The expression $\ddot{\boldsymbol{r}}_0-\sum_{j:s_j<s_i}\ddot{\boldsymbol{d}}_{ji}$ is, indeed, the absolute acceleration of the reference point P. This is seen as follows. The radius vector of P in inertial space is $\boldsymbol{r}_i+\boldsymbol{d}_{ii}$ or with Eq. (5.16) $\boldsymbol{r}_0-\sum_{j=1}^{n}\boldsymbol{d}_{ji}+\boldsymbol{d}_{ii}$. This is, indeed, identical with $\boldsymbol{r}_0-\sum_{j:s_j<s_i}\boldsymbol{d}_{ji}$ since $\boldsymbol{d}_{ji}$ is zero for $s_j\nless s_i$. Thus, the left hand side of Eq. (5.29) has the desired form. Consider now the torque $\boldsymbol{M}_i^P$. It contains, first, the resultant $\sum_{a=1}^{n}S_{ia}\boldsymbol{Y}_a$ of all internal hinge torques acting on body i and the external torque $\boldsymbol{M}_i$. The line of action of the external force $\boldsymbol{F}_i$ was said to pass through the body center of mass so that $-\boldsymbol{d}_{ii}\times\boldsymbol{F}_i$ is its torque with respect to the inboard hinge point. Also the remaining terms have the desired form in that the vectors $-\boldsymbol{d}_{ij}$ are pointing away from the inboard hinge point on body i. For values of j which satisfy the condition $s_i<s_j$ the torque $-\boldsymbol{d}_{ij}\times(M\ddot{\boldsymbol{b}}_{j0}+\boldsymbol{F}_j)$ can be interpreted as follows. Imagine that the augmented body j is isolated from the system and suspended as a pendulum in inertial space at its own inboard hinge point. Fig. 5.17 shows this pendulum as well as the bodies i and 0 and the paths between them. The augmented body j is subject to the external force $\boldsymbol{F}_j$. If now the

augmented body with mass M and center of mass B_j is rotating with its actual angular velocity and acceleration then it is exerting on its suspension the force $M\ddot{\boldsymbol{b}}_{j0}+\boldsymbol{F}_j$. This force has to be shifted until its line of action is passing through the hinge point on body i which is leading toward body j (point Q in Fig. 5.17). It then produces on body i the torque $-\boldsymbol{d}_{ij}\times(M\ddot{\boldsymbol{b}}_{j0}+\boldsymbol{F}_j)$ with respect to point P. This is the physical interpretation of the last term in $\boldsymbol{M}_i^P$.

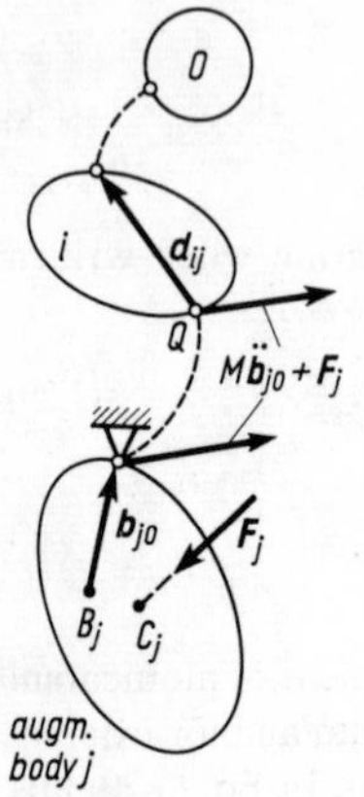

Fig. 5.17
Interpretation of $M\ddot{\boldsymbol{b}}_{j0}+\boldsymbol{F}_j$ as force applied to the suspension point of a pendulum

Eq. (5.27) will now be given a final form which is suitable for, both, numerical and non-numerical applications. Substituting Eq. (5.28) and retaining on the left hand side only terms involving angular accelerations we obtain the equation

$$\boldsymbol{K}_i\cdot\dot{\boldsymbol{\omega}}_i+M\Big[\sum_{j:s_i<s_j}\boldsymbol{d}_{ij}\times(\dot{\boldsymbol{\omega}}_j\times\boldsymbol{b}_{j0})+\boldsymbol{b}_{i0}\times\sum_{j:s_j<s_i}\dot{\boldsymbol{\omega}}_j\times\boldsymbol{d}_{ji}\Big]$$
$$=\boldsymbol{M}_i'+\boldsymbol{M}_i+\sum_{a=1}^{n}S_{ia}\boldsymbol{Y}_a\qquad i=1\dots n\qquad(5.30)$$

with
$$\boldsymbol{M}_i'=-\boldsymbol{\omega}_i\times\boldsymbol{K}_i\cdot\boldsymbol{\omega}_i-M\Big\{\sum_{j:s_i<s_j}\boldsymbol{d}_{ij}\times[\boldsymbol{\omega}_j\times(\boldsymbol{\omega}_j\times\boldsymbol{b}_{j0})]+$$
$$+\boldsymbol{b}_{i0}\times\Big[-\ddot{\boldsymbol{r}}_0+\sum_{j:s_j<s_i}\boldsymbol{\omega}_j\times(\boldsymbol{\omega}_j\times\boldsymbol{d}_{ji})\Big]\Big\}-\sum_{j:s_i\leq s_j}\boldsymbol{d}_{ij}\times\boldsymbol{F}_j\qquad i=1\dots n.\qquad(5.31)$$

The double vector-cross products on the left hand side are formulated as scalar products of a tensor and a vector, for instance

$$\boldsymbol{d}_{ij}\times(\dot{\boldsymbol{\omega}}_j\times\boldsymbol{b}_{j0})=(\boldsymbol{b}_{j0}\cdot\boldsymbol{d}_{ij}\boldsymbol{E}-\boldsymbol{b}_{j0}\boldsymbol{d}_{ij})\cdot\dot{\boldsymbol{\omega}}_j\,.$$

With the new tensors

$$\boldsymbol{K}_{ij}=\begin{cases}\boldsymbol{K}_i & i=j\\ M(\boldsymbol{b}_{j0}\cdot\boldsymbol{d}_{ij}\boldsymbol{E}-\boldsymbol{b}_{j0}\boldsymbol{d}_{ij}) & s_i<s_j\\ M(\boldsymbol{d}_{ji}\cdot\boldsymbol{b}_{i0}\boldsymbol{E}-\boldsymbol{d}_{ji}\boldsymbol{b}_{i0}) & s_j<s_i\\ \boldsymbol{0} & \text{otherwise}\end{cases}\qquad i,j=1\dots n\qquad(5.32)$$

which satisfy the relationship

$$\boldsymbol{K}_{ji}=\bar{\boldsymbol{K}}_{ij} \quad \text{(conjugate of } \boldsymbol{K}_{ij}) \qquad i,j=1\ldots n. \tag{5.33}$$

Eq. (5.30) can be written in the final form

$$\sum_{j=1}^{n} \boldsymbol{K}_{ij}\cdot\dot{\boldsymbol{\omega}}_j=\boldsymbol{M}'_i+\boldsymbol{M}_i+\sum_{a=1}^{n} S_{ia}\boldsymbol{Y}_a \qquad i=1\ldots n. \tag{5.34}$$

These n first-order differential equations (equivalent to $3n$ scalar equations) have to be supplemented by another $3n$ scalar differential equations which relate the angular velocities $\boldsymbol{\omega}_i$ to the time derivatives of generalized coordinates. In the choice of such coordinates one has complete freedom. Suppose that it is decided to use Euler angles ψ_i, θ_i and ϕ_i for the description of the angular orientation of body i ($i=1\ldots n$) relative to inertial space. Because of the nature of ball-and-socket joints there are no kinematic constraints between the Euler angles of different bodies. The kinematic differential equations are, therefore, n sets of three equations, each having the form of Eq. (2.29) with an index i attached to $\omega_{1,2,3}$ and to ψ, θ and ϕ.

Problems

5.4 Draw in Fig. 5.10 all vectors $\boldsymbol{d}_{2j}$ ($j=1\ldots 7$) and $\boldsymbol{b}_{2j}$ ($j=0\ldots 7$).

5.5 Identify in Fig. 5.10 the vectors $\boldsymbol{d}_{ij}$, $\boldsymbol{b}_{i0}$ and $\boldsymbol{b}_{ij}$ of Eq. (5.21) for the following sets of indices (i,j): (1,2), (1,3), (2,1), (2,2) and (2,6).

5.6 Write a FORTRAN or ALGOL program for the calculation of the constant coordinates of the vectors $\boldsymbol{d}_{ij}$ and of the tensors $\boldsymbol{K}_i$ ($i,j=1\ldots n$) in the respective body-fixed bases $\underline{e}^{(i)}$. Use as input data the masses and inertia components of the individual bodies and the constant coordinates of the vectors $\boldsymbol{c}_{ia}$ ($i,a=1\ldots n$). For the calculation of the vectors $\boldsymbol{b}_{ij}$ see Sec. 5.2.4.

5.2.3 The special case of plane motions

Given is, again, a multi-body system with tree structure. One of its bodies is coupled to a body 0 whose motion is prescribed. In contrast to the previous section the following conditions prevail. First, all hinges are pin joints, and the axes of all pin joints are parallel to each other (two bodies coupled by a pin joint have one degree of freedom of rotation relative to each other. A translation relative to each other along the hinge axis is not possible). Second, the prescribed absolute angular velocity of body 0 has no component normal to the pin joint axes. A three-body system of this kind is illustrated in Fig. 5.18. Because of the constraints in the hinges each body is in plane motion relative to all other bodies. The motions relative to inertial space are plane motions only if the translation of body 0 has no component along the hinge axes. For the moment, this is not assumed to be the case. Equations of motion for such systems are easily developed from the equations of motion for systems with ball-and-socket joints (Eq. (5.34)). There is only one difference between the two kinds of systems. This is explained as follows. Imagine that in each pin joint an arbitrary point on the axis is chosen and that a ball-and-socket joint is inserted at this point. The result of this procedure is a system of the kind considered in the previous section. From this system the original system with pin joints can be recovered, again, if in each ball-and-socket joint a constraint torque is introduced which eliminates the degrees of freedom of rotation normal to the pin joint axis. These constraint torques are directed normal to the pin joint axes. Consequently, the equations of motion for the system with pin joints are identical with Eq. (5.34) except that the said constraint torques have to be added to the already existing internal hinge torques $\boldsymbol{Y}_1 \dots \boldsymbol{Y}_n$. When these equations are scalar multiplied by a unit vector $\boldsymbol{p}$ which is parallel to the pin joint axes the additional constraint torques are eliminated, again. The set of n vectorial differential equations is thereby reduced to a set of n scalar differential equations for as many degrees of freedom of motion (one in each of the n hinges). It remains to be shown that the scalar differential equations are independent of which points on the pin joint axes are chosen for insertion of ball-and-socket joints and constraint torques. The only quantities in Eq. (5.34) which depend on the location of these points are the components of the vectors $\boldsymbol{d}_{ij}$ and $\boldsymbol{b}_{i0}$ ($i, j=1 \dots n$) along the unit vector $\boldsymbol{p}$. It is easy to verify that these components are eliminated when the equations are scalar multiplied by $\boldsymbol{p}$. As an example consider the term $\boldsymbol{p}\cdot\boldsymbol{d}_{ij}\times[\boldsymbol{\omega}_j\times(\boldsymbol{\omega}_j\times\boldsymbol{b}_{j0})]$. The absolute angular velocity $\boldsymbol{\omega}_j$ has itself the direction of $\boldsymbol{p}$. It suffices, therefore, to investigate the expression $\boldsymbol{p}\cdot\boldsymbol{d}_{ij}\times[\boldsymbol{p}\times(\boldsymbol{p}\times\boldsymbol{b}_{j0})]$. It is identical with $\boldsymbol{p}\cdot\boldsymbol{b}_{j0}\times\boldsymbol{d}_{ij}$. This form shows that, indeed, the components along $\boldsymbol{p}$ of, both, $\boldsymbol{b}_{j0}$ and $\boldsymbol{d}_{ij}$ are eliminated. In order to simplify the following calculations these components are set equal to zero at the outset. The same is done with the components along $\boldsymbol{p}$ of $\ddot{\boldsymbol{r}}_0(t)$ and of $\boldsymbol{F}_i$ ($i=1 \dots n$) and also with the components normal to $\boldsymbol{p}$ of $\boldsymbol{M}_i$ ($i=1 \dots n$) and of $\boldsymbol{Y}_a$ ($a=1 \dots n$).

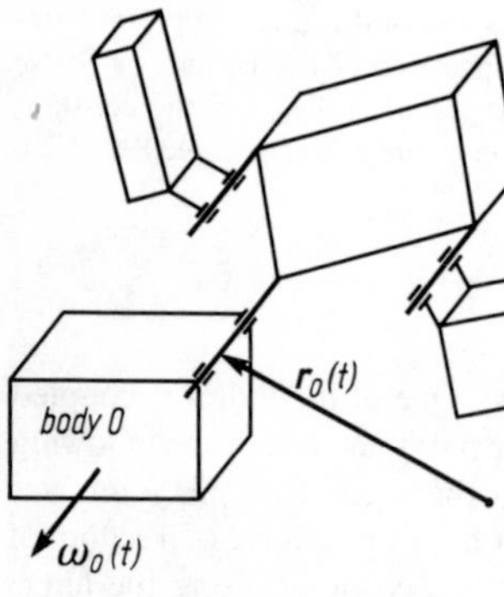

Fig. 5.18 A system with pin joints with parallel axes

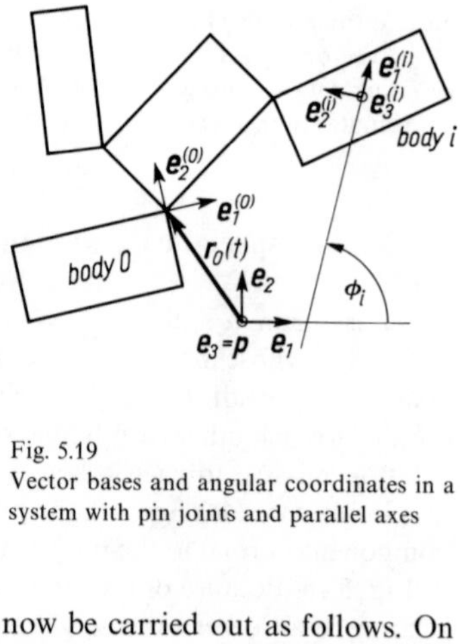

Fig. 5.19
Vector bases and angular coordinates in a system with pin joints and parallel axes

The scalar multiplication of Eq. (5.34) with $\boldsymbol{p}$ will now be carried out as follows. On each body i ($i=0 \dots n$) a vector base $\underline{\boldsymbol{e}}^{(i)}$ is fixed such that the unit vector $\boldsymbol{e}_3^{(i)}$ coincides with $\boldsymbol{p}$ (see Fig. 5.19). The orientation of $\boldsymbol{e}_1^{(i)}$ and $\boldsymbol{e}_2^{(i)}$ on the bodies is arbitrary. The origin of $\underline{\boldsymbol{e}}^{(0)}$ is located on the axis of hinge 1. Its radius vector in inertial space is the known function $\boldsymbol{r}_0(t)$. The desired scalar product of Eq. (5.34) with $\boldsymbol{p}$ is obtained when this vector equation is decomposed in the base $\underline{\boldsymbol{e}}^{(i)}$ and when only the equation for the third coordinate is retained. Fig. 5.19 shows, in addition to the bases $\underline{\boldsymbol{e}}^{(i)}$, a base $\underline{\boldsymbol{e}}$ which is fixed in inertial space. Its base vector $\boldsymbol{e}_3$ is also parallel to $\boldsymbol{p}$. As generalized coordinate for the description of the angular orientation of body i ($i=0 \dots n$) the angle ϕ_i between $\boldsymbol{e}_1$ and $\boldsymbol{e}_1^{(i)}$ is chosen. Let $\underline{A}^i$ be the transformation matrix defined by the equation

$$\underline{e}^{(i)} = \underline{A}^{i}\underline{e} \qquad i=1\dots n.$$

It has the form

$$\underline{A}^{i} = \begin{bmatrix} \cos\phi_i & \sin\phi_i & 0 \\ -\sin\phi_i & \cos\phi_i & 0 \\ 0 & 0 & 1 \end{bmatrix} \qquad i=1\dots n. \tag{5.35}$$

From this follows that the transformation matrix defined by the equation $\underline{e}^{(i)} = \underline{A}^{ij}\underline{e}^{(j)}$ $(i,j=0\dots n)$ is

$$\underline{A}^{ij} = \underline{A}^{i}\underline{A}^{j\mathrm{T}} = \begin{bmatrix} \cos(\phi_i-\phi_j) & \sin(\phi_i-\phi_j) & 0 \\ -\sin(\phi_i-\phi_j) & \cos(\phi_i-\phi_j) & 0 \\ 0 & 0 & 1 \end{bmatrix} \qquad i,j=0\dots n.$$

The coordinate matrices of $\boldsymbol{K}_i$, $\boldsymbol{d}_{ij}$, $\boldsymbol{b}_{i0}$, $\boldsymbol{\omega}_i$, $\dot{\boldsymbol{\omega}}_i$, $\boldsymbol{M}_i$ and $\boldsymbol{Y}_a$ $(i,j,a=1\dots n)$ in the base $\underline{e}^{(i)}$ are written in the form

$$\underline{K}_i = \begin{bmatrix} K_{i1} & -K_{i12} & -K_{i13} \\ -K_{i12} & K_{i2} & -K_{i23} \\ -K_{i13} & -K_{i23} & K_{i3} \end{bmatrix}, \qquad \underline{d}_{ij} = \begin{bmatrix} d_{ij}\cos\alpha_{ij} \\ d_{ij}\sin\alpha_{ij} \\ 0 \end{bmatrix}, \qquad \underline{b}_{i0} = \begin{bmatrix} b_{i0}\cos\beta_i \\ b_{i0}\sin\beta_i \\ 0 \end{bmatrix}$$

$$\underline{\omega}_i = \begin{bmatrix} 0 \\ 0 \\ \dot\phi_i \end{bmatrix}, \qquad \underline{\dot\omega}_i = \begin{bmatrix} 0 \\ 0 \\ \ddot\phi_i \end{bmatrix}, \qquad \underline{M}_i = \begin{bmatrix} 0 \\ 0 \\ M_i \end{bmatrix}, \qquad \underline{Y}_a = \begin{bmatrix} 0 \\ 0 \\ Y_a \end{bmatrix}.$$

In the expressions for $\underline{d}_{ij}$ and $\underline{b}_{i0}$ the scalars d_{ij} and b_{i0} are the absolute values of the projections of these vectors on the plane $\boldsymbol{e}_1^{(i)}-\boldsymbol{e}_2^{(i)}$. The third components have been replaced by zero for reasons that were explained earlier. The constant angles α_{ij} and β_i locate the projections of the vectors in the plane $\boldsymbol{e}_1^{(i)}-\boldsymbol{e}_2^{(i)}$. Of the external forces $\boldsymbol{F}_i$ $(i=1\dots n)$ and of the acceleration $\ddot{\boldsymbol{r}}_0(t)$ it is assumed that the coordinates F_{i1}, F_{i2} and $\ddot r_{01}$, $\ddot r_{02}$ in the base $\underline{e}$ fixed in inertial space are known. The irrelevant third coordinates are replaced by zero. The coordinate matrices of $\boldsymbol{F}_i$ and $\ddot{\boldsymbol{r}}_0$ in the base $\underline{e}^{(i)}$ are then

$$\underline{F}_i = \underline{A}^{i}[F_{i1} \quad F_{i2} \quad 0]^{\mathrm{T}} \quad i=1\dots n; \qquad \underline{\ddot r}_0 = \underline{A}^{i}[\ddot r_{01} \quad \ddot r_{02} \quad 0]^{\mathrm{T}}.$$

With these expressions Eq. (5.34) yields the scalar matrix equation

$$\sum_{j=1}^{n} \underline{K}_{ij}\underline{\dot\omega}_j = \underline{M}_i - \underline{\tilde\omega}_i\underline{K}_i\underline{\omega}_i + M\Big[\sum_{j:s_i<s_j} \dot\phi_j^2\,\underline{\tilde d}_{ij}\underline{A}^{ij}\underline{b}_{j0} + \underline{\tilde b}_{i0}\Big(\underline{\ddot r}_0 + \sum_{j:s_j<s_i} \dot\phi_j^2\,\underline{A}^{ij}\underline{d}_{ji}\Big)\Big]$$

$$-\sum_{j:s_i<s_j} \underline{\tilde d}_{ij}\underline{A}^{i}\underline{F}_j + \sum_{a=1}^{n} S_{ia}\underline{Y}_a \qquad i=1\dots n. \tag{5.36}$$

The terms involving $\dot\phi_j^2$ are explained by the identities $\boldsymbol{\omega}_j\times(\boldsymbol{\omega}_j\times\boldsymbol{b}_{j0}) = -\boldsymbol{\omega}_j^2\,\boldsymbol{b}_{j0}$ and $\boldsymbol{\omega}_j\times(\boldsymbol{\omega}_j\times\boldsymbol{d}_{ji}) = -\boldsymbol{\omega}_j^2\,\boldsymbol{d}_{ji}$. The (3×3) submatrices $\underline{K}_{ij}$ are according to Eq. (5.32)

$$\underline{K}_{ij} = \begin{cases} \underline{K}_i & i=j \\ M(\underline{b}_{j0}^{\mathrm{T}}\underline{A}^{ji}\underline{d}_{ij}\underline{E} - \underline{A}^{ij}\underline{b}_{j0}\underline{d}_{ij}^{\mathrm{T}}) & s_i<s_j \\ M(\underline{d}_{ji}^{\mathrm{T}}\underline{A}^{ji}\underline{b}_{i0}\underline{E} - \underline{A}^{ij}\underline{d}_{ji}\underline{b}_{i0}^{\mathrm{T}}) & s_j<s_i \\ \underline{0} & \text{otherwise} \end{cases} \qquad i,j=1\dots n.$$

They satisfy the identity $\underline{K}_{ji}=\underline{K}_{ij}^{\mathrm{T}}$ which is the scalar form of Eq. (5.33). Each term in Eq. (5.36) is a column matrix of three elements. Only the third element is of interest. It is found by multiplying out all products. The details are shown for two of the more complicated expressions. The first one is the product $\underline{K}_{ij}\dot{\underline{\omega}}_j$. Because of the simple form of $\dot{\underline{\omega}}_j$ only the element with indices (3,3) of $\underline{K}_{ij}$ must be evaluated. Only the case $s_i<s_j$ will be demonstrated. The product $\underline{A}^{ji}\underline{d}_{ij}$ is the matrix

$$d_{ij}\begin{bmatrix} \cos(\phi_j-\phi_i)\cos\alpha_{ij}+\sin(\phi_j-\phi_i)\sin\alpha_{ij} \\ -\sin(\phi_j-\phi_i)\cos\alpha_{ij}+\cos(\phi_j-\phi_i)\sin\alpha_{ij} \\ 0 \end{bmatrix} = d_{ij}\begin{bmatrix} \cos(\phi_i-\phi_j+\alpha_{ij}) \\ \sin(\phi_i-\phi_j+\alpha_{ij}) \\ 0 \end{bmatrix}$$

and the element (3,3) of $\underline{b}_{j0}^{\mathrm{T}}\underline{A}^{ji}\underline{d}_{ij}\underline{E}$ is

$$\begin{aligned} &b_{j0}d_{ij}[\cos\beta_j\cos(\phi_i-\phi_j+\alpha_{ij})+\sin\beta_j\sin(\phi_i-\phi_j+\alpha_{ij})] \\ &\quad = b_{j0}d_{ij}\cos(\phi_i-\phi_j+\alpha_{ij}-\beta_j). \end{aligned}$$

The element (3,3) of $\underline{A}^{ij}\underline{b}_{j0}\underline{d}_{ij}^{\mathrm{T}}$ is easily shown to be zero. Hence, in the case $s_i<s_j$ the term $\underline{K}_{ij}\dot{\underline{\omega}}_j$ contributes the expression

$$b_{j0}d_{ij}\cos(\phi_i-\phi_j+\alpha_{ij}-\beta_j)\ddot{\phi}_j$$

to the desired scalar equation of motion. The second somewhat complicated expression is $\dot{\phi}_j^2\tilde{\underline{d}}_{ij}\underline{A}^{ij}\underline{b}_{j0}$. The last two terms alone yield

$$\underline{A}^{ij}\underline{b}_{j0}=b_{j0}[\cos(\phi_i-\phi_j-\beta_j) \quad -\sin(\phi_i-\phi_j-\beta_j) \quad 0]^{\mathrm{T}}.$$

From this follows in a straightforward way that the third element of the entire expression is

$$-d_{ij}b_{j0}\sin(\phi_i-\phi_j+\alpha_{ij}-\beta_j)\dot{\phi}_j^2.$$

Evaluating also the remaining terms in Eq. (5.36) in this manner we obtain the scalar differential equations

$$\sum_{j=1}^{n}(A_{ij}\ddot{\phi}_j+B_{ij}\dot{\phi}_j^2)=R_i+\sum_{a=1}^{n}S_{ia}Y_a \qquad i=1\dots n$$

with the abbreviations

$$A_{ij} = \begin{cases} K_{i3} & i=j \\ M\,d_{ij}b_{j0}\cos(\phi_i-\phi_j+\alpha_{ij}-\beta_j) & s_i<s_j \\ M\,d_{ji}b_{i0}\cos(\phi_i-\phi_j-\alpha_{ji}+\beta_i) & s_j<s_i \\ 0 & \text{otherwise} \end{cases} \qquad i,j=1\dots n.$$

$$B_{ij} = \begin{cases} M\,d_{ij} b_{j0} \sin(\phi_i - \phi_j + \alpha_{ij} - \beta_j) & s_i < s_j \\ M\,d_{ji} b_{i0} \sin(\phi_i - \phi_j - \alpha_{ji} + \beta_i) & s_j < s_i \\ 0 & \text{otherwise} \end{cases} \qquad i,j = 1 \dots n$$

$$\begin{aligned} R_i &= M_i - M b_{i0} [\ddot{r}_{01} \sin(\phi_i + \beta_i) - \ddot{r}_{02} \cos(\phi_i + \beta_i)] + \\ &\quad + \sum_{j:s_i \leqslant s_j} d_{ij} [F_{j1} \sin(\phi_i + \alpha_{ij}) - F_{j2} \cos(\phi_i + \alpha_{ij})] \qquad i = 1 \dots n \,. \end{aligned}$$

The elements A_{ij} and B_{ij} satisfy the relationships $A_{ji} = A_{ij}$ and $B_{ji} = -B_{ij}$ $(i,j = 1 \dots n)$. The equations of motion can also be written in matrix form as

$$\underline{A} \begin{bmatrix} \ddot{\phi}_1 \\ \vdots \\ \ddot{\phi}_n \end{bmatrix} + \underline{B} \begin{bmatrix} \dot{\phi}_1^2 \\ \vdots \\ \dot{\phi}_n^2 \end{bmatrix} = \underline{S}\,\underline{Y} + \underline{R} \,. \tag{5.37}$$

Of particular interest are mechanical systems with torsional springs and dampers in the hinges. The vector bases $\underline{e}^{(i)}$ $(i = 1 \dots n)$ are now fixed on the bodies in such a way that they are all aligned parallel with each other and with the base $\underline{e}$ when the system is in a position in which all springs are unstressed. For the sake of simplicity all spring and damper coefficients are assumed constant. For hinge a $(a = 1 \dots n)$ they are called k_a and d_a, respectively. With this notation the internal torque Y_a becomes

$$Y_a = -k_a (\phi_{i^+(a)} - \phi_{i^-(a)}) - d_a (\dot{\phi}_{i^+(a)} - \dot{\phi}_{i^-(a)}) \qquad a = 1 \dots n \,.$$

The sign follows from the convention that $+Y_a$ is the torque applied to body $i^+(a)$ and that in the present case it is a torque resisting the growth of the two differences shown in brackets. The equation can also be written in the form

$$\begin{aligned} Y_a &= -k_a \sum_{i=0}^{n} S_{ia} \phi_a - d_a \sum_{i=0}^{n} S_{ia} \dot{\phi}_a \\ &= -k_a \sum_{i=1}^{n} S_{ia} \phi_a - d_a \sum_{i=1}^{n} S_{ia} \dot{\phi}_a - (k_1 \phi_0 + d_1 \dot{\phi}_0) S_{0a} \qquad a = 1 \dots n \,. \end{aligned}$$

The last term is explained by the fact that S_{0a} is different from zero for $a = 1$ only. The column matrix $\underline{Y}$ of all hinge torques is now

$$\underline{Y} = -\underline{k}\,\underline{S}^{\mathrm{T}} \underline{\phi} - \underline{d}\,\underline{S}^{\mathrm{T}} \dot{\underline{\phi}} - (k_1 \phi_0 + d_1 \dot{\phi}_0) \underline{S}_0^{\mathrm{T}}$$

where $\underline{k}$ and $\underline{d}$ are diagonal $(n \times n)$ matrices of the spring and damper coefficients, respectively. Substitution into Eq. (5.37) yields

$$\underline{A} \begin{bmatrix} \ddot{\phi}_1 \\ \vdots \\ \ddot{\phi}_n \end{bmatrix} + \underline{B} \begin{bmatrix} \dot{\phi}_1^2 \\ \vdots \\ \dot{\phi}_n^2 \end{bmatrix} + \underline{S}\,\underline{d}\,\underline{S}^{\mathrm{T}} \begin{bmatrix} \dot{\phi}_1 \\ \vdots \\ \dot{\phi}_n \end{bmatrix} + \underline{S}\,\underline{k}\,\underline{S}^{\mathrm{T}} \begin{bmatrix} \phi_1 \\ \vdots \\ \phi_n \end{bmatrix} = \underline{R} - (d_1 \dot{\phi}_0 + k_1 \phi_0) \underline{S}\,\underline{S}_0^{\mathrm{T}} . \tag{5.38}$$

The matrices $\underline{A}$, $\underline{S}\,\underline{d}\,\underline{S}^{\mathrm{T}}$ and $\underline{S}\,\underline{k}\,\underline{S}^{\mathrm{T}}$ are symmetric and $\underline{B}$ is skew-symmetric. The product $\underline{S}\,\underline{S}_0^{\mathrm{T}}$ is the column matrix $[-1 \quad 0 \quad 0 \dots 0]^{\mathrm{T}}$. The initial assumption that the springs and dampers have constant coefficients can now be dropped. The equations are obviously still valid if k_a and d_a are functions of $(\phi_{i^+(a)} - \phi_{i^-(a)})$ and $(\dot{\phi}_{i^+(a)} - \dot{\phi}_{i^-(a)})$, respectively.

The equations of motion in the special form (5.38) and in the more general form (5.37) are suitable for application to many mechanical systems of interest. One example is the problem of gait of an anthropomorphic figure. The individual links of the human body are executing motions which can with reasonable accuracy be considered plane motions. Eq. (5.37) is valid for a phase of motion in which one foot has contact with the ground. For phases of motion with no ground contact and with two-feet-contact similar equations will be established in Secs. 5.2.4 (Eq. (5.62)) and 5.3.2 (Illustrative Example 5.5).

Illustrative Example 5.1 A very special and simple system governed by Eq. (5.38) is shown in Fig. 5.20. It is a model of a cantilever beam. The beam is shown in the undeformed state and in a highly deformed state in which equations for beams known from elasticity theory are not valid. The system consists of n identical, rigid elements

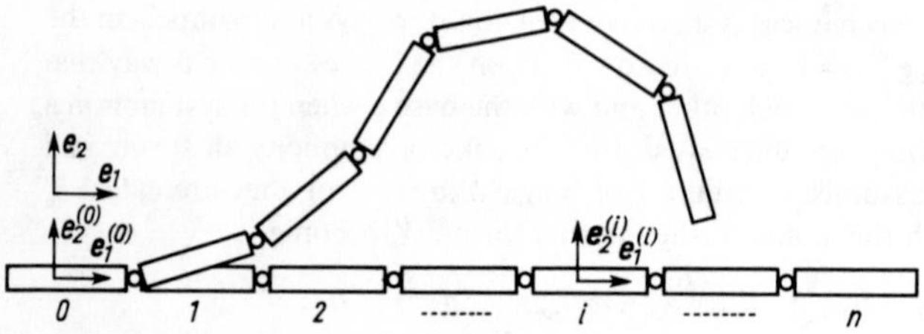

Fig. 5.20
Model of a cantilever beam with body 0 fixed in inertial space

which are coupled by pin joints with parallel hinge axes. Identical torsional springs with a constant coefficient k (and no dampers) are attached to all hinges. The base in which the beam is clamped (body 0) is fixed in inertial space. No external forces and torques are acting. The vector bases $\underline{e}$ and $\underline{e}^{(i)}$ $(i=0 \dots n)$ are oriented as shown in the figure. In the undeformed state all angles $\phi_0 \dots \phi_n$ are zero. Under these conditions the right hand side of Eq. (5.38) is identically zero. On body i $(i=1 \dots n)$ the vectors $\boldsymbol{b}_{i0}$ and $\boldsymbol{d}_{ij}$ $(j=1 \dots n)$ are parallel to $\boldsymbol{e}_1^{(i)}$. This means that $\alpha_{ij}=0$ and $\beta_i=0$ for $i,j=1 \dots n$. For all combinations of indices $i, j=1 \dots n$ $(i \neq j)$ one of the relationships $s_i < s_j$ and $s_j < s_i$ holds. If the bodies are labeled according to Fig. 5.20 then $s_i < s_j$ implies $i<j$. With these simplifications the elements of the matrices $\underline{A}$ and $\underline{B}$ are

$$A_{ij}=a_{ij}\cos(\phi_i-\phi_j), \qquad B_{ij}=a_{ij}\sin(\phi_i-\phi_j) \qquad i,j=1 \dots n$$

$$\text{with} \qquad a_{ij}=a_{ji}=\begin{cases} K_{i3} & i=j \\ M\, d_{ij} b_{j0} & i<j \end{cases} \quad i,j=1 \dots n\,.$$

The matrix $\underline{S}\underline{k}\underline{S}^{\mathrm{T}}$ becomes

$$\underline{S}\underline{k}\underline{S}^{\mathrm{T}}=k\underline{S}\underline{S}^{\mathrm{T}}=k\begin{bmatrix} 2 & -1 & 0 & 0 & \cdot & \cdot & \cdot \\ -1 & 2 & -1 & 0 & & & \\ 0 & -1 & 2 & -1 & & & \\ \cdot & & \ddots & \ddots & \ddots & & \\ \cdot & & & & -1 & 2 & -1 \\ \cdot & & & & 0 & -1 & 1 \end{bmatrix}.$$

The elements of $\underline{A}$ and $\underline{B}$ show that linearization of the equations requires that all differences $\phi_i - \phi_j$ $(i,j=1 \dots n)$ be small. It is not sufficient that the differences are small for all pairs of contiguous bodies. The linearized equations are

$$[a_{ij}]\ddot{\underline{\phi}} + k\underline{S}\,\underline{S}^{\mathrm{T}}\underline{\phi} = \underline{0}\,.$$

These equations can be used to determine the spring constant k. It should have a value which yields for the lowest eigenfrequency of the system the same result as other linearized beam equations. Once k has been determined the nonlinear equations of motion can be treated. In a Taylor series expansion of the equations the linear terms are followed by third-order terms of the form $-a_{ij}(\phi_i - \phi_j)^2\ddot{\phi}_j/2$ and $a_{ij}(\phi_i - \phi_j)\dot{\phi}_j^2$ $(i,j=1 \dots n)$.

5.2.4 Systems with ball-and-socket joints without coupling to an external body whose motion is prescribed

The systems studied in this section differ from the systems considered in Sec. 5.2.2 only in that they are not coupled with an external body whose motion is prescribed as a function of time. Typical examples for such systems are multi-body spacecraft in flight and the human body in a phase of motion without ground contact. A system of seven bodies is shown in Fig. 5.21. For the description of its motion a vector base $\underline{e}^{(0)}$

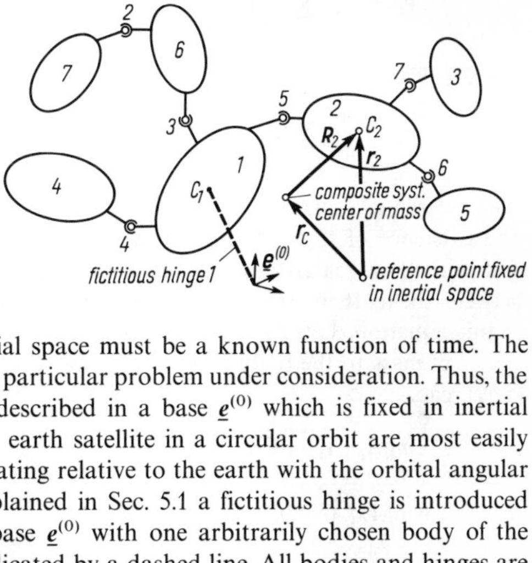

Fig. 5.21
A system with ball-and-socket joints without coupling to an external body whose motion is prescribed. Radius vectors of the system center of mass and of the body i center of mass $(i=2)$

is required whose position in inertial space must be a known function of time. The choice of this base depends upon the particular problem under consideration. Thus, the motion of a jumping man is best described in a base $\underline{e}^{(0)}$ which is fixed in inertial space. The tumbling motions of an earth satellite in a circular orbit are most easily described in a base $\underline{e}^{(0)}$ which is rotating relative to the earth with the orbital angular velocity of the satellite. As was explained in Sec. 5.1 a fictitious hinge is introduced which connects the origin of the base $\underline{e}^{(0)}$ with one arbitrarily chosen body of the system. In Fig. 5.21 this hinge is indicated by a dashed line. All bodies and hinges are then labeled according to the rules specified in Sec. 5.2.1. In Fig. 5.21 the system structure and the labeling are the same as for the systems of Fig. 5.10 and of Fig. 5.8a. Consequently, the directed graph of Fig. 5.8c can be used, again.

Equations of motion will be derived from Newton's law and from the law of moment of momentum. For this purpose, free-body diagrams are produced again by cutting all hinges of the system. The result is a set of n individual bodies with external and internal forces and torques applied to them. The same notation as in Fig. 5.11 is used for all quantities. The basic equations of motion are, therefore, identical with Eqs. (5.7) and (5.8). They read

$$m_i \ddot{\boldsymbol{r}}_i = \boldsymbol{F}_i + \sum_{a=1}^{n} S_{ia} \boldsymbol{X}_a^c \tag{5.39}$$

$$i = 1 \ldots n.$$

$$\dot{\boldsymbol{L}}_i = \boldsymbol{M}_i + \sum_{a=1}^{n} S_{ia} (\boldsymbol{c}_{ia} \times \boldsymbol{X}_a^c + \boldsymbol{Y}_a) \tag{5.40}$$

The internal force $\boldsymbol{X}_1^c$ and the internal torque $\boldsymbol{Y}_1$ in the fictitious hinge 1 are identically zero. The vector $\boldsymbol{c}_{11}$ has, therefore, no physical significance. For convenience, it is set equal to zero,

$$\boldsymbol{c}_{11} = 0 . \tag{5.41}$$

Summing all n Eqs. (5.39) and recognizing that each of the internal hinge forces $\boldsymbol{X}_2^c \ldots \boldsymbol{X}_n^c$ appears with opposite signs in two equations one obtains

$$\sum_{i=1}^{n} m_i \ddot{\boldsymbol{r}}_i = \sum_{i=1}^{n} \boldsymbol{F}_i .$$

With the total system mass M and the radius vector of the composite system center of mass

$$\boldsymbol{r}_C = \frac{1}{M} \sum_{i=1}^{n} m_i \boldsymbol{r}_i$$

this becomes

$$\ddot{\boldsymbol{r}}_C = \frac{1}{M} \sum_{i=1}^{n} \boldsymbol{F}_i . \tag{5.42}$$

This differential equation describes the motion of the composite system center of mass. The existence of this equation is the fundamental difference between systems with and without a material hinge between bodies 0 and 1. All subsequent differences between the mathematical descriptions for the two kinds of systems are a consequence of this equation. Let $\boldsymbol{R}_i$ $(i = 1 \ldots n)$ be the radius vector from the composite system center of mass to the body i center of mass (Fig. 5.21) so that $\boldsymbol{r}_i = \boldsymbol{R}_i + \boldsymbol{r}_C$. Then, Eq. (5.39) reads

$$m_i (\ddot{\boldsymbol{R}}_i + \ddot{\boldsymbol{r}}_C) = \boldsymbol{F}_i + \sum_{a=1}^{n} S_{ia} \boldsymbol{X}_a^c$$

or, in view of Eq. (5.42),

$$m_i \ddot{\boldsymbol{R}}_i = \boldsymbol{F}_i - \frac{m_i}{M} \sum_{j=1}^{n} \boldsymbol{F}_j + \sum_{a=1}^{n} S_{ia} \boldsymbol{X}_a^c$$

$$= \sum_{j=1}^{n} \mu_{ij} \boldsymbol{F}_j + \sum_{a=1}^{n} S_{ia} \boldsymbol{X}_a^c \qquad i = 1 \ldots n$$

where μ_{ij} is the dimensionless quantity

$$\mu_{ij} = \delta_{ij} - \frac{m_i}{M} \qquad i, j = 1 \ldots n . \tag{5.43}$$

The n equations of motion are combined in the single matrix equation

$$\underline{m}\,\ddot{\underline{\boldsymbol{R}}}=\underline{\mu}\,\underline{\boldsymbol{F}}+\underline{S}\,\underline{\boldsymbol{X}}^c. \tag{5.44}$$

The matrices $\underline{m}$, $\underline{\boldsymbol{F}}$, $\underline{S}$ and $\underline{\boldsymbol{X}}^c$ were defined in connection with Eq. (5.9), and $\ddot{\underline{\boldsymbol{R}}}$ and $\underline{\mu}$ are the column matrix $[\ddot{\boldsymbol{R}}_1 \dots \ddot{\boldsymbol{R}}_n]^T$ and the $(n\times n)$ matrix of the elements μ_{ij}, respectively. The n Eqs. (5.40) are written in matrix form as

$$\dot{\underline{\boldsymbol{L}}}=\underline{\boldsymbol{M}}+\underline{\boldsymbol{C}}\times\underline{\boldsymbol{X}}^c+\underline{S}\,\underline{\boldsymbol{Y}}. \tag{5.45}$$

This is identical with Eq. (5.10).

Before these equations can be further developed some important properties of the matrix $\underline{\mu}$ must be compiled. The radius vectors $\boldsymbol{R}_1 \dots \boldsymbol{R}_n$ satisfy the relationship $\sum_{j=1}^{n} m_j\boldsymbol{R}_j=\boldsymbol{0}$, from which follows

$$\boldsymbol{R}_i\equiv\boldsymbol{R}_i-\frac{1}{M}\sum_{j=1}^{n} m_j\boldsymbol{R}_j=\sum_{j=1}^{n}\left(\delta_{ij}-\frac{m_j}{M}\right)\boldsymbol{R}_j=\sum_{j=1}^{n}\mu_{ji}\boldsymbol{R}_j \qquad i=1\dots n$$

and, therefore,

$$\underline{\boldsymbol{R}}\equiv\underline{\mu}^T\underline{\boldsymbol{R}}. \tag{5.46}$$

The matrix $\underline{\mu}$ can be expressed in the form

$$\underline{\mu}=\underline{E}-\frac{1}{M}\,(\underline{m}\,\underline{1}_n\underline{1}_n^T) \tag{5.47}$$

This expression will now be used to prove the identity

$$\underline{\mu}\,\underline{m}\,\underline{\mu}^T\equiv\underline{m}\,\underline{\mu}^T. \tag{5.48}$$

The difference of the right hand side and left hand side products is

$$\underline{m}\,\underline{\mu}^T-\underline{\mu}\,\underline{m}\,\underline{\mu}^T=\frac{1}{M}\underline{m}\,\underline{1}_n\underline{1}_n^T\underline{m}\,\underline{\mu}^T=\frac{1}{M}\underline{m}\,\underline{1}_n\underline{1}_n^T\underline{m}-\frac{1}{M^2}\underline{m}\,\underline{1}_n(\underline{1}_n^T\underline{m}\,\underline{1}_n)\underline{1}_n^T\underline{m}.$$

The term in brackets equals M. From this follows at once that Eq. (5.48) is true. This implies that $\underline{\mu}$ is singular. Indeed, the sum of all rows is a row which contains only zeros. This can be expressed in the form

$$\underline{\mu}^T\underline{1}_n=\underline{0}. \tag{5.49}$$

For a proof Eq. (5.47) is substituted:

$$\underline{\mu}^T\underline{1}_n=\underline{1}_n-\frac{1}{M}\underline{1}_n(\underline{1}_n^T\underline{m}\,\underline{1}_n)=\underline{0}.$$

After these preparations Eqs. (5.44) and (5.45) are considered, again. From the first equation the constraint forces $\underline{\boldsymbol{X}}^c$ can be obtained explicitly by premultiplication with $\underline{T}$:

$$\underline{\boldsymbol{X}}^c=\underline{T}\,(\underline{m}\,\ddot{\underline{\boldsymbol{R}}}-\underline{\mu}\,\underline{\boldsymbol{F}}).$$

Substituting this into Eq. (5.45) we get

$$\dot{\underline{L}} - \underline{C} \times \underline{T}(\underline{m}\ddot{\underline{R}} - \underline{\mu}\underline{F}) = \underline{M} + \underline{S}\,\underline{Y}\,. \tag{5.50}$$

This represents $3n$ scalar differential equations. Three more are provided by Eq. (5.42) so that the total number $3n+3$ equals the total number of degrees of freedom (six in hinge 1 and three in each of the remaining hinges). Eq. (5.50) is similar to Eq. (5.12) which was further developed by expressing the radius vectors $\boldsymbol{r}_1 \ldots \boldsymbol{r}_n$ in terms of $\boldsymbol{r}_0$ and of the body-fixed vectors $\boldsymbol{d}_{ji}$ $(j,i=1 \ldots n)$. Equivalent expressions exist for the vectors $\boldsymbol{R}_1 \ldots \boldsymbol{R}_n$. The vectors $\boldsymbol{d}_{ji}$ are defined by Eq. (5.17). Their physical significance was illustrated in Fig. 5.14. In the present case it must be recognized that the vector $\boldsymbol{c}_{11}$ is zero (cf. Eq. (5.41)). This has the consequence that all vectors $\boldsymbol{d}_{1i}$ $(i=1 \ldots n)$ terminate at the body 1 center of mass. The sum of all vectors $\boldsymbol{d}_{ji}$ over $j=1 \ldots n$ for a fixed value of i is, therefore, the vector from the body i center of mass to the body 1 center of mass, i.e.

$$\boldsymbol{R}_i - \boldsymbol{R}_1 = -\sum_{j=1}^{n} \boldsymbol{d}_{ji} = -\sum_{j=1}^{n} (\underline{C}\,\underline{T})_{ji} \qquad i=1 \ldots n\,.$$

In matrix form these n equations read

$$\underline{R} - \boldsymbol{R}_1 \underline{1}_n = -(\underline{C}\,\underline{T})^{\mathrm{T}} \underline{1}_n\,.$$

When this is premultiplied by $\underline{\mu}^{\mathrm{T}}$ Eqs. (5.46) and (5.49) yield for $\underline{R}$ the explicit result

$$\underline{R} = -(\underline{C}\,\underline{T}\,\underline{\mu})^{\mathrm{T}} \underline{1}_n\,. \tag{5.51}$$

This is now substituted into Eq. (5.50):

$$\dot{\underline{L}} + \underline{C} \times \underline{T}\,\underline{m}(\ddot{\underline{C}}\,\underline{T}\,\underline{\mu})^{\mathrm{T}} \underline{1}_n + \underline{C}\,\underline{T}\,\underline{\mu} \times \underline{F} = \underline{M} + \underline{S}\,\underline{Y}\,.$$

The second term contains the product $\underline{m}\,\underline{\mu}^{\mathrm{T}}$ which is identical with $\underline{\mu}\,\underline{m}\,\underline{\mu}^{\mathrm{T}}$ according to Eq. (5.48). The additional factor $\underline{\mu}$ can, therefore, be inserted. This produces the equation

$$\dot{\underline{L}} + (\underline{C}\,\underline{T}\,\underline{\mu}) \times \underline{m}(\ddot{\underline{C}}\,\underline{T}\,\underline{\mu})^{\mathrm{T}} \underline{1}_n + (\underline{C}\,\underline{T}\,\underline{\mu}) \times \underline{F} = \underline{M} + \underline{S}\,\underline{Y}\,. \tag{5.52}$$

It corresponds to Eq. (5.18) for systems with a ball-and-socket joint 1. The essential role which was played by the vectors $(\underline{C}\,\underline{T})_{ij} = \boldsymbol{d}_{ij}$ in those systems is now played by the vectors $(\underline{C}\,\underline{T}\,\underline{\mu})_{ij}$. Like the vectors $\boldsymbol{d}_{ij}$ they have a simple interpretation. It is found by writing

$$(\underline{C}\,\underline{T}\,\underline{\mu})_{ij} = \sum_{k=1}^{n} (\underline{C}\,\underline{T})_{ik} \mu_{kj} = \sum_{k=1}^{n} \boldsymbol{d}_{ik} \mu_{kj} \qquad i,j=1 \ldots n$$

and by substituting Eq. (5.21) for $\boldsymbol{d}_{ik}$ and Eq. (5.43) for μ_{kj}. In view of Eq. (5.20) this furnishes

$$(\underline{C}\,\underline{T}\,\underline{\mu})_{ij} = \sum_{k=1}^{n} (\boldsymbol{b}_{i0} - \boldsymbol{b}_{ik}) \left(\delta_{kj} - \frac{m_k}{M} \right) = -\boldsymbol{b}_{ij} \qquad i,j=1 \ldots n\,. \tag{5.53}$$

With this relationship the radius vector $\boldsymbol{R}_i$ $(i=1 \ldots n)$ is found to be the sum

$$\boldsymbol{R}_i = \sum_{j=1}^{n} \boldsymbol{b}_{ji} \qquad i=1 \ldots n\,. \tag{5.54}$$

This follows from Eq. (5.51). The augmented body vectors $\boldsymbol{b}_{ij}$ are, thus, shown to be the dominant parameters in the equations of motion (5.52). The further development is analogous to the one of Eq. (5.18), though much simpler. First, the $(n\times n)$ matrix $(\underline{C}\,\underline{T}\,\underline{\mu})\times\underline{m}(\ddot{\underline{C}}\,\underline{T}\,\underline{\mu})^{\mathrm{T}}$ is considered. A single element with indices i,j is abbreviated $\boldsymbol{g}_{ij}$. It is

$$\boldsymbol{g}_{ij}=\sum_{k=1}^{n} m_k\,\boldsymbol{b}_{ik}\times\ddot{\boldsymbol{b}}_{jk}\qquad i,j=1\ldots n. \tag{5.55}$$

In the case $i\neq j$ this can be simplified substantially. For this purpose the directed system graph is divided into two parts by drawing a line across an arbitrary arc on the path between s_i and s_j (Fig. 5.22). Let the set of indices of all vertices of the part containing s_i be denoted by I and the set of indices of all vertices of the other part by II. Then, for all indices k belonging to I (abbreviated $k\in I$) the identity $\boldsymbol{b}_{jk}=\boldsymbol{b}_{ji}$ holds and for all $k\in II$ the identity $\boldsymbol{b}_{ik}=\boldsymbol{b}_{ij}$. With this $\boldsymbol{g}_{ij}$ becomes

$$\boldsymbol{g}_{ij}=\left(\sum_{k\in I} m_k\,\boldsymbol{b}_{ik}\right)\times\ddot{\boldsymbol{b}}_{ji}+\boldsymbol{b}_{ij}\times\sum_{k\in II} m_k\,\ddot{\boldsymbol{b}}_{jk}\qquad i\neq j.$$

The term in brackets can be written in the form

$$\sum_{k\in I} m_k\,\boldsymbol{b}_{ik}=\sum_{k=1}^{n} m_k\,\boldsymbol{b}_{ik}-\sum_{k\in II} m_k\,\boldsymbol{b}_{ik}$$

or using Eq. (5.20) and one of the identities just mentioned

$$\sum_{k\in I} m_k\,\boldsymbol{b}_{ik}=-\boldsymbol{b}_{ij}\sum_{k\in II} m_k\,.$$

In a similar manner

$$\sum_{k\in II} m_k\,\ddot{\boldsymbol{b}}_{jk}=-\ddot{\boldsymbol{b}}_{ji}\sum_{k\in I} m_k\,.$$

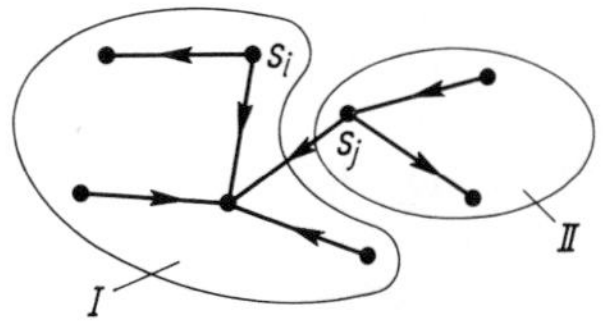

Fig. 5.22 The sets I and II of vertex indices for $s_i\neq s_j$

With this $\boldsymbol{g}_{ij}$ takes the form

$$\boldsymbol{g}_{ij}=-\left(\sum_{k\in I} m_k+\sum_{k\in II} m_k\right)\boldsymbol{b}_{ij}\times\ddot{\boldsymbol{b}}_{ji}=-M\,\boldsymbol{b}_{ij}\times\ddot{\boldsymbol{b}}_{ji}\qquad i,j=1\ldots n,\ i\neq j. \tag{5.56}$$

This simple result corresponds to the much more complicated Eq. (5.22) for systems with a ball-and-socket joint 1.

The matrix Eq. (5.52) is now split into n separate vector equations. These read

$$\dot{\boldsymbol{L}}_i+\sum_{j=1}^{n}\boldsymbol{g}_{ij}-\sum_{j=1}^{n}\boldsymbol{b}_{ij}\times\boldsymbol{F}_j=\boldsymbol{M}_i+\sum_{a=1}^{n} S_{ia}\,\boldsymbol{Y}_a\qquad i=1\ldots n$$

or with the expressions just established for $\boldsymbol{g}_{ij}$ $(i\neq j)$ and $\boldsymbol{g}_{ii}$

$$\dot{\boldsymbol{L}}_i+\sum_{k=1}^{n} m_k\,\boldsymbol{b}_{ik}\times\ddot{\boldsymbol{b}}_{ik}-M\sum_{\substack{j=1\\ \neq i}}^{n}\boldsymbol{b}_{ij}\times\ddot{\boldsymbol{b}}_{ji}-\sum_{j=1}^{n}\boldsymbol{b}_{ij}\times\boldsymbol{F}_j$$

$$=\boldsymbol{M}_i+\sum_{a=1}^{n} S_{ia}\,\boldsymbol{Y}_a\qquad i=1\ldots n. \tag{5.57}$$

This corresponds to Eq. (5.23). The two leading terms can be combined in one simpler expression. In terms of the quantities ϱ' and ϱ explained in Fig. 5.23 the absolute time derivative of the absolute angular momentum of the original (not augmented) body i with respect to the barycenter B_i is

$$\begin{aligned}\int_{m_i} \boldsymbol{\varrho}' \times \ddot{\boldsymbol{\varrho}}' \mathrm{d}m &= \int_{m_i} (\boldsymbol{\varrho}+\boldsymbol{b}_{ii}) \times (\ddot{\boldsymbol{\varrho}}+\ddot{\boldsymbol{b}}_{ii}) \mathrm{d}m \\ &= \int_{m_i} \boldsymbol{\varrho} \times \ddot{\boldsymbol{\varrho}} \mathrm{d}m + m_i \boldsymbol{b}_{ii} \times \ddot{\boldsymbol{b}}_{ii} = \dot{\boldsymbol{L}}_i + m_i \boldsymbol{b}_{ii} \times \ddot{\boldsymbol{b}}_{ii}\,.\end{aligned}$$

If to this expression the sum $\sum_{\substack{k=1\\ \neq i}}^{n} m_k \boldsymbol{b}_{ik} \times \ddot{\boldsymbol{b}}_{ik}$ is added the two leading terms on the left hand side of Eq. (5.57) are obtained. This explains their physical meaning. Together they represent the absolute time derivative of the absolute angular momentum of the augmented body i with respect to its barycenter B_i. The sum over $k \neq i$ is the contribution of the point masses m_k which are attached to body i at the tips of the vectors $\boldsymbol{b}_{ik}$ (see Fig. 5.23). Let $\boldsymbol{K}_i^*$ be the inertia tensor of the augmented body i with respect to its barycenter. In terms of the central inertia tensor $\boldsymbol{J}_i$ of the original body i it is

$$\boldsymbol{K}_i^* = \boldsymbol{J}_i + \sum_{k=1}^{n} m_k (b_{ik}^2 \boldsymbol{E} - \boldsymbol{b}_{ik} \boldsymbol{b}_{ik}) \qquad i=1 \dots n\,. \tag{5.58}$$

The two leading terms in Eq. (5.57) can now be expressed in the form

$$\boldsymbol{K}_i^* \cdot \dot{\boldsymbol{\omega}}_i + \boldsymbol{\omega}_i \times \boldsymbol{K}_i^* \cdot \boldsymbol{\omega}_i$$

where $\boldsymbol{\omega}_i$ is the absolute angular velocity of body i. With this the equations of motion read

$$\boldsymbol{K}_i^* \cdot \dot{\boldsymbol{\omega}}_i + \boldsymbol{\omega}_i \times \boldsymbol{K}_i^* \cdot \boldsymbol{\omega}_i - M \sum_{\substack{j=1\\ \neq i}}^{n} \boldsymbol{b}_{ij} \times \ddot{\boldsymbol{b}}_{ji} - \sum_{j=1}^{n} \boldsymbol{b}_{ij} \times \boldsymbol{F}_j = \boldsymbol{M}_i + \sum_{a=1}^{n} S_{ia} \boldsymbol{Y}_a \qquad i=1 \dots n\,.$$

For $\ddot{\boldsymbol{b}}_{ji}$ the expression

$$\ddot{\boldsymbol{b}}_{ji} = \dot{\boldsymbol{\omega}}_j \times \boldsymbol{b}_{ji} + \boldsymbol{\omega}_j \times (\boldsymbol{\omega}_j \times \boldsymbol{b}_{ji}) \tag{5.59}$$

must later be substituted.

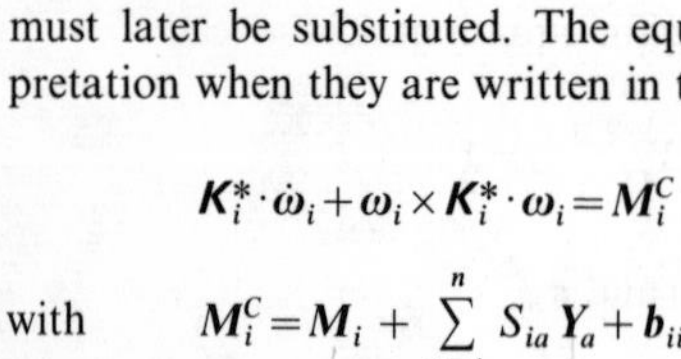

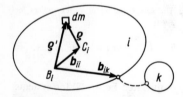

Fig. 5.23 Vectors locating a mass particle on body i. Center of mass C_i and barycenter B_i

The equations of motion allow a simple physical interpretation when they are written in the form

$$\boldsymbol{K}_i^* \cdot \dot{\boldsymbol{\omega}}_i + \boldsymbol{\omega}_i \times \boldsymbol{K}_i^* \cdot \boldsymbol{\omega}_i = \boldsymbol{M}_i^C \qquad i=1 \dots n$$

with $$\boldsymbol{M}_i^C = \boldsymbol{M}_i + \sum_{a=1}^{n} S_{ia} \boldsymbol{Y}_a + \boldsymbol{b}_{ii} \times \boldsymbol{F}_i + \sum_{\substack{j=1\\ \neq i}}^{n} \boldsymbol{b}_{ij} \times (M \ddot{\boldsymbol{b}}_{ji} + \boldsymbol{F}_j) \qquad i=1 \dots n\,.$$

This has the form of the law of moment of momentum for a single rigid body in the special case where the center of mass is used as reference point for angular momentum and external torques. The role of the rigid body is played by the augmented body i.

The reference point for $\boldsymbol{K}_i^*$ is, indeed, its center of mass, i.e. the barycenter B_i. The torque $\boldsymbol{M}_i^C$ contains, first, the external torque $\boldsymbol{M}_i$ and the resultant $\sum_{a=1}^{n} S_{ia}\boldsymbol{Y}_a$ of all internal hinge torques on the body. Second, the external force $\boldsymbol{F}_i$ whose line of action is passing through the center of mass C_i is contributing the torque $\boldsymbol{b}_{ii}\times\boldsymbol{F}_i$ (see Fig. 5.24). The last term can be interpreted as follows. Let the augmented body j ($j\neq i$) be suspended as a pendulum in inertial space at its hinge point leading toward body i and let it be subject to its external force $\boldsymbol{F}_j$. When it is then given its actual angular velocity and acceleration it exerts on the suspension point the force $M\ddot{\boldsymbol{b}}_{ji}+\boldsymbol{F}_j$. This force must be shifted so that its line of action is passing through the hinge point on body i which is leading toward body j (point Q in Fig. 5.24). It then causes the torque $\boldsymbol{b}_{ij}\times(M\ddot{\boldsymbol{b}}_{ji}+\boldsymbol{F}_j)$ about the barycenter B_i. Summing these torques over all $j\neq i$ the last term of $\boldsymbol{M}_i^C$ is obtained.

As a final preparation for numerical and non-numerical applications the equations of motion will now be brought into a form similar to Eq. (5.34). Substitution of Eq. (5.59) results in

$$\boldsymbol{K}_i^*\cdot\dot{\boldsymbol{\omega}}_i - M\sum_{\substack{j=1\\ \neq i}}^{n}\boldsymbol{b}_{ij}\times(\dot{\boldsymbol{\omega}}_j\times\boldsymbol{b}_{ji}) = \boldsymbol{M}_i' + \boldsymbol{M}_i + \sum_{a=1}^{n} S_{ia}\boldsymbol{Y}_a \qquad i=1\ldots n$$

with

$$\boldsymbol{M}_i' = -\boldsymbol{\omega}_i\times\boldsymbol{K}_i^*\cdot\boldsymbol{\omega}_i + M\sum_{\substack{j=1\\ \neq i}}^{n}\boldsymbol{b}_{ij}\times[\boldsymbol{\omega}_j\times(\boldsymbol{\omega}_j\times\boldsymbol{b}_{ji})] + \sum_{j=1}^{n}\boldsymbol{b}_{ij}\times\boldsymbol{F}_j \qquad i=1\ldots n.$$

The double vector-cross product on the left hand side is reformulated as

$$\boldsymbol{b}_{ij}\times(\dot{\boldsymbol{\omega}}_j\times\boldsymbol{b}_{ji}) = (\boldsymbol{b}_{ji}\cdot\boldsymbol{b}_{ij}\boldsymbol{E} - \boldsymbol{b}_{ji}\boldsymbol{b}_{ij})\cdot\dot{\boldsymbol{\omega}}_j\,.$$

Tensors $\boldsymbol{K}_{ij}$ are then defined as follows

$$\boldsymbol{K}_{ij} = \begin{cases}\boldsymbol{K}_i^* & i=j\\ M(\boldsymbol{b}_{ji}\boldsymbol{b}_{ij} - \boldsymbol{b}_{ji}\cdot\boldsymbol{b}_{ij}\boldsymbol{E}) & i\neq j\end{cases} \qquad i,j=1\ldots n.$$

They satisfy the relationship

$$\boldsymbol{K}_{ji} = \bar{\boldsymbol{K}}_{ij}^{T} \qquad i,j=1\ldots n. \qquad (5.60)$$

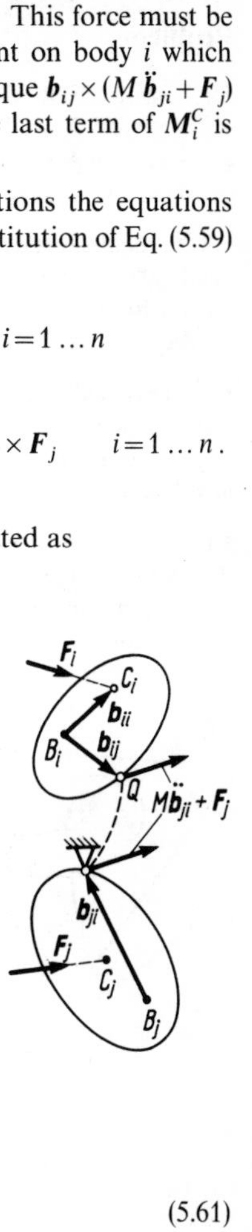

Fig. 5.24
Interpretation of $M\ddot{\boldsymbol{b}}_{ji}+\boldsymbol{F}_j$ as force applied to the suspension point of a pendulum

With them the equations of motion take the final form

$$\sum_{j=1}^{n}\boldsymbol{K}_{ij}\cdot\dot{\boldsymbol{\omega}}_j = \boldsymbol{M}_i' + \boldsymbol{M}_i + \sum_{a=1}^{n} S_{ia}\boldsymbol{Y}_a \qquad i=1\ldots n. \qquad (5.61)$$

These equations are supplemented by the single Eq. (5.42) for the motion of the composite system center of mass. In addition, kinematic differential equations have to be formulated. They are identical with the ones added to Eq. (5.34) so that no further comment is needed.

$$(\vec{\omega}_i\times\vec{b}_{ij})\cdot(\vec{b}_{ji}\times\vec{\omega}_j) \equiv \vec{\omega}_i\cdot(\vec{b}_{ji}\vec{b}_{ij} - \vec{b}_{ji}\cdot\vec{b}_{ij}\cdot E)\cdot\vec{\omega}_j$$

Eq. (5.61) is identical in form with Eq. (5.34) which was obtained for systems with a material hinge 1. The development of both equations followed the same sequence of steps (except for the formulation and re-substitution of Eq. (5.42)). For systems in which hinge 1 is not a material ball-and-socket joint all these steps are mathematically simpler. This is a consequence of the fact that in such systems no single body or hinge is playing a dominant role. In systems with a material hinge 1 this hinge together with body 0 introduces a lack of symmetry which is reflected in the mathematical description of the system.

Problems

5.7 Write a FORTRAN or ALGOL program for the calculation of the constant coordinates of the vectors $\boldsymbol{b}_{ij}$ and of the tensors $\boldsymbol{K}_i^*$ $(i=1\dots n;\, j=0\dots n)$ in the respective body-fixed bases $\underline{e}^{(i)}$. Use as input data the masses and inertia components of the individual bodies and the constant coordinates of the vectors $\boldsymbol{c}_{ia}$ $(i,a=1\dots n)$.

5.8 n identical rods of length l, mass m and central moment of inertia J (about an axis perpendicular to the rod) are connected with one another at their end points to form a chain. Give a formula for the central moment of inertia (about an axis perpendicular to the rod) for the i-th augmented body $(i=1\dots n)$.

5.9 Two bodies are coupled by a ball-and-socket joint. In a torque-free invironment the system can be in a state of permanent rotation in which the bodies have equal and constant angular velocities $\boldsymbol{\omega}_1=\boldsymbol{\omega}_2\equiv\boldsymbol{\omega}=\text{const.}$ Show that in such a state of motion the vectors $\boldsymbol{\omega}$, $\boldsymbol{K}_1^*\cdot\boldsymbol{\omega}$, $\boldsymbol{K}_2^*\cdot\boldsymbol{\omega}$, $\boldsymbol{J}_1\cdot\boldsymbol{\omega}$, $\boldsymbol{J}_2\cdot\boldsymbol{\omega}$, $\boldsymbol{b}_{12}$ and $\boldsymbol{b}_{21}$ lie in one common plane.

5.10 A system of interconnected flat discs is moving without friction on an inclined plane (Fig. 5.25). Springs and dampers are mounted in the pin joints between the bodies. The constant spring and damper coefficients in hinge a $(a=1\dots n)$ are called k_a and d_a, respectively. The base vectors of an inertial frame of reference $\underline{e}$ are directed as shown in the figure. On body i

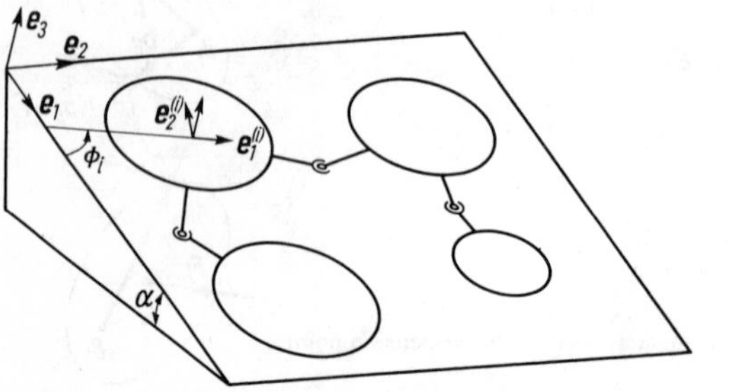

Fig. 5.25
A system of discs on an inclined plane

$(i=1\dots n)$ a base $\underline{e}^{(i)}$ is fixed in such a way that $\boldsymbol{e}_3^{(i)}$ is parallel to $\boldsymbol{e}_3$. Furthermore, the bases $\underline{e}^{(i)}$ $(i=1\dots n)$ are aligned parallel to one another when all springs are unstressed. The angle of rotation of body i $(i=1\dots n)$ about $\boldsymbol{e}_3$ is called ϕ_i. It is measured between $\boldsymbol{e}_1$ and $\boldsymbol{e}_1^{(i)}$. Apply the methods of Sec. 5.2.3 to Eq. (5.61) and show that these equations can be reduced to

$$\underline{A}\begin{bmatrix}\ddot{\phi}_1\\ \vdots\\ \ddot{\phi}_n\end{bmatrix}+\underline{B}\begin{bmatrix}\dot{\phi}_1^2\\ \vdots\\ \dot{\phi}_n^2\end{bmatrix}+\underline{S}\,\underline{d}\,\underline{S}^{\mathrm{T}}\begin{bmatrix}\dot{\phi}_1\\ \vdots\\ \dot{\phi}_n\end{bmatrix}+\underline{S}\,\underline{k}\,\underline{S}^{\mathrm{T}}\begin{bmatrix}\phi_1\\ \vdots\\ \phi_n\end{bmatrix}=\underline{R} \tag{5.62}$$

where the matrices $\underline{A}$, $\underline{B}$, $\underline{d}$, $\underline{k}$ and $\underline{R}$ have the elements

$$\left.\begin{aligned} A_{ij} &= \begin{cases} K^*_{i3} & i=j \\ -M\, b_{ij} b_{ji} \cos(\phi_i - \phi_j + \beta_{ij} - \beta_{ji}) & i \neq j \end{cases} \\ B_{ij} &= -M\, b_{ij} b_{ji} \sin(\phi_i - \phi_j + \beta_{ij} - \beta_{ji}) \\ d_{ij} &= d_i \delta_{ij}, \qquad k_{ij} = k_i \delta_{ij} \end{aligned}\right\} i,j = 1 \dots n$$

$$R_i = M_i + \sum_{j=1}^{n} b_{ij}[-F_{j1} \sin(\phi_i + \beta_{ij}) + F_{j2} \cos(\phi_i + \beta_{ij})] \quad i = 1 \dots n.$$

The angle β_{ij} $(i,j=1 \dots n)$ is measured on body i between $e_1^{(i)}$ and the projection of $\boldsymbol{b}_{ij}$ onto the inclined plane. The quantities M_i, F_{i1} and F_{i2} $(i=1 \dots n)$ represent the resultant external torque on body i about $\boldsymbol{e}_3$ (with respect to the body center of mass) and the coordinates of the resultant external force on body i in the directions of $\boldsymbol{e}_1$ and $\boldsymbol{e}_2$. In the particular case of motions on an inclined plane these quantities are $M_i = 0$, $F_{i1} = m_i g \sin\alpha$ and $F_{i2} = 0$ $(i = 1 \dots n)$.

5.2.5 The special case of a multi-body satellite in a circular orbit

In this section one of the exceptional cases is studied in which for a multi-body system results of, both, theoretical and practical interest can be found by non-numerical methods. The physical phenomenon to be investigated is explained, first, for the simple case where instead of a multi-body system a single rigid body is considered. The body is moving as a satellite in a circular orbit about the Earth. The gravitational force is given by Newton's law. This means that a mass particle dm of the satellite at a radius vector $\boldsymbol{r}$ from the center of the Earth is attracted by the force

$$\mathrm{d}\boldsymbol{F} = -\varkappa \frac{\mathrm{d}m\,\boldsymbol{r}}{r^3}$$

where $\varkappa$ denotes the product of the universal gravitational constant and the mass of the Earth. The relationship points out the physical phenomenon to be examined. Particles of identical mass but at different locations within the satellite experience different gravitational attraction forces. If a typical length of the satellite is on the order of several meters and the radius of the orbit trajectory on the order of 6.500 km then the ratio between the two lengths is approximately 10^{-6}. The difference between the gravitational attaction forces acting on two particles of identical mass is, therefore, exceedingly small compared with the attraction force itself. The difference can safely be neglected when the orbit trajectory is determined. In this part of the problem, therefore, for all mass particles the radius vector $\boldsymbol{r}$ is replaced by the radius vector $\boldsymbol{r}_C$ of the satellite center of mass. This simplification results in a Keplerian orbit for the center of mass. In the present case, in particular, it is assumed that the orbit is circular so that the magnitude of $\boldsymbol{r}_C$ is independent of time. The satellite is moving along its trajectory with a constant orbital angular velocity $\boldsymbol{\omega}_0$ whose magnitude depends on the orbit radius r_C. The relationship is given by Kepler's third law

$$\omega_0^2 = \frac{\varkappa}{r_C^3}. \tag{5.63}$$

The nonhomogeneity of the gravitational field over the volume occupied by the satellite must not be neglected, however, when rotational motions of the satellite are of concern. The force d$\boldsymbol{F}$ applied to a mass particle dm causes a torque about the body center of mass. When this is integrated over the entire mass a resultant gravitational torque is obtained which, in general, is not zero. Although this torque is extremely small it must not be neglected for the simple reason that it is the only external torque on the body (it is assumed here that there are no other torques such as those caused by solar pressure on the satellite surface or by interaction with the Earth's magnetic field, for instance). From the explanation given for the resultant gravitational torque it follows at once that it is a function of the angular orientation of the body relative to the Earth. For the description of this orientation the orbital reference frame $\underline{e}$ shown in Fig. 5.26 is used. Its origin coincides at all times with the satellite center of mass, and it is rotating relative to the Earth with the orbital angular velocity $\boldsymbol{\omega}_0$. The base vector $\boldsymbol{e}_3$ is directed along the local outward vertical and the vector $\boldsymbol{e}_2$ along $\boldsymbol{\omega}_0$. Let $\boldsymbol{J}$ and $\boldsymbol{\omega}$ be the central inertia tensor of the body and its absolute angular velocity, respectively. If $\boldsymbol{M}_{\text{grav}}$ denotes the resultant gravitational torque then rotational motions of the body are governed by the equation

$$\boldsymbol{J}\cdot\dot{\boldsymbol{\omega}}+\boldsymbol{\omega}\times\boldsymbol{J}\cdot\boldsymbol{\omega}=\boldsymbol{M}_{\text{grav}}\,.$$

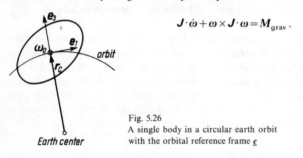

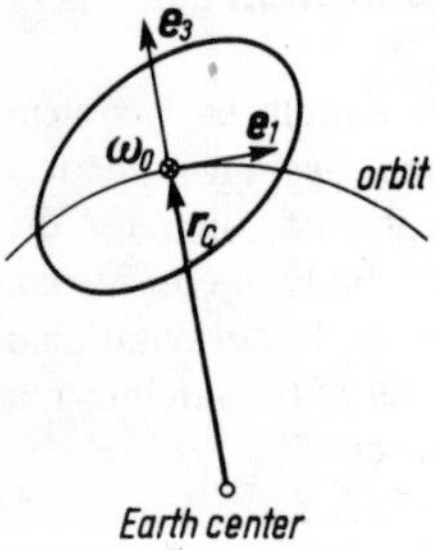

Fig. 5.26
A single body in a circular earth orbit with the orbital reference frame $\underline{e}$

Relative to the orbital reference frame the body is rotating with an angular velocity called $\boldsymbol{\omega}_{\text{rel}}$. Hence, $\boldsymbol{\omega}=\boldsymbol{\omega}_0+\boldsymbol{\omega}_{\text{rel}}$ and $\dot{\boldsymbol{\omega}}=\dot{\boldsymbol{\omega}}_{\text{rel}}$. The rotation relative to the base $\underline{e}$ is, therefore, governed by the equation

$$\boldsymbol{J}\cdot\dot{\boldsymbol{\omega}}_{\text{rel}}+(\boldsymbol{\omega}_0+\boldsymbol{\omega}_{\text{rel}})\times\boldsymbol{J}\cdot(\boldsymbol{\omega}_0+\boldsymbol{\omega}_{\text{rel}})=\boldsymbol{M}_{\text{grav}}\,. \tag{5.64}$$

It is interesting to consider whether it is possible that this equation has the solution $\boldsymbol{\omega}_{\text{rel}}\equiv\boldsymbol{0}$, i.e. whether the satellite can remain stationary relative to the rotating reference frame $\underline{e}$. Such a state exists and is called relative equilibrium position. From Eq. (5.64) follows as condition for relative equilibrium the equation

$$\boldsymbol{\omega}_0\times\boldsymbol{J}\cdot\boldsymbol{\omega}_0=\boldsymbol{M}_{\text{grav}}\,. \tag{5.65}$$

The quantity on the right hand side was shown to be a function of the angular orientation of the body relative to the base $\underline{e}$. The same is true for the term on the left hand side since $\boldsymbol{\omega}_0$ has constant coordinates in $\underline{e}$ whereas $\boldsymbol{J}$ has constant inertia components in a body-fixed frame of reference. The equation is, therefore, determining the unknown angular orientation in the state of relative equilibrium. Relative equilibrium positions of this kind can be observed in nature. The moon is in relative equilibrium in its Earth orbit and the planet Mercury in its orbit about the sun. Relative equilibrium positions have considerable practical importance for the performance of orbiting spacecraft. In the design phase of orbiting artificial satellites for observation and signal transmission purposes the relative equilibrium positions must be known in advance.

Only then can cameras and antennas be mounted in such a way that during flight they are always pointing vertically toward the Earth.

After these introductory remarks the general problem to be treated here can be formulated. Given is a multi-body system with tree structure and with ball-and-socket joints without any internal hinge torques. Each individual body represents a gyrostat with rotors whose angular velocities relative to the carrier body are kept constant by control devices. The entire system is moving as a satellite in a circular orbit about the Earth. The questions to be answered are: Does the system possess relative equilibrium positions in the sense that all carriers of the system are simultaneously in a state of relative equilibrium with the rotors rotating relative to the carriers? If so, how do the relative equilibrium positions depend upon the parameters of the system, in particular upon the angular momenta of the rotors relative to the carriers? Mutual gravitational attraction forces between bodies of the system can be neglected.

The solution will be found by following the line of arguments described above for the single-body satellite. First, the equations of rotational motions of the system will be formulated. For the external gravitational forces and torques explicit expressions will be developed. From the resulting equations equilibrium conditions will be obtained by introducing the identities $\boldsymbol{\omega}_i \equiv \boldsymbol{\omega}_0$ $(i=1 \dots n)$ for the absolute angular velocities of all carriers of the system. The equations of rotational motions are developed from Eq. (5.61). The internal hinge torques $\boldsymbol{Y}_a$ $(a=1 \dots n)$ are zero by assumption. The external forces $\boldsymbol{F}_i$ and $\boldsymbol{M}_i$ $(i=1 \dots n)$ caused by the Earth's gravitational field will be examined later. The only other point requiring attention is the presence of rotors on the bodies. Eq. (5.61) governs a system without rotors. It is a simple matter, however, to add terms which render the equations applicable to the present case. For this purpose it must be remembered that, except for the formulation, Eqs. (5.61) and (5.57) are identical. In the latter $\dot{\boldsymbol{L}}_i$ denotes the time derivative of the absolute angular momentum of body i with respect to its center of mass. If body i is a gyrostat consisting of a carrier and of rotors with constant angular velocities relative to the carrier then the absolute angular momentum is composed of two parts (see Chap. 4.6). The first part is the angular momentum of the body (carrier plus rotors) when all rotors are "frozen". This part is the quantity called $\boldsymbol{L}_i$ in Eq. (5.57). The second part is the resultant angular momentum relative to the carrier of all rotors mounted on the carrier. It is a vector $\boldsymbol{h}_i$ whose coordinates in a carrier-fixed reference frame are constant. If $\boldsymbol{\omega}_i$ denotes the absolute angular velocity of the carrier i then $\dot{\boldsymbol{L}}_i$ has to be replaced by the expression $\dot{\boldsymbol{L}}_i + \boldsymbol{\omega}_i \times \boldsymbol{h}_i$. The vector-cross product is the only additional term caused by the rotors. It follows that the rotational equations (5.61) have to be replaced by

$$\sum_{j=1}^{n} \boldsymbol{K}_{ij} \cdot \dot{\boldsymbol{\omega}}_j = -\boldsymbol{\omega}_i \times (\boldsymbol{K}_i^* \cdot \boldsymbol{\omega}_i + \boldsymbol{h}_i) + M \sum_{\substack{j=1 \\ \neq i}}^{n} \boldsymbol{b}_{ij} \times [\boldsymbol{\omega}_j \times (\boldsymbol{\omega}_j \times \boldsymbol{b}_{ji})]$$

$$+ \sum_{j=1}^{n} \boldsymbol{b}_{ij} \times \boldsymbol{F}_j + \boldsymbol{M}_i \qquad i=1 \dots n. \tag{5.66}$$

Next, expressions will be developed for $\boldsymbol{F}_i$ and $\boldsymbol{M}_i$ $(i=1 \dots n)$. They are independent of the rotations of the rotors relative to the carriers since only the mass distribution of the system is relevant. Therefore, the rotors are assumed, for the moment, to be "frozen". Fig. 5.27 shows the system together with the orbital reference frame $\underline{e}$ whose origin is at the composite system center of mass at the radius vector $\boldsymbol{r}_C$ from the Earth's center. The magnitude r_C of $\boldsymbol{r}_C$ is constant by assumption. The vector from the system center of mass to the center of mass of body i ($i=1 \dots n$) is called $\boldsymbol{R}_i$ as in Fig. 5.21. The location of the mass particle dm on body i is defined by the body-fixed radius vector $\boldsymbol{\varrho}$. The gravitational force acting on the mass element is

$$\mathrm{d}\boldsymbol{F}_i = -\varkappa \frac{\boldsymbol{r}_C+\boldsymbol{R}_i+\boldsymbol{\varrho}}{|\boldsymbol{r}_C+\boldsymbol{R}_i+\boldsymbol{\varrho}|^3}\,\mathrm{d}m.$$

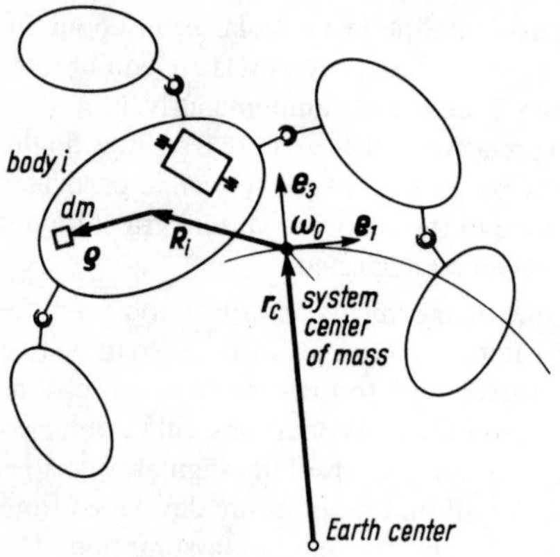

Fig. 5.27
Multi-body satellite in a circular orbit with vectors locating a mass particle on body i

The denominator is developed into the Taylor series

$$|\boldsymbol{r}_C+\boldsymbol{R}_i+\boldsymbol{\varrho}|^3 = [(\boldsymbol{r}_C+\boldsymbol{R}_i+\boldsymbol{\varrho})^2]^{3/2} = r_C^3\left[1+\frac{2\boldsymbol{e}_3\cdot(\boldsymbol{R}_i+\boldsymbol{\varrho})}{r_C}+\cdots\right]^{3/2}$$
$$= r_C^3\left[1+\frac{3\boldsymbol{e}_3\cdot(\boldsymbol{R}_i+\boldsymbol{\varrho})}{r_C}+\cdots\right].$$

Dots indicate terms of second and higher order in $|\boldsymbol{R}_i+\boldsymbol{\varrho}|/r_C$ which can be neglected (in the numerical example given earlier the second order terms are on the order of 10^{-12}). With this expression d$\boldsymbol{F}_i$ takes the form

$$\mathrm{d}\boldsymbol{F}_i = -\frac{\varkappa}{r_C^3}(\boldsymbol{r}_C+\boldsymbol{R}_i+\boldsymbol{\varrho})\left[1-\frac{3\boldsymbol{e}_3\cdot(\boldsymbol{R}_i+\boldsymbol{\varrho})}{r_C}\right]\mathrm{d}m+\cdots. \tag{5.67}$$

When this is multiplied out the term $(\boldsymbol{R}_i+\boldsymbol{\varrho})^2/r_C$ can be neglected against $|\boldsymbol{R}_i+\boldsymbol{\varrho}|$ as a second-order term. The factor in front of the first bracket is $-\omega_0^2$ (see Eq. (5.63)). Hence,

$$\mathrm{d}\boldsymbol{F}_i = -\omega_0^2[\boldsymbol{r}_C-3\boldsymbol{e}_3\boldsymbol{e}_3\cdot(\boldsymbol{R}_i+\boldsymbol{\varrho})+\boldsymbol{R}_i+\boldsymbol{\varrho}]\,\mathrm{d}m+\cdots.$$

Now the integration can be performed over the total mass of body i. Recognizing the identities $\sum_{i=1}^{n} m_i\boldsymbol{R}_i=\boldsymbol{0}$ and $\int_{m_i}\boldsymbol{\varrho}\,\mathrm{d}m=\boldsymbol{0}$ we get

$$\boldsymbol{F}_i = -\omega_0^2 m_i(\boldsymbol{r}_C+\boldsymbol{R}_i-3\boldsymbol{e}_3\boldsymbol{R}_i\cdot\boldsymbol{e}_3) \qquad i=1\dots n. \tag{5.68}$$

The force would simply be $-\omega_0^2 m_i\boldsymbol{r}_C = -\varkappa m_i\boldsymbol{r}_C/r_C^3$ if the total mass of body i were to be concentrated at the composite system center of mass. The remaining terms which are smaller by a factor on the order of 10^{-6} in the example given earlier are caused by the finite dimensions of the system.

Next, the torque $\boldsymbol{M}_i$ with respect to the body i center of mass is evaluated. It is represented by the integral

$$\boldsymbol{M}_i = \int_{m_i} \boldsymbol{\varrho} \times \mathrm{d}\boldsymbol{F}_i \mathrm{d}m \qquad i=1\dots n$$

or with Eq. (5.67)

$$\boldsymbol{M}_i = -\omega_0^2 \int_{m_i} \boldsymbol{\varrho} \times [\boldsymbol{r}_C - 3\boldsymbol{e}_3\boldsymbol{e}_3\cdot(\boldsymbol{R}_i+\boldsymbol{\varrho}) + \boldsymbol{R}_i + \boldsymbol{\varrho}]\mathrm{d}m\,.$$

Because of the identity $\int_{m_i} \boldsymbol{\varrho}\mathrm{d}m = \boldsymbol{0}$ this is equal to

$$\boldsymbol{M}_i = 3\omega_0^2 \int_{m_i} \boldsymbol{\varrho} \times \boldsymbol{e}_3\boldsymbol{e}_3\cdot\boldsymbol{\varrho}\mathrm{d}m = -3\omega_0^2\boldsymbol{e}_3 \times \int_{m_i} \boldsymbol{\varrho}\boldsymbol{\varrho}\mathrm{d}m\cdot\boldsymbol{e}_3\,.$$

This can also be written in the form

$$\boldsymbol{M}_i = 3\omega_0^2\boldsymbol{e}_3 \times \int_{m_i} (\varrho^2\boldsymbol{E} - \boldsymbol{\varrho}\boldsymbol{\varrho})\mathrm{d}m\cdot\boldsymbol{e}_3$$

since $\boldsymbol{e}_3 \times \boldsymbol{E}\cdot\boldsymbol{e}_3 = \boldsymbol{e}_3 \times \boldsymbol{e}_3$ is equal to zero. The integral represents the central inertia tensor $\boldsymbol{J}_i$ of body i. Thus, the final result is obtained

$$\boldsymbol{M}_i = 3\omega_0^2\boldsymbol{e}_3 \times \boldsymbol{J}_i\cdot\boldsymbol{e}_3 \qquad i=1\dots n\,. \tag{5.69}$$

It confirms the statement made earlier that in a nonhomogeneous gravitational field a body of finite dimensions is, in general, subject to a very small torque which is a function of the angular orientation of the body relative to the orbital reference frame. The magnitude of the torque is mainly determined by ω_0 which for near-Earth orbits is on the order of $2\pi/(100\,\mathrm{min}) \approx 10^{-3}/\mathrm{s}$.

Before substituting the expressions for $\boldsymbol{F}_i$ and $\boldsymbol{M}_i$ into the equations of motion let us briefly return to the special case of a single rigid body in orbit. Its equation of motion is now

$$\boldsymbol{J}\cdot\dot{\boldsymbol{\omega}} + \boldsymbol{\omega} \times \boldsymbol{J}\cdot\boldsymbol{\omega} = 3\omega_0^2\boldsymbol{e}_3 \times \boldsymbol{J}\cdot\boldsymbol{e}_3\,.$$

From this follows the remarkable result that two different bodies with identical ratios $J_1{:}J_2{:}J_3$ of the principal moments of inertia move with identical angular velocities $\boldsymbol{\omega}(t)$ provided ω_0 and the initial conditions are also identical. The size of the bodies has no influence. With $\boldsymbol{\omega} = \omega_0\boldsymbol{e}_2$ the equilibrium condition (5.65) for the single rigid body becomes

$$\boldsymbol{e}_2 \times \boldsymbol{J}\cdot\boldsymbol{e}_2 = 3\boldsymbol{e}_3 \times \boldsymbol{J}\cdot\boldsymbol{e}_3\,. \tag{5.70}$$

Either side of the equation is zero if $\boldsymbol{e}_2$ as well as $\boldsymbol{e}_3$ are parallel to principal axes of inertia of the body. Then, all three principal axes of inertia are parallel to the base vectors $\boldsymbol{e}_{1,2,3}$. It can be shown that these positions of relative equilibrium are the only solutions of the equilibrium condition. For this purpose the equation is decomposed into three scalar equations for coordinates in the orbital reference frame. The coordinate matrices of $\boldsymbol{e}_2$, $\boldsymbol{e}_3$ and $\boldsymbol{J}$ are

$$\underline{e}_2 = \begin{bmatrix} 0 \\ 1 \\ 0 \end{bmatrix} \qquad \underline{e}_3 = \begin{bmatrix} 0 \\ 0 \\ 1 \end{bmatrix} \qquad \underline{J} = \begin{bmatrix} J_{11} & -J_{12} & -J_{13} \\ -J_{12} & J_{22} & -J_{23} \\ -J_{13} & -J_{23} & J_{33} \end{bmatrix}.$$

The elements of $\underline{J}$ are, of course, still unknown because the equilibrium position is unknown. Eq. (5.70) yields $\tilde{\underline{e}}_2 \underline{J} \underline{e}_2 = 3\tilde{\underline{e}}_3 \underline{J} \underline{e}_3$ or after multiplying out both sides

$$\begin{bmatrix} -J_{23} \\ 0 \\ J_{12} \end{bmatrix} = 3 \begin{bmatrix} J_{23} \\ -J_{13} \\ 0 \end{bmatrix}.$$

Hence, in a position of relative equilibrium all three products of inertia J_{12}, J_{13} and J_{23} are zero. This proves that in positions of relative equilibrium all principal axes of inertia are parallel to the base vectors $\boldsymbol{e}_{1,2,3}$. It would now be necessary to investigate the stability of these equilibrium positions. This will not be done here. The reader is referred to Magnus [6] and Belezki [16].

We return now to the equations of motion (5.66) with $\boldsymbol{F}_i$ and $\boldsymbol{M}_i$ given by Eqs. (5.68) and (5.69), respectively. First, the sum $\sum_{k=1}^{n} \boldsymbol{b}_{ik} \times \boldsymbol{F}_k$ will be examined. For $\boldsymbol{R}_k$ the expression $\sum_{j=1}^{n} \boldsymbol{b}_{jk}$ is substituted (see Eq. (5.54)). This yields

$$\sum_{k=1}^{n} \boldsymbol{b}_{ik} \times \boldsymbol{F}_k = -\omega_0^2 \sum_{k=1}^{n} \boldsymbol{b}_{ik} \times \left[\boldsymbol{r}_C + \sum_{j=1}^{n} \boldsymbol{b}_{jk} - 3\boldsymbol{e}_3 \sum_{j=1}^{n} \boldsymbol{b}_{jk} \cdot \boldsymbol{e}_3 \right] m_k .$$

The contribution of the term involving $\boldsymbol{r}_C$ is zero because of the relationship $\sum_{k=1}^{n} \boldsymbol{b}_{ik} m_k = \boldsymbol{0}$ (see Eq. (5.20)). The remaining expression can be rewritten in the form

$$\sum_{k=1}^{n} \boldsymbol{b}_{ik} \times \boldsymbol{F}_k = -\omega_0^2 \sum_{j=1}^{n} \sum_{k=1}^{n} m_k \boldsymbol{b}_{ik} \times \boldsymbol{b}_{jk} - 3\omega_0^2 \boldsymbol{e}_3 \times \sum_{j=1}^{n} \sum_{k=1}^{n} m_k \boldsymbol{b}_{ik} \boldsymbol{b}_{jk} \cdot \boldsymbol{e}_3 . \quad (5.71)$$

The first term contains the sum $\sum_{k=1}^{n} m_k \boldsymbol{b}_{ik} \times \boldsymbol{b}_{jk}$. Except for the absence of differentiation dots, it is identical with $\boldsymbol{g}_{ij}$ in Eq. (5.55). In the case $j \neq i$ this was reduced to Eq. (5.56). The same line of arguments is applicable here. Hence,

$$\sum_{k=1}^{n} m_k \boldsymbol{b}_{ik} \times \boldsymbol{b}_{jk} = -M \boldsymbol{b}_{ij} \times \boldsymbol{b}_{ji} \qquad i,j = 1 \dots n;\ i \neq j. \quad (5.72)$$

In the case $j = i$ the sum is zero. The second term in Eq. (5.71) contains the sum $\sum_{k=1}^{n} m_k \boldsymbol{b}_{ik} \boldsymbol{b}_{jk}$ which differs from the one just considered only by the multiplication symbol. The arguments leading from Eq. (5.55) to Eq. (5.56) can, again, be used analogously. This yields

$$\sum_{k=1}^{n} m_k \boldsymbol{b}_{ik} \boldsymbol{b}_{jk} = \begin{cases} \sum_{k=1}^{n} m_k \boldsymbol{b}_{ik} \boldsymbol{b}_{ik} & i = j \\ -M \boldsymbol{b}_{ij} \boldsymbol{b}_{ji} & i \neq j \end{cases} \qquad i,j = 1 \dots n .$$

Combining this and Eq. (5.72) with Eq. (5.71) one obtains the expression

$$\sum_{k=1}^{n} \boldsymbol{b}_{ik}\times\boldsymbol{F}_k = -3\omega_0^2\boldsymbol{e}_3\times\sum_{k=1}^{n} m_k\boldsymbol{b}_{ik}\boldsymbol{b}_{ik}\cdot\boldsymbol{e}_3 + \\ +\omega_0^2 M\sum_{\substack{j=1\\ \neq i}}^{n}(\boldsymbol{b}_{ij}\times\boldsymbol{b}_{ji}+3\boldsymbol{e}_3\times\boldsymbol{b}_{ij}\boldsymbol{b}_{ji}\cdot\boldsymbol{e}_3) \qquad i=1\ldots n. \tag{5.73}$$

The first term and the torque $\boldsymbol{M}_i$ from Eq. (5.69) can be combined to yield

$$3\omega_0^2\boldsymbol{e}_3\times\left[\boldsymbol{J}_i-\sum_{k=1}^{n} m_k\boldsymbol{b}_{ik}\boldsymbol{b}_{ik}\right]\cdot\boldsymbol{e}_3.$$

This is identical with

$$3\omega_0^2\boldsymbol{e}_3\times\left[\boldsymbol{J}_i+\sum_{k=1}^{n} m_k(b_{ik}^2\boldsymbol{E}-\boldsymbol{b}_{ik}\boldsymbol{b}_{ik})\right]\cdot\boldsymbol{e}_3$$

since $\boldsymbol{e}_3\times\boldsymbol{E}\cdot\boldsymbol{e}_3=\boldsymbol{e}_3\times\boldsymbol{e}_3$ is equal to zero. Comparison with Eq. (5.58) shows that the expression in square brackets represents the central inertia tensor of the augmented body i. The entire expression is, therefore, simply $3\omega_0^2\boldsymbol{e}_3\times\boldsymbol{K}_i^*\cdot\boldsymbol{e}_3$. Substituting this and the second term of Eq. (5.73) into the equation of motion (5.66) we get

$$\sum_{j=1}^{n}\boldsymbol{K}_{ij}\cdot\dot{\boldsymbol{\omega}}_j = -\boldsymbol{\omega}_i\times(\boldsymbol{K}_i^*\cdot\boldsymbol{\omega}_i+\boldsymbol{h}_i)+M\sum_{\substack{j=1\\ \neq i}}^{n}\boldsymbol{b}_{ij}\times[\boldsymbol{\omega}_j\times(\boldsymbol{\omega}_j\times\boldsymbol{b}_{ji})]+ \\ +3\omega_0^2\boldsymbol{e}_3\times\boldsymbol{K}_i^*\cdot\boldsymbol{e}_3+\omega_0^2 M\sum_{\substack{j=1\\ \neq i}}^{n}(\boldsymbol{b}_{ij}\times\boldsymbol{b}_{ji}+3\boldsymbol{e}_3\times\boldsymbol{b}_{ij}\boldsymbol{b}_{ji}\cdot\boldsymbol{e}_3) \qquad i=1\ldots n.$$

Conditions for relative equilibrium can now be obtained by substituting $\dot{\boldsymbol{\omega}}_i\equiv\boldsymbol{0}$, $\boldsymbol{\omega}_i\equiv\omega_0\boldsymbol{e}_2$ for $i=1\ldots n$. This yields

$$\boldsymbol{e}_2\times\left(\boldsymbol{K}_i^*\cdot\boldsymbol{e}_2+\frac{\boldsymbol{h}_i}{\omega_0}\right)-M\sum_{\substack{j=1\\ \neq i}}^{n}\boldsymbol{b}_{ij}\times[\boldsymbol{e}_2\times(\boldsymbol{e}_2\times\boldsymbol{b}_{ji})] \\ =3\boldsymbol{e}_3\times\boldsymbol{K}_i^*\cdot\boldsymbol{e}_3+M\sum_{\substack{j=1\\ \neq i}}^{n}(\boldsymbol{b}_{ij}\times\boldsymbol{b}_{ji}+3\boldsymbol{e}_3\times\boldsymbol{b}_{ij}\boldsymbol{b}_{ji}\cdot\boldsymbol{e}_3).$$

A further simplification is achieved when the triple vector-cross product on the left hand side is rewritten in the form

$$\boldsymbol{b}_{ij}\times[\boldsymbol{e}_2\times(\boldsymbol{e}_2\times\boldsymbol{b}_{ji})] = \boldsymbol{b}_{ij}\times(\boldsymbol{e}_2\boldsymbol{b}_{ji}\cdot\boldsymbol{e}_2-\boldsymbol{b}_{ji}) \\ = -\boldsymbol{e}_2\times\boldsymbol{b}_{ij}\boldsymbol{b}_{ji}\cdot\boldsymbol{e}_2-\boldsymbol{b}_{ij}\times\boldsymbol{b}_{ji}.$$

The second term in this expression produces the sum $M\sum_{\substack{j=1\\ \neq i}}^{n}\boldsymbol{b}_{ij}\times\boldsymbol{b}_{ji}$ on the left hand side of the equilibrium conditions. The same expression appears also on the right hand side. These can be canceled out leaving the equation

$$\boldsymbol{e}_2 \times \boldsymbol{B}_i \cdot \boldsymbol{e}_2 - 3\boldsymbol{e}_3 \times \boldsymbol{B}_i \cdot \boldsymbol{e}_3 + \boldsymbol{e}_2 \times \frac{\boldsymbol{h}_i}{\omega_0} = \boldsymbol{0} \qquad i = 1 \dots n \tag{5.74}$$

in which $\boldsymbol{B}_i$ is an abbreviation for the tensor

$$\boldsymbol{B}_i = \boldsymbol{K}_i^* + M \sum_{\substack{j=1 \\ \neq i}}^{n} \boldsymbol{b}_{ij} \boldsymbol{b}_{ji} \qquad i = 1 \dots n\,. \tag{5.75}$$

It should be noticed that $\boldsymbol{B}_i$ contains vectors which are fixed on different bodies. The coordinates of the tensor in a vector base fixed on body i are, consequently, not constant. The equilibrium conditions just found can be considered from the following point of view. As was mentioned earlier the relative angular momentum vector $\boldsymbol{h}_i$ on carrier i has constant coordinates in a vector base fixed on this carrier. It is assumed that there are at least three rotors on each carrier whose axes are not all parallel to one plane. Then, it is possible to give $\boldsymbol{h}_i$ any desired magnitude and direction in the carrier-fixed vector base by a proper choice of rotor angular velocities relative to the carrier. The vectors $\boldsymbol{h}_1 \dots \boldsymbol{h}_n$ influence the relative equilibrium positions. This suggests the following question. Is it possible to choose $\boldsymbol{h}_1 \dots \boldsymbol{h}_n$ in such a way that the system possesses a relative equilibrium position with certain prescribed characteristics? The practical significance of this problem is illustrated by the following example. Suppose a satellite with a camera mounted on one of the bodies is in orbit. As a result of a change in the mass distribution of the system caused by fuel consumption the original relative equilibrium position is disturbed. This causes the camera axis to leave its nominal vertical orientation. It would be desirable to change the system parameters in such a way that a new relative equilibrium position is created in which the camera axis is, again, in its nominal orientation. The only system parameters that can be changed on command from Earth are the relative rotor angular velocities. In order to find an answer to this problem the equilibrium conditions are resolved explicitly for $\boldsymbol{h}_1 \dots \boldsymbol{h}_n$. With the abbreviations

$$\boldsymbol{c}_i = \boldsymbol{B}_i \cdot \boldsymbol{e}_3; \qquad \boldsymbol{d}_i = \omega_0(\boldsymbol{e}_2 \times \boldsymbol{B}_i \cdot \boldsymbol{e}_2 - 3\boldsymbol{e}_3 \times \boldsymbol{B}_i \cdot \boldsymbol{e}_3) \qquad i = 1 \dots n \tag{5.76}$$

Eq. (5.74) reads

$$\boldsymbol{d}_i + \boldsymbol{e}_2 \times \boldsymbol{h}_i = \boldsymbol{0} \qquad i = 1 \dots n\,. \tag{5.77}$$

It states that $\boldsymbol{h}_i$ as well as $\boldsymbol{e}_2$ are orthogonal to $\boldsymbol{d}_i$. Hence, $\boldsymbol{h}_i$ has the general form

$$\boldsymbol{h}_i = \lambda_i \boldsymbol{e}_2 + \mu_i \boldsymbol{e}_2 \times \boldsymbol{d}_i \qquad i = 1 \dots n \tag{5.78}$$

with still unknown factors λ_i and μ_i. Furthermore, the equation $\boldsymbol{e}_2 \cdot \boldsymbol{d}_i = 0$ $(i = 1 \dots n)$ holds which in view of Eq. (5.76) reads

$$\boldsymbol{e}_2 \cdot \boldsymbol{e}_2 \times \boldsymbol{B}_i \cdot \boldsymbol{e}_2 - 3\boldsymbol{e}_2 \cdot \boldsymbol{e}_3 \times \boldsymbol{B}_i \cdot \boldsymbol{e}_3 = 0 \qquad i = 1 \dots n\,.$$

The first term is zero because two factors of the product are identical. The remainder can be rewritten in the form

$$\boldsymbol{e}_2 \times \boldsymbol{e}_3 \cdot \boldsymbol{B}_i \cdot \boldsymbol{e}_3 = \boldsymbol{e}_1 \cdot \boldsymbol{B}_i \cdot \boldsymbol{e}_3 = 0 \qquad i = 1 \dots n$$

or using Eq. (5.76)

$$e_1 \cdot c_i = 0 \qquad i = 1 \dots n . \tag{5.79}$$

Substitution of Eq. (5.78) into Eq. (5.77) results in

$$d_i + \mu_i e_2 \times (e_2 \times d_i) = (1 - \mu_i)\, d_i = 0 .$$

Hence, μ_i equals one for $i = 1 \dots n$, and h_i becomes

$$h_i = \lambda_i e_2 + e_2 \times d_i \qquad i = 1 \dots n . \tag{5.80}$$

This is the desired explicit formulation. It contains a free parameter λ_i. The vector $\lambda_i e_2$ is the component of h_i in the direction of the orbital angular velocity ω_0. This explains why λ_i has no influence on the relative equilibrium positions. The n equations (5.79) together with the n equations (5.80) are equivalent to the original equilibrium conditions (5.74). Each relative equilibrium position is a solution of Eq. (5.74). Conversely, each solution of Eq. (5.79) determines a relative equilibrium position. To each such solution belong vectors $h_1 \dots h_n$ which are determined by Eq. (5.80). Consequently, only Eq. (5.79) needs to be solved. With Eqs. (5.76) and (5.75) the vector c_i is

$$c_i = K_i^* \cdot e_3 + M \sum_{\substack{j=1 \\ \neq i}}^{n} b_{ij} (b_{ji} \cdot e_3) \qquad i = 1 \dots n .$$

Let us now suppose that on each carrier an axis of arbitrary direction is defined by a unit vector fixed on the carrier and that all n unit vectors are aligned parallel to the base vector e_3 in the local vertical. This does not yet uniquely determine the position of the system in the orbital reference frame. Each carrier can still be rotated about e_3 through an arbitrary angle. These n angles have no influence on the projections $b_{ji} \cdot e_3$ of the body-fixed vectors b_{ji} $(i, j = 1 \dots n)$ on the direction of e_3 and no influence on the coordinates of the vector $K_i^* \cdot e_3$ $(i = 1 \dots n)$ in a vector base fixed on body i. From this follows that during such rotations about e_3 the vector c_i is fixed on carrier i. On each carrier there is one such vector. In the following, first, the general case is assumed that none of these vectors is zero or parallel to e_3. Then, all n equations (5.79) can be satisfied by choosing the still undetermined rotation angles in such a way that on each carrier $i = 1 \dots n$ the vector c_i is normal to e_1. This determines two positions for each body and, hence, 2^n different positions for the entire system. Each one of them is a relative equilibrium position provided the vectors $h_1 \dots h_n$ are chosen in accordance with Eq. (5.80). For the degenerate case in which the vector c_i is either zero or parallel to e_3 for one or several bodies the solution is obvious. For these particular bodies Eq. (5.79) is satisfied independently of the rotation angle about e_3. Each such body has an infinite number of relative equilibrium positions. For each of the remaining bodies two relative equilibrium positions exist as before. The results just obtained can be summarized as follows. There exist, in general, 2^n different relative equilibrium positions (an infinite number in degenerate cases) which satisfy the requirement that n arbitrarily chosen axes, one on each carrier, are parallel to the local vertical. For each such position the necessary vectors $h_1 \dots h_n$ are determined explicitly from Eq. (5.80). The necessary stability analysis is not presented here. For this the reader is referred to Wittenburg/Lilov [17]. Let it only be mentioned that in the stability analysis the free parameters $\lambda_1 \dots \lambda_n$ play an essential role.

Problem

5.11 Formulate expressions for the kinetic and potential energy of the satellite considered in this section.

5.2.6 Systems with ball-and-socket, universal and pin joints

The equations of motion of Secs. 5.2.2 and 5.2.4 for systems with ball-and-socket joints can be generalized to be valid also for systems with universal and pin joints. In practice these three types of joints are very common, in particular the pin joint. This justifies the devotion of a separate section to such systems. All three types of joints have one essential property in common. The motion of two contiguous bodies relative to each other is a pure rotation. Only the number of degrees of freedom is different. In a universal joint (Fig. 5.28) one geometric constraint is active which forces the angular velocity of the two bodies relative to one another to lie in the plane of the two axes (the axes do not have to be orthogonal to each other; they must intersect each other, however). This requires an internal constraint torque orthogonal to this plane. In a pin joint (Fig. 5.29) two geometric constraints are active. These force the relative angu-

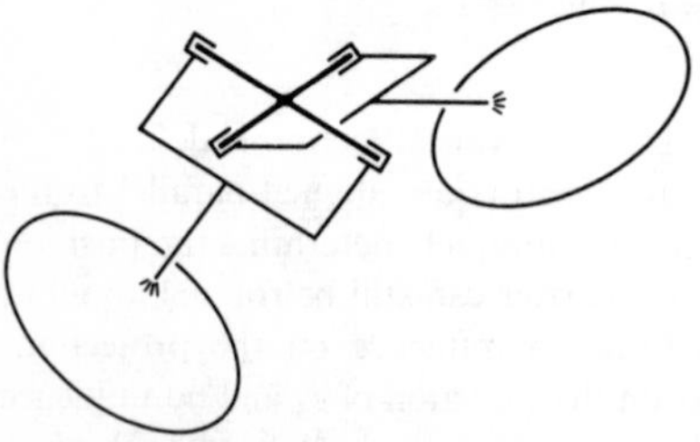

Fig. 5.28 Universal joint

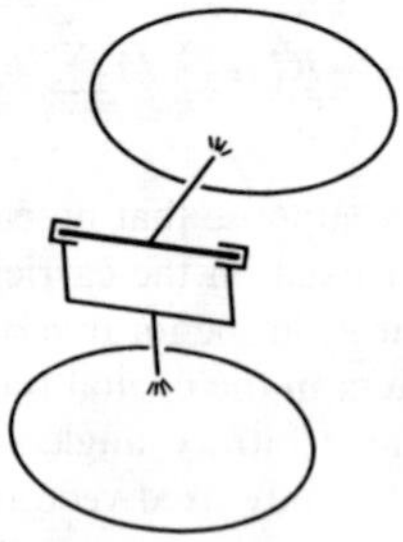

Fig. 5.29 Pin joint

lar velocity vector to lie in the axis. This requires an internal constraint torque with two mutually perpendicular components normal to the axis. Equations of motion will be established on the basis of the following model. Both, universal and pin joints are modeled as ball-and-socket joint plus constraint torque. In the case of a universal joint the geometric center of the substitute ball-and-socket joint is the intersection point of the two axes. In the case of a pin joint it is an arbitrary point on the axis. With this model the system is reduced to one for which the equations of motion are known. They are Eq. (5.34) for systems with a material ball-and-socket joint 1 and Eqs. (5.42) and (5.61) for systems without a material hinge 1. The tensors $\boldsymbol{K}_{ij}$ and the vectors $\boldsymbol{M}'_i$ have different definitions in the two cases. For the present system Eqs. (5.34) and (5.61) must be replaced by

$$\sum_{j=1}^{n} \boldsymbol{K}_{ij} \cdot \dot{\boldsymbol{\omega}}_j = \boldsymbol{M}'_i + \boldsymbol{M}_i + \sum_{a=1}^{n} S_{ia} (\boldsymbol{Y}_a + \boldsymbol{Y}^c_a) \qquad i = 1 \dots n .$$

The terms $\boldsymbol{Y}^c_a$ represent the additional constraint torques in universal and pin joints whereas $\boldsymbol{Y}_a$ $(a = 1 \dots n)$ denotes as before internal hinge torques which do not represent constraint torques. If body 1 is not coupled by a material hinge 1 with an external body whose motion is prescribed then Y_1 and Y_1^c are identically zero. All n equations can be combined in the single matrix equation

$$\underline{K}\cdot\dot{\underline{\omega}}=\underline{M}'+\underline{M}+\underline{S}(\underline{Y}+\underline{Y}^c) \tag{5.81}$$

in which $\dot{\underline{\omega}}$, $\underline{M}$, $\underline{S}$ and $\underline{Y}$ are familiar from previous sections. New are the column matrices $\underline{M}'=[M_1'\dots M_n']^T$ and $\underline{Y}^c=[Y_1^c\dots Y_n^c]^T$ and the $(n\times n)$ matrix $\underline{K}$ whose elements are the tensors K_{ij} $(i,j=1\dots n)$. The scalar product of such a matrix with a matrix containing vectors as elements was explained in Chap. 1. There remain only two steps to be taken in order to prodeed from Eq. (5.81) to a final form for the equations of motion. One is the description of the system kinematics and the other is the elimination of the constraint torques $\underline{Y}^c$.

5.2.6.1 Kinematics of motion of contiguous bodies relative to one another In the presence of universal and pin joints it is no longer possible to treat the absolute angular velocities ω_i of the individual bodies as independent quantities (independent in the sense that they are unconstrained). On the contrary, there exist constraints of the kind already described between the absolute angular velocities of contiguous bodies. It is useful to introduce as generalized coordinates angles of rotation about the axes of pin joints and universal joints which describe the position of the bodies relative to each other. For the sake of uniformity of description such angular variables are introduced also for ball-and-socket joints although this is not necessary because of the absence of constraints. One angle ϕ_{a1} is required if joint a $(a=1\dots n)$ is a pin joint. Two angles ϕ_{a1}, ϕ_{a2} are required if it is a universal joint and three angles $\phi_{a1}, \phi_{a2}, \phi_{a3}$ if it is a ball-and-socket joint. The hinge number one requires an extra comment. If the multi-body system is not coupled materially with an external body 0 undergoing a prescribed motion then hinge 1 is a fictitious hinge with six degrees of freedom of motion of body 1 relative to some well-defined vector base $\underline{e}^{(0)}$. The fact that the center of mass of body 1 can translate relative to $\underline{e}^{(0)}$ without any constraint led to Eq. (5.42) for the motion of the composite system center of mass. The rotation of body 1 relative to $\underline{e}^{(0)}$ is uncoupled from its translation. From the point of view of kinematics it can be treated as if the center of mass of body 1 and the origin of $\underline{e}^{(0)}$ were coupled by a ball-and-socket joint. In the mathematical description, therefore, no distinction need be made from a true, material ball-and-socket joint. On each body i $(i=1\dots n)$ a vector base $\underline{e}^{(i)}$ is fixed in an arbitrary orientation. Let Ω_a $(a=1\dots n)$ be the angular velocity of body $i^-(a)$ relative to body $i^+(a)$. This definition contains a basic s i g n c o n v e n t i o n comparable in importance to the convention that the internal hinge torque $+Y_a$ is acting on body $i^+(a)$. Whatever the actual meaning of the angular variables ϕ_{ai}, the quantity Ω_a has the form

$$\Omega_a=\sum_{i=1}^{n_a} p_{ai}\dot{\phi}_{ai} \qquad a=1\dots n \tag{5.82}$$

where n_a is the number of degrees of freedom in the hinge $(n_a=1, 2 \text{ or } 3)$ and p_{ai} are unit vectors along axes of rotation. Two examples illustrate this:

In a pin joint (Fig. 5.30) n_a is equal to one. The unit vector p_{a1} is fixed in the hinge axis with an arbitrarily chosen sense of direction. The coordinate ϕ_{a1} is defined as the angle through which body $i^-(a)$ is rotated relative to body $i^+(a)$ from some defined position $\phi_{a1}=0$ in a clockwise direction about $\mathbf{p}_{a1}$. In this particular case $\mathbf{p}_{a1}$ is fixed on both bodies. In the second example a ball-and-socket joint with $n_a=3$ is considered. As coordinates Bryant angles are used. They are defined as described in Chap. 2.1.2 (see Fig. 2.3) with $\underline{e}^{(i^+(a))}$ and $\underline{e}^{(i^-(a))}$ playing the roles of $\underline{e}^{(1)}$ and $\underline{e}^{(2)}$, respectively. Eq. (5.82) is obtained by adapting Eq. (2.30) to the new notation. This yields

$$\boldsymbol{\Omega}_a=\dot{\phi}_{a1}\,\boldsymbol{e}_1^{(i^+(a))}+\dot{\phi}_{a2}\,\boldsymbol{e}_2^{(i^-(a))'}+\dot{\phi}_{a3}\,\boldsymbol{e}_3^{(i^-(a))}.$$

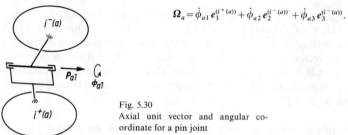

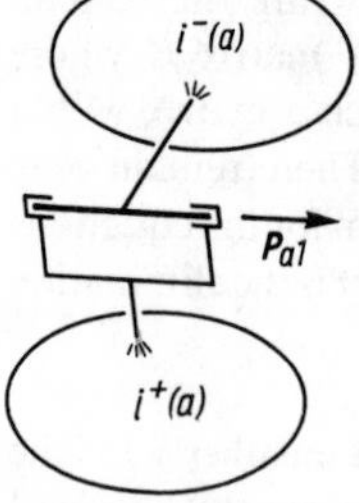

Fig. 5.30
Axial unit vector and angular coordinate for a pin joint

In this case $\mathbf{p}_{a1}$ is fixed on body $i^+(a)$, $\mathbf{p}_{a3}$ on body $i^-(a)$ and $\mathbf{p}_{a2}$ is not fixed on any of the two bodies. The coordinates of $\mathbf{p}_{a2}$ in both body-fixed bases are, however, known as functions of the angles.

Let $\mathring{\boldsymbol{\Omega}}_a$ $(a=1\dots n)$ be defined as the angular acceleration of body $i^-(a)$ relative to body $i^+(a)$. It is the time derivative of $\boldsymbol{\Omega}_a$ in the vector base $\underline{e}^{(i^-(a))}$ (and also in the base $\underline{e}^{(i^+(a))}$). From Eq. (5.82) follows the expression

$$\mathring{\boldsymbol{\Omega}}_a=\sum_{i=1}^{n_a}\left(\mathbf{p}_{ai}\ddot{\phi}_{ai}+\sum_{j=1}^{n_a}\frac{\partial\mathbf{p}_{ai}}{\partial\phi_{aj}}\dot{\phi}_{ai}\dot{\phi}_{aj}\right)\qquad a=1\dots n.$$

In the second term, partial differentiation has to be applied to the coordinates of $\mathbf{p}_{ai}$ either in the base $\underline{e}^{(i^-(a))}$ or in $\underline{e}^{(i^+(a))}$. An example is treated in Illustrative Example 5.2. With the abbreviation

$$\mathbf{w}_a=\sum_{i=1}^{n_a}\sum_{j=1}^{n_a}\frac{\partial\mathbf{p}_{ai}}{\partial\phi_{aj}}\dot{\phi}_{ai}\dot{\phi}_{aj}\qquad a=1\dots n \tag{5.83}$$

one obtains

$$\mathring{\boldsymbol{\Omega}}_a=\sum_{i=1}^{n_a}\mathbf{p}_{ai}\ddot{\phi}_{ai}+\mathbf{w}_a\qquad a=1\dots n. \tag{5.84}$$

The kinematics for the individual hinges is completed with a formula for the transformation matrix between the two body-fixed vector bases $\underline{e}^{(i^+(a))}$ and $\underline{e}^{(i^-(a))}$. This matrix is called $\underline{G}_a$ and defined by the equation

$$\underline{e}^{(i^-(a))}=\underline{G}_a\underline{e}^{(i^+(a))}\qquad a=1\dots n. \tag{5.85}$$

For all three types of hinges and for any choice of generalized coordinates $\underline{G}_a$ is a known function of these coordinates. For details see Illustrative Example 5.2.

5.2.6.2 Kinematics of motion of bodies relative to inertial space From the kinematics of motion of contiguous bodies relative to each other we proceed now to the kinematics of motion of the bodies relative to inertial space. The link between the two is the definition of $\boldsymbol{\Omega}_a$ as the difference

$$\boldsymbol{\Omega}_a = \boldsymbol{\omega}_{i^-(a)} - \boldsymbol{\omega}_{i^+(a)} \qquad a=1\ldots n. \tag{5.86}$$

This can be written in the form

$$\boldsymbol{\Omega}_a = -\sum_{i=0}^{n} S_{ia}\boldsymbol{\omega}_i = -S_{0a}\boldsymbol{\omega}_0 - \sum_{i=1}^{n} S_{ia}\boldsymbol{\omega}_i \qquad a=1\ldots n.$$

All n equations are combined in the single matrix equation

$$\underline{\boldsymbol{\Omega}} = -\boldsymbol{\omega}_0 \underline{S}_0^{\mathrm{T}} - \underline{S}^{\mathrm{T}}\underline{\boldsymbol{\omega}}$$

with column matrices $\underline{\boldsymbol{\Omega}} = [\boldsymbol{\Omega}_1 \ldots \boldsymbol{\Omega}_n]^{\mathrm{T}}$ and $\underline{\boldsymbol{\omega}} = [\boldsymbol{\omega}_1 \ldots \boldsymbol{\omega}_n]^{\mathrm{T}}$. Multiplication from the left by $\underline{T}^{\mathrm{T}}$ yields with Eqs. (5.5) and (5.6)

$$\underline{\boldsymbol{\omega}} = -\underline{T}^{\mathrm{T}}\underline{\boldsymbol{\Omega}} + \boldsymbol{\omega}_0 \underline{1}_n. \tag{5.87}$$

For a single absolute angular velocity this yields

$$\boldsymbol{\omega}_i = -\sum_{a=1}^{n} T_{ai}\boldsymbol{\Omega}_a + \boldsymbol{\omega}_0 \qquad i=1\ldots n.$$

From this the absolute angular acceleration is obtained:

$$\dot{\boldsymbol{\omega}}_i = -\sum_{a=1}^{n} T_{ai}(\mathring{\boldsymbol{\Omega}}_a + \boldsymbol{w}_a^*) + \dot{\boldsymbol{\omega}}_0 \qquad i=1\ldots n \tag{5.88}$$

where $\boldsymbol{w}_a^*$ is an abbreviation for

$$\boldsymbol{w}_a^* = \boldsymbol{\omega}_{i^-(a)} \times \boldsymbol{\Omega}_a \qquad a=1\ldots n.$$

The angular velocity $\boldsymbol{\omega}_{i^-(a)}$ in this expression could be expressed, in turn, in terms of $\boldsymbol{\Omega}_1 \ldots \boldsymbol{\Omega}_n$ with the help of Eq. (5.87). This substitution, however, should not be carried out in symbolic form. The equations of motion to be developed can be used only for numerical computations. In a computer program the quantities $\boldsymbol{\Omega}_1 \ldots \boldsymbol{\Omega}_n$ and $\boldsymbol{\omega}_1 \ldots \boldsymbol{\omega}_n$ would be calculated first, and then the products $\boldsymbol{\omega}_{i^-(a)} \times \boldsymbol{\Omega}_a$ would be evaluated. All n equations (5.88) can be combined in the single matrix equation

$$\dot{\underline{\boldsymbol{\omega}}} = -\underline{T}^{\mathrm{T}}(\mathring{\underline{\boldsymbol{\Omega}}} + \underline{\boldsymbol{w}}^*) + \dot{\boldsymbol{\omega}}_0 \underline{1}_n \tag{5.89}$$

in which $\mathring{\underline{\boldsymbol{\Omega}}}$ and $\underline{\boldsymbol{w}}^*$ are the column matrices $[\mathring{\boldsymbol{\Omega}}_1 \ldots \mathring{\boldsymbol{\Omega}}_n]^{\mathrm{T}}$ and $[\boldsymbol{w}_1^* \ldots \boldsymbol{w}_n^*]^{\mathrm{T}}$, respectively. It remains to express $\mathring{\underline{\boldsymbol{\Omega}}}$ explicitly in terms of generalized coordinates and of their time derivatives. This is done by combining the n equations (5.84) in the single matrix equation

$$\mathring{\underline{\boldsymbol{\Omega}}} = \underline{\boldsymbol{p}}^{\mathrm{T}}\ddot{\underline{\phi}} + \underline{\boldsymbol{w}}. \tag{5.90}$$

On the right hand side stand the column matrix $\underline{\boldsymbol{w}} = [\boldsymbol{w}_1 \ldots \boldsymbol{w}_n]^{\mathrm{T}}$, the column matrix

$$\ddot{\underline{\phi}} = [\ddot{\phi}_{11} \ldots \ddot{\phi}_{1n_1} \quad \ddot{\phi}_{21} \ldots \ddot{\phi}_{2n_2} \;\cdots\cdots\; \ddot{\phi}_{n1} \ldots \ddot{\phi}_{nn_n}]^{\mathrm{T}}$$

and the quasi-diagonal matrix $\underline{\boldsymbol{p}}$ whose transpose is

$$\underline{\boldsymbol{p}}^{\mathrm{T}} = \begin{bmatrix} \boldsymbol{p}_{11} \cdots \boldsymbol{p}_{1n_1} & & & \underline{0} \\ & \boldsymbol{p}_{21} \cdots \boldsymbol{p}_{2n_2} & & \\ & & \ddots & \\ \underline{0} & & & \boldsymbol{p}_{n1} \cdots \boldsymbol{p}_{nn_n} \end{bmatrix}. \tag{5.91}$$

This matrix $\underline{p}$ has n columns, each corresponding to one hinge, and as many rows as there are angular variables in the entire system. Eq. (5.90) is now substituted into Eq. (5.89):

$$\dot{\underline{\omega}} = -\underline{T}^{\mathrm{T}}(\underline{p}^{\mathrm{T}}\ddot{\underline{\phi}} + \underline{f}) + \dot{\omega}_0 \underline{1}_n . \tag{5.92}$$

In this final result for the angular accelerations $\underline{f}$ is the column matrix

$$\underline{f} = \underline{w} + \underline{w}^* . \tag{5.93}$$

On the right hand side of the equation for $\dot{\underline{\omega}}$ the angular acceleration $\dot{\omega}_0$ is a known function of time, the matrix $\underline{p}$ is a known function of the generalized coordinates and $\underline{f}$ is a known function of the generalized coordinates and of their first time derivatives. The angular orientation of body i ($i=1 \dots n$) in the reference frame $\underline{e}^{(0)}$ is a function of the generalized coordinates ϕ_{ai} ($a=1 \dots n$; $i=1 \dots n_a$) of the system. This function can be expressed as follows. A transformation matrix $\underline{A}_i$ is defined by the equation

$$\underline{e}^{(0)} = \underline{A}_i \underline{e}^{(i)} \qquad i=1 \dots n . \tag{5.94}$$

These matrices are related to the transformation matrices $\underline{G}_a$ ($a=1 \dots n$) from Eq. (5.85) through the alternative equations

$$\underline{A}_{i^-(a)} = \underline{A}_{i^+(a)} \underline{G}_a^{\mathrm{T}}, \qquad \underline{A}_{i^+(a)} = \underline{A}_{i^-(a)} \underline{G}_a \qquad a=1 \dots n . \tag{5.95}$$

The practical application of these formulas to a recursive calculation of the matrices $\underline{A}_1 \dots \underline{A}_n$ from $\underline{G}_1 \dots \underline{G}_n$ is demonstrated for a system whose directed system graph has the form shown in Fig. 5.8c. The matrices are calculated in the order

$$\begin{array}{llll} \underline{A}_1 = \underline{G}_1^{\mathrm{T}}, & \underline{A}_4 = \underline{A}_1 \underline{G}_4, & \underline{A}_6 = \underline{A}_1 \underline{G}_3, & \underline{A}_7 = \underline{A}_6 \underline{G}_2^{\mathrm{T}}, \\ \underline{A}_2 = \underline{A}_1 \underline{G}_5, & \underline{A}_3 = \underline{A}_2 \underline{G}_7, & \underline{A}_5 = \underline{A}_2 \underline{G}_6^{\mathrm{T}}. & \end{array}$$

The mathematical description of the system kinematics is now complete. The angular orientation as well as the absolute angular velocities and accelerations of all bodies have been expressed as functions of ϕ_{ai}, $\dot{\phi}_{ai}$ and $\ddot{\phi}_{ai}$ ($a=1 \dots n$; $i=1 \dots n_a$). The equations (5.87) and (5.92) for the angular velocities and accelerations are substituted into the equations of motion (5.81). This leads to

$$\underline{K} \cdot [-\underline{T}^{\mathrm{T}}(\underline{p}^{\mathrm{T}}\ddot{\underline{\phi}} + \underline{f}) + \dot{\omega}_0 \underline{1}_n] = \underline{M}' + \underline{M} + \underline{S}(\underline{Y} + \underline{Y}^c)$$

where $\underline{M}'$ is now a known function of time and of ϕ_{ai} and of $\dot{\phi}_{ai}$ ($a=1 \dots n$; $i=1 \dots n_a$).

5.2.6.3 Elimination of constraint torques It was mentioned earlier that the last step in the development of equations of motion is the elimination of the constraint torques $\underline{Y}^c$. Like the constraint forces which were eliminated in Secs. 5.2.2 and 5.2.4 the constraint torques are of interest only for an analysis of stresses in the hinge mechanisms during motion. They are obtained explicitly by premultiplying the last equation with $\underline{T}$:

$$\underline{Y}^c = \underline{T}\{\underline{K} \cdot [-\underline{T}^{\mathrm{T}}(\underline{p}^{\mathrm{T}}\ddot{\underline{\phi}} + \underline{f}) + \dot{\omega}_0 \underline{1}_n] - \underline{M}' - \underline{M}\} - \underline{Y} . \tag{5.96}$$

The elimination of $\underline{Y}^c$ from these equations is now a straight-forward procedure. Each constraint torque appears in only one single vectorial equation. Suppose that hinge a ($a=1 \dots n$) is a pin joint. Then, $\boldsymbol{Y}_a^c$ is perpendicular to the unit vector $\boldsymbol{p}_{a1}$ in the hinge axis. It disappears when the a-th equation is scalar-multiplied by $\boldsymbol{p}_{a1}$. This results in one single scalar differential equation for each hinge with $n_a=1$ degree of freedom. The equation contains, among other terms, the product $\boldsymbol{p}_{a1} \cdot \boldsymbol{Y}_a$. This is a torque about the hinge axis which might be caused, for instance, by a torsional spring or damper. It is a known function of ϕ_{a1} and $\dot{\phi}_{a1}$. Suppose, next, that hinge a ($a=1 \dots n$) is a universal joint. Then, $\boldsymbol{Y}_a^c$ is perpendicular to both unit vectors $\boldsymbol{p}_{a1}$ and $\boldsymbol{p}_{a2}$ in the hinge axes. It disappears when the a-th equation of Eq. (5.96) is scalar-multiplied by either $\boldsymbol{p}_{a1}$ or $\boldsymbol{p}_{a2}$. Carrying out both multiplications one gets two linearly independent scalar differential equations for each hinge with $n_a=2$ degrees of freedom. In these two equations appear the products $\boldsymbol{p}_{a1} \cdot \boldsymbol{Y}_a$ and $\boldsymbol{p}_{a2} \cdot \boldsymbol{Y}_a$. They are interpreted as torques about the hinge axes which could be caused, for instance, by torsional springs and dampers. As in the case of a pin joint they are known functions of the hinge variables ϕ_{ai} and $\dot{\phi}_{ai}$ ($i=1,2$). Finally, suppose that hinge a ($a=1 \dots n$) is a ball-and-socket joint. In this case the constraint torque $\boldsymbol{Y}_a^c$ is identically zero. We obtain three linearly independent scalar differential equations when the a-th equation of Eq. (5.96) is scalar-multiplied by $\boldsymbol{p}_{a1}, \boldsymbol{p}_{a2}$ and $\boldsymbol{p}_{a3}$ separately. The products $\boldsymbol{p}_{ai} \cdot \boldsymbol{Y}_a$ for $i=1,2,3$ are interpreted the same way as in the previous cases. It is easily verified that all scalar multiplications just explained are carried out in one stroke if Eq. (5.96) is scalar-premultiplied by the matrix $\underline{\boldsymbol{p}}$ whose transpose is given by Eq. (5.91). Rearranging the remaining right hand side terms the equations of motion are, thus, obtained in the final form

$$\underline{A}\,\ddot{\underline{\phi}} = \underline{B} \tag{5.97}$$

with

$$\underline{A} = (\underline{\boldsymbol{p}}\,\underline{T}) \cdot \underline{\boldsymbol{K}} \cdot (\underline{\boldsymbol{p}}\,\underline{T})^{\mathrm{T}} \tag{5.98}$$

$$\underline{B} = -(\underline{\boldsymbol{p}}\,\underline{T}) \cdot [\underline{\boldsymbol{K}} \cdot (\underline{T}^{\mathrm{T}} \underline{\boldsymbol{f}} - \dot{\boldsymbol{\omega}}_0 \underline{1}_n) + \underline{\boldsymbol{M}}' + \underline{\boldsymbol{M}}] - \underline{\boldsymbol{p}} \cdot \underline{\boldsymbol{Y}}\,. \tag{5.99}$$

The coefficient matrix $\underline{A}$ is, obviously, not constant. Its elements contain scalar products of vectors and tensors whose coordinates are defined in vector bases which are fixed on different bodies. In numerical calculations the matrix $\underline{A}$ must, therefore, be evaluated and inverted anew every time the column matrix $\ddot{\underline{\phi}}$ is computed. This requires the bulk of the total computation time necessary for numerical integrations. Both, the evaluation and inversion of $\underline{A}$ are substantially simplified by the fact that $\underline{A}$ is symmetric. This is a consequence of the relationship $\boldsymbol{K}_{ji} = \bar{\boldsymbol{K}}_{ij}$ which is satisfied by the elements of the matrix $\underline{\boldsymbol{K}}$ (see Eqs. (5.33) and (5.60)). The formal proof that $\underline{A}$ is positive definite will be obtained as a by-product in Sec. 5.2.8.

Illustrative Example 5.2 Fig. 5.31a shows an arm prothesis with a ball-and-socket shoulder joint, a pin joint as elbow and a universal joint at the wrist. The two hinge axes fixed on the forearm are perpendicular to one another. The motion of the vector base $\underline{e}^{(0)}$ which is fixed on the shoulder is prescribed as a function of time. For this particular system some of the relevant quantities which enter Eq. (5.97) will be explained in detail. It should be noted that the following decisions about the system graph as well as about the choice of generalized coordinates represent just one out of many possibilities. The bodies and hinges are labeled as shown in Fig. 5.31a. This is a regular labeling in the sense of Sec. 5.2.1. For the arcs in the corresponding system graph the directions shown in Fig. 5.31b are chosen. This choice of arc directions offers the advantage that the generalized coordinates to be defined later describe the position and motion of the hand relative to the forearm, of the forearm relative to the upper arm and of the upper arm relative to the base $\underline{e}^{(0)}$, respectively. This seems more

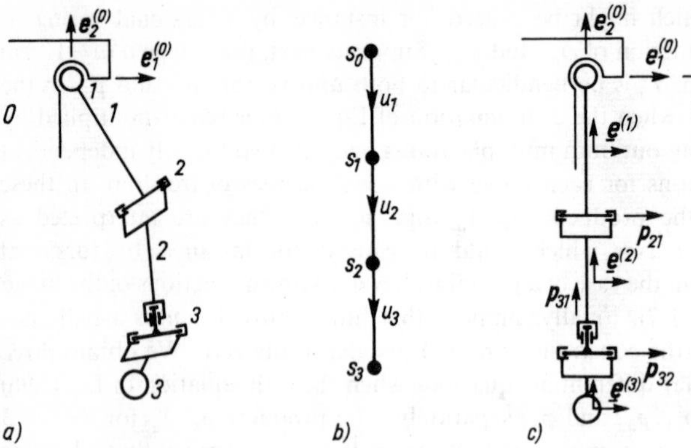

Fig. 5.31 a) An arm prothesis with six degrees of freedom, b) its directed system graph and c) the null position in which all six angular coordinates are zero

natural than to describe, for instance, the motion of the forearm relative to the hand. The regular labeling and the particular choice of arc directions offer still other advantages, however, which are more important. In the matrix $\underline{T}$ for the graph,

$$\underline{T} = \begin{bmatrix} -1 & -1 & -1 \\ 0 & -1 & -1 \\ 0 & 0 & -1 \end{bmatrix},$$

all nonzero elements are in the upper triangle and, furthermore, the signs of all nonzero elements are identical. These two properties can be used to shorten computer running times. Two examples illustrate this. In Eq. (5.87) the absolute angular velocities can be calculated from the recursion formula

$$\boldsymbol{\omega}_i = \boldsymbol{\Omega}_i + \boldsymbol{\omega}_{i-1} \qquad i=1,2,3,$$

and Eq. (5.95) for the transformation matrices $\underline{A}_1 \ldots \underline{A}_n$ reduces to the recursion formula

$$\underline{A}_i = \begin{cases} \underline{G}_1^{\mathrm{T}} & i=1 \\ \underline{A}_{i-1}\underline{G}_i^{\mathrm{T}} & i=2,3. \end{cases}$$

The particular form of $\underline{T}$ simplifies also the execution of all multiplications by $\underline{T}$ and $\underline{T}^{\mathrm{T}}$ in Eqs. (5.98) and (5.99). These advantages of a regular labeling in combination with uniform arc directions are particularly important for numerical calculations on systems with large numbers of bodies and hinges.

In Fig. 5.31c the arm is depicted in what will be called the null position because all angular variables to be defined later are zero in this position. In the null position the arm is hanging vertically with the elbow stretched and with one wrist axis and with the elbow axis being parallel to the line passing through both shoulder joints. In this position the body-fixed vector bases $\underline{e}^{(i)}$ $(i=0\dots3)$ are parallel to one another and oriented as shown. As generalized coordinates Bryant angles ϕ_{11}, ϕ_{12}, ϕ_{13} are chosen for hinge 1 and angles ϕ_{21} for the elbow and ϕ_{31}, ϕ_{32} for the wrist. For the definition of Bryant angles see Fig. 2.3 which is adapted to the present case by replacing $\underline{e}^{(1)}$ by $\underline{e}^{(0)}$, $\underline{e}^{(2)}$ by $\underline{e}^{(1)}$ and ϕ_i by ϕ_{1i} for $i=1,2,3$. For the wrist the axes of rotation are indicated in Fig. 5.31 c. The unit vectors $\boldsymbol{p}_{ai}$ in Eq. (5.82) have the following coordinate matrices:

$$\left.\begin{aligned}\underline{p}_{11}&=[\cos\phi_{12}\cos\phi_{13} \quad -\cos\phi_{12}\sin\phi_{13} \quad \sin\phi_{12}]^{\mathrm{T}}\\ \underline{p}_{12}&=[\sin\phi_{13} \quad \cos\phi_{13} \quad 0]^{\mathrm{T}}\\ \underline{p}_{13}&=[0 \quad 0 \quad 1]^{\mathrm{T}}\end{aligned}\right\}\text{ in }\underline{e}^{(1)}$$

$$\underline{p}_{21}=[1 \quad 0 \quad 0]^{\mathrm{T}} \quad \text{in }\underline{e}^{(2)}$$

$$\left.\begin{aligned}\underline{p}_{31}&=[0 \quad \cos\phi_{32} \quad -\sin\phi_{32}]^{\mathrm{T}}\\ \underline{p}_{32}&=[1 \quad 0 \quad 0]^{\mathrm{T}}\end{aligned}\right\}\text{ in }\underline{e}^{(3)}$$

The transformation matrices $\underline{G}_a$ defined by Eq. (5.85) are (with the abbreviations $c_{ai}=\cos\phi_{ai}$ and $s_{ai}=\sin\phi_{ai}$)

$$\underline{G}_1=\begin{bmatrix} c_{12}c_{13} & c_{11}s_{13}+s_{11}s_{12}c_{13} & s_{11}s_{13}-c_{11}s_{12}c_{13}\\ -c_{12}s_{13} & c_{11}c_{13}-s_{11}s_{12}s_{13} & s_{11}c_{13}+c_{11}s_{12}s_{13}\\ s_{12} & -s_{11}c_{12} & c_{11}c_{12}\end{bmatrix}$$

$$\underline{G}_2=\begin{bmatrix}1&0&0\\0&c_{21}&s_{21}\\0&-s_{21}&c_{21}\end{bmatrix},\qquad \underline{G}_3=\begin{bmatrix}c_{31}&0&-s_{31}\\ s_{31}s_{32}&c_{32}&c_{31}s_{32}\\ s_{31}c_{32}&-s_{32}&c_{31}c_{32}\end{bmatrix}.$$

The matrix $\underline{G}_1$ is copied from Eq. (2.5). In order to find expressions for the vectors $\boldsymbol{w}_a$ of Eq. (5.83) the coordinates of $\boldsymbol{p}_{ai}$ in the base $\underline{e}^{(i^-(a))}$ are differentiated with respect to time. The results are $\underline{w}_2=\underline{0}$ and

$$\underline{w}_1=\dot\phi_{11}\begin{bmatrix}-\dot\phi_{12}s_{12}c_{13}-\dot\phi_{13}c_{12}s_{13}\\ \dot\phi_{12}s_{12}s_{13}-\dot\phi_{13}c_{12}c_{13}\\ \dot\phi_{12}c_{12}\end{bmatrix}+\dot\phi_{12}\begin{bmatrix}\dot\phi_{13}c_{13}\\ -\dot\phi_{13}s_{13}\\ 0\end{bmatrix}\text{ in }\underline{e}^{(1)},$$

$$\underline{w}_3=\dot\phi_{31}\begin{bmatrix}0\\ -\dot\phi_{32}s_{32}\\ -\dot\phi_{32}c_{32}\end{bmatrix}\text{ in }\underline{e}^{(3)}.$$

In a computer program for numerical integration of Eq. (5.97) the column matrix $\ddot{\underline{\phi}}$ must be calculated for given values of the generalized coordinates, the generalized velocities and of time. First, the matrices $\underline{p}_{ai}$, $\underline{G}_a$ and $\underline{w}_a$ are determined. From $\underline{G}_a(a=1,2,3)$ the transformation matrices $\underline{A}_i(i=1,2,3)$ are obtained by the recursion formula explained earlier. With these matrices the coordinates of all vectors and tensors can then be calculated in the common reference base $\underline{e}^{(0)}$. Following this all dot and cross multiplications of vectors and tensors are carried out which lead to the matrices $\underline{A}$ and $\underline{B}$ of Eq. (5.97). In programming these details matrix formulations of the kind shown in Chap. 1 are used. More detailed programming instructions will be given in Sec. 5.2.7. ■

5.2.6.4 The case of controlled variables It frequently happens that in a technical system governed by Eq. (5.97) one or more of the angular variables are controlled by built-in motors to be prescribed functions of time. In such cases it is desirable to reduce Eq. (5.97) to a new matrix differential equation for the smaller number of uncontrolled variables. In addition, explicit expressions are desired for the unknown motor control torques. How this problem can be solved will now be shown for the case in which a single variable is controlled. The generalization to an arbitrary number of controlled variables will then be obvious. For the sake of simplicity the elements of $\ddot{\underline{\phi}}$ in Eq. (5.97) are now identified by a single index which is running from 1 to N, N being the total number of variables in the system, i.e. $\ddot{\underline{\phi}}=[\ddot{\phi}_1 \dots \ddot{\phi}_N]^T$. The controlled variable is called $\phi_k(t)$. The elements of $\underline{B}$ in Eq. (5.99) are $B_1 \dots B_N$. The k-th element, in particular, is written as $B_k = B_k^* - M_k^{\text{mot}}$ with M_k^{mot} being the k-th element of $\underline{p} \cdot \underline{Y}$. This element represents the desired motor control torque. Eq. (5.97) can now be written in the form

$$\begin{bmatrix} A_{11} \dots A_{1k} \dots A_{1N} \\ \vdots \\ A_{k1} \quad A_{kk} \quad A_{kN} \\ \vdots \\ A_{N1} \quad A_{Nk} \quad A_{NN} \end{bmatrix} \begin{bmatrix} \ddot{\phi}_1 \\ \vdots \\ \ddot{\phi}_k(t) \\ \vdots \\ \ddot{\phi}_N \end{bmatrix} = \begin{bmatrix} B_1 \\ \vdots \\ B_k^* - M_k^{\text{mot}} \\ \vdots \\ B_N \end{bmatrix}.$$

The k-th equation is extracted from this matrix equation and solved for the unknown motor control torque:

$$M_k^{\text{mot}} = B_k^* - [A_{k1} \dots A_{kk} \dots A_{kN}]\ddot{\underline{\phi}}. \tag{5.100}$$

In the remaining $N-1$ equations all terms involving $\ddot{\phi}_k(t)$ are shifted to the other side. This produces for the uncontrolled variables the matrix equation

$$\begin{bmatrix} A_{11} \dots A_{1,k-1} A_{1,k+1} \dots A_{1N} \\ \vdots \\ A_{k-1,1} \\ A_{k+1,1} \\ \vdots \\ A_{N1} \quad \dots \quad A_{NN} \end{bmatrix} \begin{bmatrix} \ddot{\phi}_1 \\ \vdots \\ \ddot{\phi}_{k-1} \\ \ddot{\phi}_{k+1} \\ \vdots \\ \ddot{\phi}_N \end{bmatrix} = \begin{bmatrix} B_1 \\ \vdots \\ B_{k-1} \\ B_{k+1} \\ \vdots \\ B_N \end{bmatrix} - \ddot{\phi}_k(t) \begin{bmatrix} A_{1k} \\ \vdots \\ A_{k-1,k} \\ A_{k+1,k} \\ \vdots \\ A_{Nk} \end{bmatrix}. \tag{5.101}$$

It has the same standard form as Eq. (5.97). Its smaller coefficient matrix is symmetric, again. Since this matrix is a function of all variables including $\phi_k(t)$ it is an explicit function of time. Once numerical solutions have been determined for the uncontrolled variables all quantities are known to calculate from Eq. (5.100) the motor control torque M_k^{mot}.

Problems

5.12 In Fig. 5.32 a hinge is shown which differs from a universal joint in that the axes on the intermediate body do not intersect each other. The intermediate body is assumed to be massless. Discuss the equations of motion for systems which have such hinges in addition to universal and pin joints.

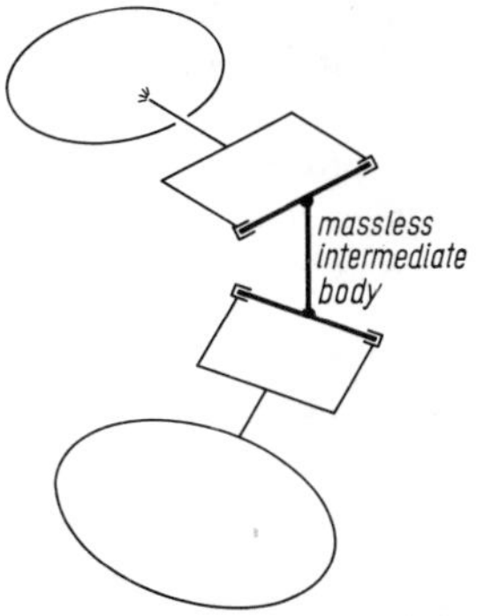

Fig. 5.32
Two bodies coupled by an intermediate massless body

5.13 The transformation matrix $\underline{G}_a$ for a pin joint is simplest if each of the two vector bases $\underline{e}^{(i^+(a))}$ and $\underline{e}^{(i^-(a))}$ on the bodies coupled by this hinge has one base vector parallel to the hinge axis (see $\underline{G}_2$ in Illustrative Example 5.2). Such an orientation is not possible for all pin joints if, for example, on each body several pin joints are located whose axes are not all parallel. What is the general form of $\underline{G}_a$ for arbitrary orientations of the bases?

5.2.7 Programming instructions

In the course of numerical integrations of Eq. (5.97) the matrices $\underline{A}$ and $\underline{B}$ must be evaluated many times. It is, therefore, necessary to carry out these calculations in a way which requires as little time as possible. The instructions given in this section will enable the reader to write a computer programm which satisfies this requirement and which, at the same time, economizes on storage space. To be specific, multi-body systems will be considered which are coupled with an external body whose motion is prescribed as a function of time. The matrices $\underline{\boldsymbol{K}}$ and $\underline{\boldsymbol{M}}'$ are then given by Eqs. (5.32) and (5.31), respectively.

Attention! Throughout this section it will be assumed that the labeling of vertices and arcs in the system graph is regular and that, in addition, all arcs are directed away from s_0. This greatly simplifies the program structure and shortens computer running times. The reasons were explained in Illustrative Example 5.2.

Every time numerical values for $\underline{A}$ and $\underline{B}$ are to be determined the following data must be available:

(i) Constant parameters describing the system. These are the quantities listed (1) to (5) in Sec. 5.1.

(ii) Constant parameters which are derived from the quantities just mentioned.

(iii) Values for ϕ_{ai} and $\dot{\phi}_{ai}$ $(a=1 \dots n;\ i=1 \dots n_a)$ and for time t.

(iv) The coordinates of $\ddot{r}_0(t)$, $\omega_0(t)$ and $\dot{\omega}_0(t)$ in the base $\underline{e}^{(0)}$.

(v) Functions for the external forces $\boldsymbol{F}_i$ and external torques $\boldsymbol{M}_i$ $(i=1 \dots n)$ in terms of ϕ_{ai}, $\dot{\phi}_{ai}$ $(a=1 \dots n; i=1 \dots n_a)$ and of t, i.e. functions for the coordinates of these forces and torques in specified frames of reference.

(vi) Functions $\boldsymbol{p}_{ai} \cdot \boldsymbol{Y}_a$ $(a=1 \dots n; i=1 \dots n_a)$ in terms of ϕ_{ai}, $\dot{\phi}_{ai}$ and t.

The derived constant parameters mentioned as second item in the list are body-fixed coordinates of the vectors $\boldsymbol{d}_{ij}$ and $\boldsymbol{b}_{i0}$ $(i,j=1 \dots n)$ and inertia components of the augmented bodies. These quantities are calculated once in a separate Parameter Program. They are then stored on magnetic disc in order to be read in again by the main program for numerical integrations. In the Parameter Program computer running time and storage requirements are of no concern. This program requires the following input data:

a) The number n of bodies; program name N.

b) The matrix $\underline{T}$; program name $T(i,j)$; $i,j=1 \dots n$.

c) The integer function $i^+(a)$; program name $IPLUS\ (i)$, $i=1 \dots n$. Because of the regular labeling in combination with the chosen arc directions $i^-(a)$ is identical with a for $a=1 \dots n$. For this reason it is no longer necessary to distinguish between vertex indices $i=1 \dots n$ and arc indices $a=1 \dots n$.

d) The masses $m_1 \dots m_n$ of the bodies; program name $AM(i)$, $i=1 \dots n$.

e) The central inertia components of the bodies in body-fixed frames of reference; program name $AJ(i,j,k)$ with $i=1 \dots n$ (body index) and $j,k=1,2,3$.

f) The coordinates of all $2n-1$ vectors $\boldsymbol{c}_{ia}$. Each data card contains the integers i and a and the three coordinates of $\boldsymbol{c}_{ia}$ in the base $\underline{e}^{(i)}$. See item (C) below.

g) The total number of pin joints, universal joints and ball-and-socket joints; program names NPJ, NUJ and NSJ, respectively.

h) The indices of the arcs representing

(i) pin joints;	program name $IPJ(i)$,	$i=1 \dots NPJ$
(ii) universal joints;	program name $IUJ(i)$,	$i=1 \dots NUJ$
(iii) ball-a.-s. joints;	program name $ISJ(i)$,	$i=1 \dots NSJ$.

i) The unit vector $\boldsymbol{p}_{a1}$ for pin joints, the unit vectors $\boldsymbol{p}_{a1}$ and $\boldsymbol{p}_{a2}$ for universal joints and the unit vectors $\boldsymbol{p}_{a1}$ and $\boldsymbol{p}_{a3}$ for ball-and-socket joints are body-fixed vectors. Their constant coordinates are input data (in pin joints $\boldsymbol{p}_{a1}$ is decomposed in the reference frame fixed on body $i^-(a)$!). Each data card carries the number of a row of the matrix $\underline{\boldsymbol{p}}$ and the coordinates of the vector $\boldsymbol{p}_{ai}$ in that row. Program name $P(i,j)$, $i=$row number of $\underline{\boldsymbol{p}}$, $j=1,2,3$.

k) The coordinate transformation matrix $\underline{G}_a$ defined by Eq. (5.85) has the general form

$$\underline{G}_a = \underline{G}_a^1 \underline{G}_a^2 \underline{G}_a^3 \qquad a=1 \dots n \tag{5.102}$$

where $\underline{G}_a^1$ and $\underline{G}_a^3$ are constant matrices whereas $\underline{G}_a^2$ depends on the angular variables in hinge a (see Problem 5.13). The matrices $\underline{G}_a^1$ and $\underline{G}_a^3$ are input data; program names $G1(i,j,k)$ and $G3(i,j,k)$ with $i=1 \dots n$ (arc index) and $j,k=1,2,3$.

From these input data the Parameter Program generates the following quantities:

A) The total number of generalized coordinates; program name NE. It is $NE = NPJ + 2*NUJ + 3*NSJ$.

B) From Eq. (5.1) the submatrix $\underline{S}$ of the incidence matrix; program name $S(i,j)$; $i,j = 1 \dots n$.

C) From Eq. (5.11) the $(3n \times n)$ scalar matrix associated with $\underline{C}$ (see Prob. 5.6); program name $CM(i,j)$; $i = 1 \dots 3n$, $j = 1 \dots n$. This section of the program has the following form. It reads in the data cards described in item (f).

```
      N3=3*N
      N2=2*N-1
      DO 10 I=1,N3
      DO 10 J=1,N
   10 CM(I,J)=0.
      DO 40 K=1,N2
      READ 20,I,J,AB(1),AB(2),AB(3)
   20 FORMAT (2I5,3F10.5)
      IN=3*(I-1)
      DO 30 L=1,3
   30 CM(IN+L,J)=S(I,J)*AB(L)
   40 CONTINUE
```

D) The total system mass M; program name $CAPM$.

E) The coordinate matrix associated with the matrix $\underline{C}\,\underline{T}\,\underline{\mu}$ whose elements are, according to Eq. (5.53), the vectors $-\boldsymbol{b}_{ij}$. Program name $CTM(i,j)$ with $i = 1 \dots 3n$, $j = 1 \dots n$.

F) The constant coordinates of the body-fixed vectors $\boldsymbol{b}_{i0}$ $(i = 1 \dots n)$; program name $BN(i,j)$ with $i = 1 \dots n$ (body index) and $j = 1, 2, 3$. For details see Problem 5.7.

G) Auxiliary integer arrays $NT1(i)$ and $NT(i,j)$, $i = 1 \dots n$, $j = 1 \dots NT1(i)$. They are calculated from the array $T(i,j)$ and are defined as follows. For $i = 1 \dots n$ $NT1(i)$ is the number of nonzero elements in the i-th row of $\underline{T}$. For $i = 1 \dots n$ and $j = 1 \dots NT1(i)$ $NT(i,j)$ is the column index of the j-th nonzero element in the i-th row of $\underline{T}$. Example: For the directed system graph shown in Fig. 5.33 the arrays $T(i,j)$, $NT1(i)$ and $NT(i,j)$ are

$$T(i,j) = \begin{bmatrix} -1 & -1 & -1 & -1 & -1 \\ 0 & -1 & -1 & 0 & -1 \\ 0 & 0 & -1 & 0 & -1 \\ 0 & 0 & 0 & -1 & 0 \\ 0 & 0 & 0 & 0 & -1 \end{bmatrix}, \qquad NT1(i) = \begin{bmatrix} 5 \\ 3 \\ 2 \\ 1 \\ 1 \end{bmatrix}, \qquad NT(i,j) = \begin{bmatrix} 1 & 2 & 3 & 4 & 5 \\ 2 & 3 & 5 & & \\ 3 & 5 & & & \\ 4 & & \text{not} & & \\ 5 & & \text{defined} & & \end{bmatrix}.$$

The purpose of these arrays is to reduce computer running time.

H) Auxiliary integer arrays $NV1(i)$ and $NV(i,j)$, $i = 1 \dots n$, $j = 1 \dots NV1(i)$. For $i = 1 \dots n$ $NV1(i)$ is the number of vertices for which s_i is the inboard vertex. The

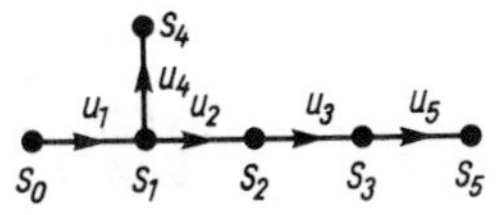

Fig. 5.33
Directed system graph

indices of these vertices are called $NV(i,j), j=1 \dots NV1(i)$. For the graph of Fig. 5.33 the two arrays are

$$NV1(i) = \begin{bmatrix} 2 \\ 1 \\ 1 \\ 0 \\ 0 \end{bmatrix}, \qquad NV(i,j) = \begin{bmatrix} 2 & 4 \\ 3 & \\ 5 & \\ \text{not} & \\ \text{defined} & \end{bmatrix}.$$

The arrays are constructed from $T(i,j)$.

I) Auxiliary integer arrays $NR(i,j)$ and $NB(k)$, an integer NI and an array $D(k,l)$ for $i,j=1 \dots n$, $k=1 \dots NI$, $l=1,2,3$. They allow reduction of computer running time and of memory space for the storage of the vectors $\boldsymbol{d}_{ij}$. These vectors are calculated from Eq. (5.21) by means of the arrays $CTM(i,j)$ and $BN(i,j)$ from items (E) and (F) above. The vectors $\boldsymbol{d}_{ij}$ are zero for many combinations of indices i and j. Among the nonzero vectors many are identical. Consider, for example, the system represented by Fig. 5.33. Its matrix $\underline{C}\,\underline{T}$ has the form

$$\underline{C}\,\underline{T} = \begin{bmatrix} \boldsymbol{d}_{11} & \boldsymbol{d}_{12} & \boldsymbol{d}_{12} & \boldsymbol{d}_{14} & \boldsymbol{d}_{12} \\ \boldsymbol{0} & \boldsymbol{d}_{22} & \boldsymbol{d}_{23} & \boldsymbol{0} & \boldsymbol{d}_{23} \\ \boldsymbol{0} & \boldsymbol{0} & \boldsymbol{d}_{33} & \boldsymbol{0} & \boldsymbol{d}_{35} \\ \boldsymbol{0} & \boldsymbol{0} & \boldsymbol{0} & \boldsymbol{d}_{44} & \boldsymbol{0} \\ \boldsymbol{0} & \boldsymbol{0} & \boldsymbol{0} & \boldsymbol{0} & \boldsymbol{d}_{55} \end{bmatrix}.$$

It contains only $2n-1$ different vectors. This number is called NI. The coordinates of the NI different vectors are arranged in the new array $D(k,l)$, $k=1 \dots NI$, $l=1,2,3$. Its first index k is a function of the two indices i,j of $\boldsymbol{d}_{ij}$. This integer function is called $NR(i,j)$. The index of the body on which a vector stored as $D(k,l)$ is fixed must be identifiable. This index is a function of the first index k of $D(k,l)$. The function is called $NB(k)$. For the graph in Fig. 5.33 the arrays $NR(i,j)$ and $NB(k)$ could be, for example,

$$NR(i,j) = \begin{bmatrix} 1 & 2 & 2 & 3 & 2 \\ - & 4 & 5 & - & 5 \\ - & - & 6 & - & 7 \\ - & - & - & 8 & - \\ - & - & - & - & 9 \end{bmatrix}, \qquad NB(k) = [1\ \ 1\ \ 1\ \ 2\ \ 2\ \ 3\ \ 3\ \ 4\ \ 5].$$

(— : not defined)

The integer NI and the arrays $NR(i,j)$, $NB(i)$ and $D(i,j)$ can be constructed from the arrays $T(i,j)$, $CTM(i,j)$, $BN(i,j)$, $NT1(i)$, $NT(i,j)$, $NV1(i)$ and $NV(i,j)$ in the FORTRAN loop

```
NI=0
DO 120 I=1,N
NI=NI+1
NR(I,I)=NI
DO 60 K=1,3
```

```
60  D(NI,K)=CTM(3*(I-1)+K,I)+BN(I,K)
    NB(NI)=I
    JMAX=NV1 (I)
    IF (JMAX) 120,120,70
70  DO 110 J=1,JMAX
    L=NV(I,J)
    NI=NI+1
    NR(I,L)=NI
    DO 80 K=1,3
80  D(NI,K)=CTM(3*(I-1)+K,L)+BN(I,K)
    NB(NI)=I
    MMAX=NT1(L)-1
    IF (MMAX) 110,110,90
90  DO 100 M=1,MMAX
100 NR(I,NT (L,M+1))=NI
110 CONTINUE
120 CONTINUE
```

K) From Eq. (5.24) the inertia components of the augmented bodies measured in body-fixed frames of reference. They are stored in the same array $AJ(i,j,k)$ which initially contains the inertia components of the original bodies. This section of the program is shown in detail in order to demonstrate the use of the arrays $NT1(i)$, $NT(i,j)$, $NR(i,j)$ and $D(i,j)$:

```
    DO 160 I=1,N
    LMAX=NT1(I)
    DO 150 L=1,LMAX
    J=NT(I,L)
    K=NR(I,J)
    B=AM(J)
    S1=B*(D(K,1)*D(K,1)+D(K,2)*D(K,2)+D(K,3)*D(K,3))
    DO 140 KK=1,3
    DO 130 JJ=1,3
130 AJ(I,KK,JJ)=-B*D(K,KK)*D(K,JJ)+AJ(I,KK,JJ)
140 AJ(I,KK,KK)=AJ(I,KK,KK)+S1
150 CONTINUE
160 CONTINUE
```

The cards $LMAX=NT1(I)$ until $J=NT(I,L)$ have the effect that all terms in the sum in Eq. (5.24) in which $\boldsymbol{d}_{ik}$ equals zero are disregarded.

L) The auxiliary integer arrays $NP(j)$ and $NF(j)$, $j=1 \dots n$. In the matrix $\underline{p}$ defined by Eq. (5.91) each column has either one or two or three nonzero elements. For the column j this number is called $NP(j)$. The index of the uppermost row with a nonzero element in the j-th column is called $NF(j)$. Both arrays are generated from NPJ, NUJ, NSJ, $IPJ(i)$, $IUJ(i)$ and $ISJ(i)$.

At the end of the Parameter Program the following data are stored on magnetic disc:

N, $IPLUS(i)$, $T(i,j)$, $AM(i)$, NPJ, NUJ, NSJ, $IPJ(i)$, $IUJ(i)$, $ISJ(i)$, $G1(i,j,k)$, $G3(i,j,k)$, NE, $CAPM$, $BN(i,j)$, $NT1(i)$, $NT(i,j)$, $NV1(i)$, $NV(i,j)$, $NR(i,j)$, $NB(i)$, NI, $D(i,j)$, $AJ(i,j,k)$, $NP(i)$, $NF(i)$, $P(i,j)$.

In what follows the main program will be outlined in which the matrices $\underline{A}$ and $\underline{B}$ of Eq. (5.97) are computed. At the start of this program all data listed above is read in from magnetic disc. The angular coordinates ϕ_{ai} and their first derivatives $\dot{\phi}_{ai}$ $(a=1 \dots n, i=1 \dots n_a)$ are stored in the one-dimensional array $Y(i), i=1 \dots 2*NE$, in the order

$$Y(i)=[\dot{\phi}_{11} \dots \dot{\phi}_{1n_1} \dots \dot{\phi}_{n1} \dots \dot{\phi}_{nn_n} \quad \phi_{11} \dots \phi_{1n_1} \dots \phi_{n1} \dots \phi_{nn_n}] .$$

The program starts with a section on kinematic quantities and on transformations. This section has the following subsections:

(1) Coordinates of the unit vectors $\boldsymbol{p}_{a2}$ for all ball-and-socket joints in the respective bases $\underline{\boldsymbol{e}}^{(i^-(a))}$. To be specific it is assumed that Bryant angles are used and that they are defined the same way as for hinge 1 in Illustrative Example 5.2. The FORTRAN statements read

```
   DO 10 I=1,NSJ
   J=NF(ISJ(I))+1
   FI=Y(J+1+NE)
   P(J,1)=SIN(FI)
   P(J,2)=COS(FI)
10 P(J,3)=0.
```

(2) Coordinates of the vectors $\boldsymbol{w}_a$ of Eq. (5.83) for all universal and ball-and-socket joints in the respective bases $\underline{\boldsymbol{e}}^{(i^-(a))}$; program name $WE(i,j), i=1 \dots n$ (hinge index), $j=1,2,3$. If for all universal joints the geometry is the same as for hinge 3 in Illustrative Example 5.2 then the program section for these joints has the simple form

```
   DO 20 I=1,NUJ
   J=IUJ(I)
   K=NF(J)
   FI=Y(K+1+NE)
   FIDOT1=Y(K)
   FIDOT2=Y(K+1)
   A=FIDOT1*FIDOT2
   WE(J,1)=0.
   WE(J,2)=-A*SIN(FI)
20 WE(J,3)=A*COS(FI)
```

It is left to the reader to develop the general form for arbitrary matrices $\underline{G}_a^1$ and $\underline{G}_a^3$ and for arbitrary directions of the vectors $\boldsymbol{p}_{a1}$ and $\boldsymbol{p}_{a2}$ on the bodies.

(3) Transformation matrices $\underline{G}_a$ for all hinges. The matrices called $\underline{G}_a^2$ in Eq. (5.102) are calculated first (cf. Illustrative Example 5.2). Then, the matrix multiplications with $\underline{G}_a^1$ and $\underline{G}_a^3$ are carried out. The result is the three-dimensional array $GA(i,j,k)$ with $i=1 \dots n$ (hinge index) and $j,k=1,2,3$.

(4) Transformation matrices $\underline{A}_i$ defined by Eq. (5.94). They are calculated from the recursion formula derived from Eq. (5.95)

$$\underline{A}_j = \begin{cases} \underline{G}_1^{\mathrm{T}} & j=1 \\ \underline{A}_{i^+(j)} \underline{G}_j^{\mathrm{T}} & j=2 \dots n . \end{cases}$$

The result is stored in the array $TRM(i,j,k)$ with $i=1 \dots n$ (body index) and $j,k=1,2,3$.

(5) Transformation from body-fixed reference frames into the common reference frame $\underline{e}^{(0)}$ of all vectors $\boldsymbol{p}_{ai}$, $\boldsymbol{d}_{ij}$, $\boldsymbol{w}_a$ and $\boldsymbol{b}_{i0}$ as well as of all augmented body inertia tensors. Details are shown for the vectors $\boldsymbol{p}_{ai}$ and $\boldsymbol{d}_{ij}$ only. To the coordinates $P(i,j)$ and $D(k,l)$ before transformation belong the coordinates $PS(i,j)$ and $DS(k,l)$, respectively, after transformation.

```
C  TRANSFORMATION FROM P(I,J) TO PS(I,J)
   DO 70 J=1,N
   L=NP(J)
   IN=NF(J)-1
   DO 65 II=1,L
   IF (2*II-L) 25,25,30
25 K=IPLUS (J)
   GO TO 35
30 K=J
35 I=IN+II
   IF (K) 65,40,50
40 DO 45 J1=1,3
45 PS(I,J1)=P(I,J1)
   GO TO 65
50 DO 60 J1=1,3
   PS(I,J1)=0.
   DO 55 K1=1,3
55 PS(I,J1)=PS(I,J1)+TRM(K,J1,K1)*P(I,K1)
60 CONTINUE
65 CONTINUE
70 CONTINUE
C  TRANSFORMATION FROM D(I,J) TO DS(I,J)
   DO 85 I=1,NI
   K=NB(I)
   DO 80 J=1,3
   DS(I,J)=0.
   DO 75 L=1,3
75 DS(I,J)=DS(I,J)+TRM(K,J,L)*D(I,L)
80 CONTINUE
85 CONTINUE
```

In a similar manner coordinates in the base $\underline{e}^{(0)}$ are calculated for the vectors $\boldsymbol{w}_a$ and $\boldsymbol{b}_{i0}$ (from the arrays $WE(i,j)$ and $BN(i,j)$, respectively) and for the augmented body inertia tensors (from $AJ(i,j,k)$). They are called $WES(i,j)$, $BNS(i,j)$ and $AJS(i,j,k)$, respectively. From this point on all manipulations on vectors and tensors will be carried out in terms of coordinates in the common reference frame $\underline{e}^{(0)}$. The first example is the calculation of

(6) the angular velocities $\boldsymbol{\Omega}_i$, $i=1 \ldots n$; program name $RAV(i,j)$, $i=1 \ldots n$, $j=1,2,3$. For ball-and-socket joints, for instance, the FORTRAN statements read (cf. Eq. (5.82))

```
   DO 110 I=1,NSJ
   J=ISJ(I)
   KMIN=NF(J)
   KMAX=KMIN+2
   DO 100 L=1,3
```

```
      RAV(J,L)=0.
      DO 90 K=KMIN,KMAX
   90 RAV(J,L)=RAV(J,L)+Y(K)*PS(K,L)
  100 CONTINUE
  110 CONTINUE
```

(7) Absolute angular velocities ω_i, $i=1\ldots n$; program name $AAV(i,j)$, $i=1\ldots n$, $j=1,2,3$. The program is based on the recursion formula derived from Eq. (5.87)

$$\omega_j = \begin{cases} \Omega_1 + \omega_0(t) & j=1 \\ \Omega_j + \omega_{i^+(j)} & j=2\ldots n\,. \end{cases}$$

```
      CALL REFBASE (TIME, AAVNUL)
      DO 120 J=1,3
  120 AAV(1,J)=RAV(1,J)+AAVNUL(J)
      DO 140 I=2,N
      K=IPLUS(I)
      DO 130 J=1,3
  130 AAV(I,J)=RAV(I,J)+AAV(K,J)
  140 CONTINUE
```

REFBASE(TIME,AAVNUL) is a subroutine for the coordinates of $\omega_0(t)$ in $\underline{e}^{(0)}$ (program name $AAVNUL(i), i=1,2,3$) as functions of *TIME*. This concludes the section on kinematic quantities and transformations.

In the next section of the main program the scalar coordinate matrix associated with the matrix $\underline{\boldsymbol{K}}$ in Eq. (5.98) is calculated. Each element of $\underline{\boldsymbol{K}}$ is a tensor $\boldsymbol{K}_{ij}$ with nine coordinates in the base $\underline{e}^{(0)}$ which form a (3×3) matrix $\underline{K}_{ij}$. All n^2 matrices together yield a symmetric $(3n\times3n)$ matrix whose program name is $DK(k,l)$. Only the submatrices $\underline{K}_{ij}$ with $j\geqslant i$ are calculated. They are zero for those combinations of i and j for which $T(i,j)$ equals zero. Zero elements are not assigned because they will not be used later. The program reads

```
    C SUBMATRICES ALONG THE DIAGONAL
      DO 180 I=1,N
      IN=3*(I-1)
      DO 170 J=1,3
      II=J+IN
      DO 160 K=1,3
  160 DK(II,K+IN)=AJS(I,J,K)
  170 CONTINUE
  180 CONTINUE
    C SUBMATRICES ABOVE THE DIAGONAL
      N1=N-1
      DO 230 II=1,N1
      IN=3*(II-1)
      KMAX=NT1(II)
      IF (KMAX-1) 230,230,190
  190 DO 220 K=2,KMAX
      JJ=NT(II,K)
      JN=3*(JJ-1)
      L=NR(II,JJ)
```

```
      S=CAPM*(BNS(JJ,1)*DS(L,1)+BNS(JJ,2)*DS(L,2)
                                +BNS(JJ,3)*DS(L,3))
      DO 210 I=1,3
      I1=IN+I
      DO 200 J=1,3
200   DK(I1,JN+J)=-CAPM*BNS(JJ,I)*DS(L,J)
210   DK(I1,JN+I)=DK(I1,JN+I)+S
220   CONTINUE
230   CONTINUE
```

In the following section of the program the matrix $\underline{B}$ of Eq. (5.99) is constructed. As a first step the coordinates in the base $\underline{e}^{(0)}$ are calculated for the vectors $\boldsymbol{f}_1 \dots \boldsymbol{f}_n$ of Eq. (5.93). They form the array *EF (i,j)*, $i=1 \dots n, j=1,2,3$. The term $\underline{T}^{\mathrm{T}}\underline{f}-\dot{\omega}_0 \underline{1}_n$ has the same form as the right hand side of Eq. (5.87) for $\underline{\omega}$. Hence, FORTRAN statements similar to those shown in item (7) can be used. This time the results are not stored in a two-dimensional array (like *AAV (i,j)* in item (7)) but in the one-dimensional array *TF (i)*, $i=1 \dots 3n$. In the next step the scalar form of $\underline{\boldsymbol{K}}\cdot(\underline{T}^{\mathrm{T}}\underline{f}-\dot{\omega}_0 \underline{1}_n)$ is obtained as the product of *DK (i,j)* with the array *TF (j)*. To the resulting one-dimensional array called *DKTF (i)*, $i=1 \dots 3n$, two other one-dimensional arrays are added which are the scalar coordinate matrices associated with $\underline{\boldsymbol{M}}'$ and $\underline{\boldsymbol{M}}$, respectively. According to Eq. (5.31) each of the n vectors in the matrix $\underline{\boldsymbol{M}}'$ is the sum of five terms. Only the more complicated terms

$$\sum_{j:s_i<s_j} \boldsymbol{d}_{ij}\times \underbrace{[\boldsymbol{\omega}_j\times(\boldsymbol{\omega}_j\times \boldsymbol{b}_{j0})]}_{\boldsymbol{a}_j} \quad \text{and} \quad \boldsymbol{\sigma}_i = \sum_{j:s_j<s_i} \underbrace{\boldsymbol{\omega}_j\times(\boldsymbol{\omega}_j\times \boldsymbol{d}_{ji})}_{\boldsymbol{p}_{ji}}$$

are discussed here in some detail. The vector abbreviated $\boldsymbol{a}_j$ depends on the index j alone. Because of the condition $s_i<s_j$ only $\boldsymbol{a}_2 \dots \boldsymbol{a}_n$ are needed. The coordinates of these vectors in the base $\underline{e}^{(0)}$ are calculated first. Then, the sums over $\boldsymbol{d}_{ij}\times \boldsymbol{a}_j$ are constructed. Consider now the second sum. Since there are fewer different vectors $\boldsymbol{d}_{ji}$ then there are index combinations (j,i) the same is true also for the vectors abbreviated $\boldsymbol{p}_{ji}$. First, all different vectors $\boldsymbol{p}_{ji}$ are calculated; program name *PE (j,i,l)* with $l=1,2,3$. This is done in the FORTRAN loop

```
      DO 255 J=1,N
      LMAX=NV1(J)
      IF (LMAX) 255, 255, 240
240   DO 250 L=1,LMAX
      I=NV(J,L)
      K=NR (J,I)
      B1=-AAV (J,3)*DS (K,2)+AAV (J,2)*DS (K,3)
      B2=AAV (J,3)*DS(K,1)-AAV (J,1)*DS(K,3)
      B3=-AAV (J,2)*DS(K,1)+AAV (J,1)*DS(K,2)
      PE(J,I,1)=-AAV (J,3)*B2+AAV (J,2)*B3
      PE(J,I,2)=AAV (J,3)*B1-AAV (J,1)*B3
250   PE(J,I,3)=-AAV (J,2)*B1+AAV (J,1)*B2
255   CONTINUE
```

To the vector $\boldsymbol{\sigma}_i$ only those values of j contribute which satisfy the conditions $T\,(j,i)\neq 0$ and $j\neq i$. This explains the FORTRAN loop

```
    DO 280 I=2,N
    DO 260 L=1,3
260 SIGMA (I,L)=0.
    DO 275 J=1,I-1
    IF (T(J,I)) 265,275,275
265 KMAX=NV1(J)
    DO 266 K=1,KMAX
    M=NV(J,K)
    IF (T(M,I)) 267,266,266
266 CONTINUE
267 DO 270 L=1,3
270 SIGMA(I,L)=SIGMA(I,L)+PE(J,M,L)
275 CONTINUE
280 CONTINUE
```

The final result of these calculations is a one-dimensional array $BR(i)$, $i=1\ldots 3n$, which represents the scalar coordinate matrix associated with the expression in square brackets in Eq. (5.99). From this the scalar coordinate matrix associated with $\underline{T}[\underline{\boldsymbol{K}}\cdot(\underline{T}^{\mathrm{T}}\,\underline{\boldsymbol{f}}-\dot{\omega}_0\underline{1}_n)+\underline{\boldsymbol{M}}'+\underline{\boldsymbol{M}}]$—called $TBR(i)$, $i=1\ldots 3n$—is calculated as follows

```
    N3=3*N
    DO 285 I=1,N3
285 TBR(I)=0.
    DO 300 I=1,N
    IN=3*(I-1)
    KMAX=NT1(I)
    DO 295 K=1,KMAX
    J=NT(I,K)
    JN=3*(J-1)
    DO 290 II=1,3
    L=IN+II
290 TBR(L)=TBR(L)-BR(JN+II)
295 CONTINUE
300 CONTINUE
```

The last step in the calculation of the leading term in $\underline{B}$ is the scalar premultiplication by $\underline{\boldsymbol{p}}$. This is done as follows.

```
    DO 315 J=1,N
    JN=3*(J-1)
    L=NP(J)
    IN=NF(J)-1
    DO 310 II=1,L
    I=IN+II
    PTBR(I)=0.
    DO 305 K=1,3
305 PTBR(I)=PTBR(I)+PS(I,K)*TBR(JN+K)
310 CONTINUE
315 CONTINUE
```

In order to obtain the matrix $\underline{B}$ we must subtract from minus $PTBR(i)$ the column matrix $\underline{\boldsymbol{p}}\cdot\underline{\boldsymbol{Y}}$. The elements of this matrix are given functions of $\phi_{ai}, \dot{\phi}_{ai}$ ($a=1\ldots n$, $i=1\ldots n_a$) and of time t. These functions must be provided in a subroutine.

In the next section of the program the matrix $\underline{A}$ of Eq. (5.98) is constructed. As compared with the program for the matrix $\underline{B}$ this part of the program is rather short since the expresion for $\underline{A}$ is simpler than the one for $\underline{B}$. It requires more computer running time, however, because $\underline{A}$ is a square matrix whereas $\underline{B}$ is a column matrix. First, the scalar coordinate matrix associated with $\underline{T}\underline{\boldsymbol{K}}\underline{T}^{\mathrm{T}}$ is constructed. A single element of $\underline{T}\underline{\boldsymbol{K}}\underline{T}^{\mathrm{T}}$ is the double sum $\sum_{i=1}^{n}\sum_{j=1}^{n} T_{ai}T_{bj}\boldsymbol{K}_{ij}\,(a,b=1\dots n)$. The reader may verify for himself that for a directed system graph with regular labeling and with all arcs pointing away from s_0 the following relationships are valid.

$$(\underline{T}\underline{\boldsymbol{K}}\underline{T}^{\mathrm{T}})_{ki}=\left.\begin{cases}\sum\limits_{l\in P_k}(\underline{T}\underline{\boldsymbol{K}}\underline{T}^{\mathrm{T}})_{lk}+\sum\limits_{j:s_k\leqslant s_j}\boldsymbol{K}_{kj} & \text{if } i=k\\ 0 & \left.\begin{matrix}\text{if } T_{ik}=0\\ \text{if } T_{ik}\neq 0\end{matrix}\right\} i<k\\ \sum\limits_{l\in Q_k}(\underline{T}\underline{\boldsymbol{K}}\underline{T}^{\mathrm{T}})_{kl}+\sum\limits_{j:s_k\leqslant s_j}\bar{\boldsymbol{K}}_{ij} & \end{cases}\right\}\ k=1\dots n.$$

In these formulas P_k and Q_k are defined as follows. P_k is the set of indices of all vertices for which s_k is the inboard vertex. These indices are $NV(k,j), j=1\dots NV1(k)$. For all $l\in P_k$ the inequality $l>k$ holds. Q_k is the set of indices of all vertices s_j whose inboard vertex is s_i and for which, in addition, $T(j,k)=1$. The set is either empty or it has one element l for which the inequality $i<l\leqslant k$ holds. The formula allows a recursive calculation of $(\underline{T}\underline{\boldsymbol{K}}\underline{T}^{\mathrm{T}})_{ki}$ in the order indicated by the FORTRAN statements:

```
      DO 480 K=N,1,-1
      calculate (TKT^T)_kk
      DO 470 I=K-1,1,-1
      calculate (TKT^T)_ki
  470 CONTINUE
  480 CONTINUE
```

The equation for the elements of $\underline{T}\underline{\boldsymbol{K}}\underline{T}^{\mathrm{T}}$ has the property that the tensor $\boldsymbol{K}_{kk}\,(k=1\dots n)$ appears only once, namely in the calculation of $(\underline{T}\underline{\boldsymbol{K}}\underline{T}^{\mathrm{T}})_{kk}$. This allows storing the results in the same array $DK(i,j)$ which contains the coordinates of $\boldsymbol{K}_{ij}\,(j\geqslant i)$. At the end of the following program the (3×3) submatrices along and under the main diagonal of the array $DK(i,j)$ are occupied by the coordinates of $(\underline{T}\underline{\boldsymbol{K}}\underline{T}^{\mathrm{T}})_{ki}$ with $k=1\dots n, i\leqslant k$ (this implies that the coordinates of $\boldsymbol{K}_{kk}\,(k=1\dots n)$ are no longer available).

```
      DO 480 K=N,1,-1
C     SUBMATRICES ALONG THE MAIN DIAGONAL
      KK3=3*K
      KK1=KK3-2
      N1=NV1(K)
      IF (N1) 340,340,320
  320 DO 330 J=1,N1
      KP=3*(NV(K,J)-K)
      DO 325 II=KK1,KK3
      DO 325 KK=KK1,KK3
  325 DK(KK,II)=DK(KK,II)+DK(KK+KP,II)
  330 CONTINUE
  340 N2=NT1(K)
      IF (N2-1) 380,380,350
```

```
350 DO 370 J=2,N2
    KP=3*(NT(K,J)-K)
    DO 360 II=KK1,KK3
    DO 360 KK=KK1,KK3
360 DK(KK,II)=DK(KK,II)+DK(KK,II+KP)
370 CONTINUE
380 CONTINUE
  C SUBMATRICES BELOW THE MAIN DIAGONAL
    IF (K-1) 480,480,390
390 DO 470 I=K-1,1,-1
    IF (T(I,K)) 400,470,400
400 II3=3*I
    II1=II3-2
    ND=NV1(I)
    IF(ND) 435,435,405
405 DO 410 J=1,ND
    I1=NV(I,J)
    IF (T(I1,K)) 420,410,420
410 CONTINUE
    GO TO 435
420 IP=3*(I1-I)
    DO 430 II=II1,II3
    DO 430 KK=KK1,KK3
430 DK(KK,II)=DK(II,KK)+DK(KK,II+IP)
435 IF (N2-1) 470,470,440
440 DO 460 J=2,N2
    KP=3*(NT(K,J)-K)
    DO 450 II=II1,II3
    DO 450 KK=KK1,KK3
450 DK(KK,II)=DK(KK,II)+DK(II,KK+KP)
460 CONTINUE
470 CONTINUE
480 CONTINUE
```

The final step in the computation of the matrix $\underline{A}$ is the scalar pre- and postmultiplication of $\underline{T}\underline{\boldsymbol{K}}\underline{T}^{\mathrm{T}}$ by $\underline{\boldsymbol{p}}$ and $\underline{\boldsymbol{p}}^{\mathrm{T}}$, respectively. In the following program the upper half of the symmetric matrix is calculated. It is stored in the array *DK(i,j)* in the locations *DK(1,4) ... DK(1,3+NE) ... DK(NE,3+NE)* (this causes an overflow if $3+NE>3n$ or $NE>3(n-1)$, i.e. if the system contains more than $n-2$ ball-and-socket joints. The overflow is avoided if the dimension declaration for *DK* is *DK(3n,3n+3)*). In the statements following "*590 CONTINUE*" the upper half of $\underline{A}$ is shifted to the locations *DK(1,1) ... DK(1,NE) ... DK(NE,NE)*.

```
    DO 590 L=1,N
    LP=3*L-2
    NL=NP(L)
    J1=NF(L)
    J2=J1-1+NL
    DO 580 J=J1,J2
    DO 570 K=L,N
    KP=3*(K-1)
```

```
      NK=NP(K)
      I1=NF(K)
      I2=I1-1+NK
      IF (T(L,K)) 530,500,530
500   DO 520 I=I1,I2
      IF (J-I) 510,510,520
510   DK(J,I+3)=0.
520   CONTINUE
      GO TO 570
530   DO 540 L1=1,3
      KK=KP+L1
540   B(L1)=DK(KK,LP)*PS(J,1)+DK(KK,LP+1)*PS(J,2)
                               +DK(KK,LP+2)*PS(J,3)
      DO 560 I=I1,I2
      IF (J-I) 550,550,560
550   DK(J,I+3)=PS(I,1)*B(1)+PS(I,2)*B(2)+PS(I,3)*B(3)
560   CONTINUE
570   CONTINUE
580   CONTINUE
590   CONTINUE
      DO 600 I=1,NE
      DO 600 J=1,NE
600   DK(I,J)=DK(I,J+3)
```

This concludes the programming instructions for the matrices $\underline{A}$ and $\underline{B}$ in Eq. (5.97).

5.2.8 Systems with arbitrary holonomic constraints in the hinges

In the previous sections multi-body systems with special types of hinges were investigated. In the present section hinges with arbitrary holonomic constraints are taken into consideration (the case of nonholonomic constraints will be

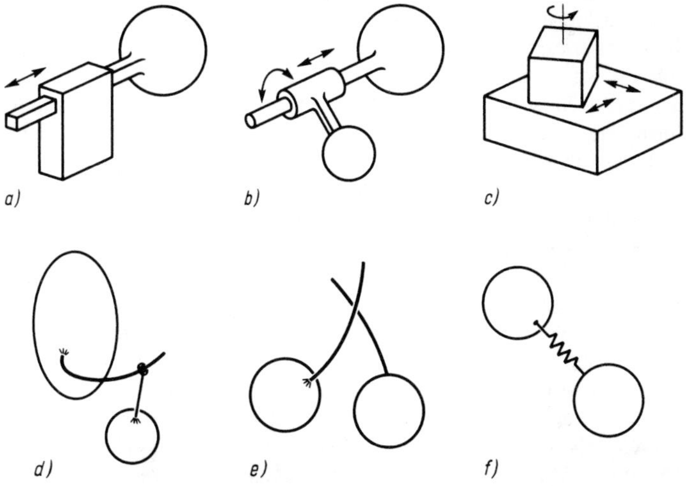

Fig. 5.34 Six hinges with 1, 2, 3, 4, 5 and 6 degrees of freedom

treated in Sec. 5.3.2). The constraints may be either scleronomic or rheonomic. In Fig. 5.34a to f six examples of hinges with scleronomic constraints are shown. Bodies coupled by these hinges have—in this order—one, two, three, four, five and six degrees of freedom of motion relative to each other. In Fig. 5.34c two plane surfaces, one on each body, are in constant touch. In Fig. 5.34d one of the bodies is a pendulum whose suspension point is free to move along a guide which is fixed on the other body. In Fig. 5.34e each body has its own guide. The guides are constrained to touch each other but they are free to slip along each other. In Fig. 5.34f the only internal hinge force is provided by a spring. The degenerate case that even this spring is missing is not ruled out. Hinges with rheonomic constraints are obtained if, for instance, the shape of the guides in Figs. 5.34d and e changes according to some prescribed function of time.

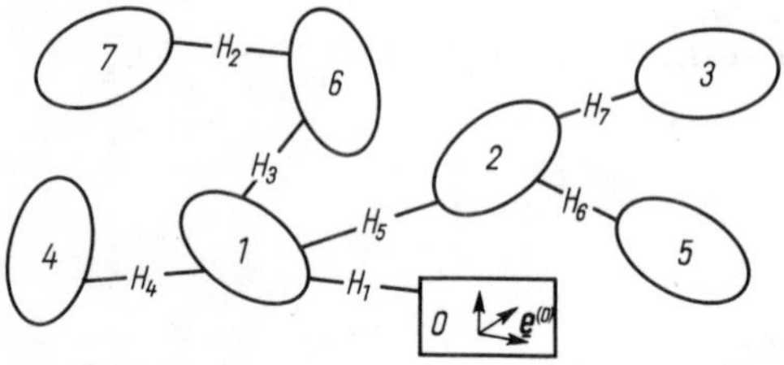

Fig. 5.35
A system with hinges with arbitrary holonomic constraints. Hinges are indicated by the letter "H"

In the seven-body system which is depicted in Fig. 5.35 for illustrative purposes the hinges are simply marked by the symbol $—H_a—$ without any indication of their specific properties. These properties are assumed to be listed separately. The motion of body 0 relative to inertial space is prescribed as a function of time. The labeling of bodies and hinges is the same as in Fig. 5.8a so that the directed system graph of Fig. 5.8c and the associated matrices $\underline{S}_0$, $\underline{S}$ and $\underline{T}$ can be used, again. On body 0 a vector base $\underline{e}^{(0)}$ is fixed. The hinge number 1 between the bodies 0 and 1 may be a material hinge. It may also be a fictitious hinge in the sense familiar from Sec. 5.1. In the latter case body 0 need not be a material body. Only the reference frame $\underline{e}^{(0)}$ is of interest whose motion relative to inertial space is prescribed as some conveniently chosen function of time. In contrast to previous sections it is no longer necessary to formulate different equations of motion for systems with a material hinge 1 and for systems with a fictitious hinge 1. The reason is that any hinge may be free of constraint forces as well as of other internal forces so that the fictitious hinge is not an exceptional case any more. It is still possible, however, to formulate a special set of equations of motion for systems in which hinge 1 is fictitious. As in Sec. 5.2.4 this set consists of one single equation describing the motion of the composite system center of mass and of a set of equations describing motions relative to this center of mass. In practical applications the use of these special equations is often advantageous. From the point of view of numerical accuracy it may even be necessary. Both sets of equations will, therefore, be developed. For both sets the procedure is for the most part identical. The paragraphs which refer to the special equations only will be marked Special Case. Only two such paragraphs will be necessary. A very short one is inserted near the beginning. The other follows after the equations of general applicability will have been formulated.

5.2.8.1 Formulation of d'Alembert's principle In the previous sections equations of motion were obtained on the basis of Newton's law and of the law of moment of momentum formulated for individual bodies which had been isolated by cuts through the hinges. This approach is, in principle, practicable also in the case of hinges other than ball-and-socket, universal and pin joints. It requires, however, some information about the direction of constraint forces and torques in the hinges, on the one hand, and about the kinematics of rotational and translational motions of contiguous bodies relative to each other, on the other. Even this minimum of information is not available here since the properties of the hinges are not specified at all. For this reason, methods of analytical mechanics are chosen. In Chap. 3.5 d'Alembert's principle was formulated for a single rigid body (see Eq. (3.25)). The same approach is used in the present case. For a system of n rigid bodies d'Alembert's principle can be written in the form

$$\sum_{i=1}^{n} [\delta \boldsymbol{r}_i \cdot (\boldsymbol{F}_i - m_i \ddot{\boldsymbol{r}}_i) + \boldsymbol{\delta\pi}_i \cdot (\boldsymbol{M}_i - \dot{\boldsymbol{L}}_i)] + \delta W = 0 \tag{5.103}$$

with $\quad \dot{\boldsymbol{L}}_i = \boldsymbol{J}_i \cdot \dot{\boldsymbol{\omega}}_i + \boldsymbol{\omega}_i \times \boldsymbol{J}_i \cdot \boldsymbol{\omega}_i \qquad i = 1 \dots n\,.$

In this expression m_i, $\boldsymbol{r}_i$, $\boldsymbol{L}_i$, $\boldsymbol{J}_i$ and $\boldsymbol{\omega}_i$ denote the same quantities as in the previous sections, namely the mass of body i, the radius vector of its center of mass from a reference point fixed in inertial space, its absolute angular momentum with respect to its center of mass, its central inertia tensor and its absolute angular velocity, respectively. Also $\boldsymbol{F}_i$ and $\boldsymbol{M}_i$ are the same quantities as before, namely the resultant external force and the resultant external torque on body i. The line of action of $\boldsymbol{F}_i$ is passing through the body i center of mass. The vector $\delta \boldsymbol{r}_i$ is the variation of $\boldsymbol{r}_i$, and $\boldsymbol{\delta\pi}_i$ is the product of an arbitrary unit vector with an infinitesimal angle. It is necessary to emphasize here that the dots on $\ddot{\boldsymbol{r}}_i$ and $\dot{\boldsymbol{\omega}}_i$ denote differentiation with respect to time in an inertial reference frame and that $\delta \boldsymbol{r}_i$ and $\boldsymbol{\delta\pi}_i$ describe variations of the position and orientation of body i relative to this inertial reference frame. The summation over i ranges only from 1 to n. Body 0 is excluded since in d'Alembert's principle variation means variation with time t held fixed. From this follows that the variation of any quantity associated with the motion of body 0 is zero since this motion is prescribed as a function of time. The term δW represents the total virtual work done in the hinges of the system. The constraint forces do not contribute to it because they are assumed to be ideal. But virtual work is done by springs, dampers and similar elements in the hinges. In the presence of constraints in the hinges the variations $\delta \boldsymbol{r}_i$ and $\boldsymbol{\delta\pi}_i$ ($i = 1 \dots n$) are not independent of each other. From this follows the necessity to express the variations in terms of variations of other quantities—generalized coordinates—whose variations are independent. This is a problem of pure kinematics and will be handled in the following subsections. The dynamics part of the problem is fully contained in Eq. (5.103). For later use this equation is rewritten in matrix form as

$$\delta \underline{\boldsymbol{r}}^{\mathrm{T}} \cdot (\underline{\boldsymbol{F}} - \underline{m}\ddot{\underline{\boldsymbol{r}}}) + \underline{\boldsymbol{\delta\pi}}^{\mathrm{T}} \cdot (\underline{\boldsymbol{M}} - \underline{\boldsymbol{J}} \cdot \dot{\underline{\boldsymbol{\omega}}} - \underline{\boldsymbol{V}}) + \delta W = 0 \tag{5.104}$$

with the following matrices

$$\left.\begin{array}{ll}\underline{\ddot{r}} = [\ddot{r}_1 \dots \ddot{r}_n]^T & \underline{\delta\pi} = [\delta\pi_1 \dots \delta\pi_n]^T \\ \delta\underline{r} = [\delta r_1 \dots \delta r_n]^T & \underline{F} = [F_1 \dots F_n]^T \\ \underline{M} = [M_1 \dots M_n]^T & \underline{\dot{\omega}} = [\dot{\omega}_1 \dots \dot{\omega}_n]^T \\ \underline{V} = [V_1 \dots V_n]^T \quad \text{with} & V_i = \omega_i \times J_i \cdot \omega_i \\ \underline{J} = \begin{bmatrix} J_1 & & \\ & \ddots & \\ & & J_n \end{bmatrix} & \underline{m} = \begin{bmatrix} m_1 & & \\ & \ddots & \\ & & m_n \end{bmatrix} \end{array}\right\} \tag{5.105}$$

The matrices $\underline{\ddot{r}}$, $\underline{\dot{\omega}}$ and $\underline{m}$ were used in Secs. 5.2.2 and 5.2.4 with the same definitions.

Special Case Before turning to kinematics d'Alembert's principle is formulated for the special case of a system with a fictitious hinge 1. The radius vector $\boldsymbol{r}_i$ is expressed as the sum $\boldsymbol{r}_i = \boldsymbol{r}_C + \boldsymbol{R}_i$ where $\boldsymbol{r}_C$ is the radius vector of the composite system center of mass from the reference point fixed in inertial space and $\boldsymbol{R}_i$ is the vector from the system center of mass to the body i center of mass. The quantities $\boldsymbol{r}_C$ and $\boldsymbol{R}_i$ have the same definitions as in Fig. 5.21. Substitution of

$$\delta \boldsymbol{r}_i = \delta \boldsymbol{r}_C + \delta \boldsymbol{R}_i, \qquad \ddot{\boldsymbol{r}}_i = \ddot{\boldsymbol{r}}_C + \ddot{\boldsymbol{R}}_i \qquad i = 1 \dots n \tag{5.106}$$

into Eq. (5.103) yields

$$\sum_{i=1}^{n} \{(\delta \boldsymbol{r}_C + \delta \boldsymbol{R}_i) \cdot [\boldsymbol{F}_i - m_i(\ddot{\boldsymbol{r}}_C + \ddot{\boldsymbol{R}}_i)] + \boldsymbol{\delta\pi}_i \cdot (\boldsymbol{M}_i - \dot{\boldsymbol{L}}_i)\} + \delta W = 0.$$

The quantities $\boldsymbol{R}_1 \dots \boldsymbol{R}_n$ satisfy the relationship

$$\sum_{i=1}^{n} m_i \boldsymbol{R}_i = \boldsymbol{0}. \tag{5.107}$$

This reduces d'Alembert's principle to the form

$$\delta \boldsymbol{r}_C \cdot \sum_{i=1}^{n} (\boldsymbol{F}_i - m_i \ddot{\boldsymbol{r}}_C) + \sum_{i=1}^{n} [\delta \boldsymbol{R}_i \cdot (\boldsymbol{F}_i - m_i \ddot{\boldsymbol{R}}_i) + \boldsymbol{\delta\pi}_i \cdot (\boldsymbol{M}_i - \dot{\boldsymbol{L}}_i)] + \delta W = 0.$$

In hinge 1 no internal forces and torques are acting. Hence, $\delta \boldsymbol{R}_i$ and $\boldsymbol{\delta\pi}_i$ $(i = 1 \dots n)$ as well as δW are independent of $\delta \boldsymbol{r}_C$. From this follows that d'Alembert's principle breaks up into the equations

$$M \ddot{\boldsymbol{r}}_C = \sum_{i=1}^{n} \boldsymbol{F}_i \tag{5.108}$$

and
$$\sum_{i=1}^{n} [\delta \boldsymbol{R}_i \cdot (\boldsymbol{F}_i - m_i \ddot{\boldsymbol{R}}_i) + \boldsymbol{\delta\pi}_i \cdot (\boldsymbol{M}_i - \dot{\boldsymbol{L}}_i)] + \delta W = 0.$$

The first equation in which M denotes the total system mass determines the motion of the composite system center of mass. The second equation has the same structure as Eq. (5.103). Because of the constraints in the hinges and because of the additional constraint represented by Eq. (5.107) the variations $\delta \boldsymbol{R}_i$ and $\boldsymbol{\delta\pi}_i$ $(i = 1 \dots n)$ are not independent of each other. They have to be expressed in terms of appropriately chosen generalized coordinates. In matrix form the equation reads

$$\delta \underline{R}^{\mathrm{T}} \cdot (\underline{F} - \underline{m}\, \ddot{\underline{R}}) + \underline{\delta\pi}^{\mathrm{T}} \cdot (\underline{M} - \underline{J} \cdot \dot{\underline{\omega}} - \underline{V}) + \delta W = 0 . \tag{5.109}$$

This differs from Eq. (5.104) only in the column matrices $\ddot{\underline{R}} = [\ddot{R}_1 \ldots \ddot{R}_n]^{\mathrm{T}}$ and $\delta\underline{R} = [\delta R_1 \ldots \delta R_n]^{\mathrm{T}}$. The development of equations of motion for the Special Case is left in this state. It will be taken up, again, after the formulation of generally applicable equations of motion.

The following two subsections are devoted to system kinematics. The final goal is to formulate expressions for $\delta\underline{r}$, $\ddot{\underline{r}}$, $\underline{\delta\pi}$, $\dot{\underline{\omega}}$ and other kinematic quantities in Eq. (5.104) in terms of generalized coordinates and of their time derivatives and variations. This will be achieved as follows. In the first subsection the kinematics of motion of two contiguous bodies relative to one another is treated. This part is concerned with individual hinges. In the second subsection the kinematics of motion of individual bodies relative to inertial space is described. This is done by combining formulas developed in the first subsection with concepts of graph theory. The entire kinematics formulation is similar to the one in Sec. 5.2.6. It is more difficult, though, because of the more general character of the hinges. All mathematical quantities are defined in such a way that the two formulations are fully identical in the particular case that all hinges are ball-and-socket or universal or pin joints. The reader is advised to verify this at each step of the development.

5.2.8.2 Kinematics of motion of contiguous bodies relative to one another In Fig. 5.36 two bodies $i^+(a)$ and $i^-(a)$ coupled by hinge a $(a = 1 \ldots n)$ are shown. On the bodies, vector bases $\underline{e}^{(i^+(a))}$ and $\underline{e}^{(i^-(a))}$, respectively, are fixed as follows. On body 0 the origin of the base $\underline{e}^{(0)}$ is called C_0. This is an arbitrarily chosen point on the body (note the difference in comparison with Sec. 5.2.2 where C_0 was chosen to be the hinge point of hinge 1 as is shown in Fig. 5.10). On all other bodies $i = 1 \ldots n$ the origin of $\underline{e}^{(i)}$ is fixed at the body center of mass C_i. The orientation of the bases on the bodies is arbitrary. The number of degrees of freedom of motion of the bodies relative to each other is called n_a for hinge a $(a = 1 \ldots n)$. It is in the range $1 \leqslant n_a \leqslant 6$ depending on the properties of the hinge. An equal number of generalized coordinates $q_{a1} \ldots q_{an_a}$ is required to describe the motion of the bodies relative to one another.

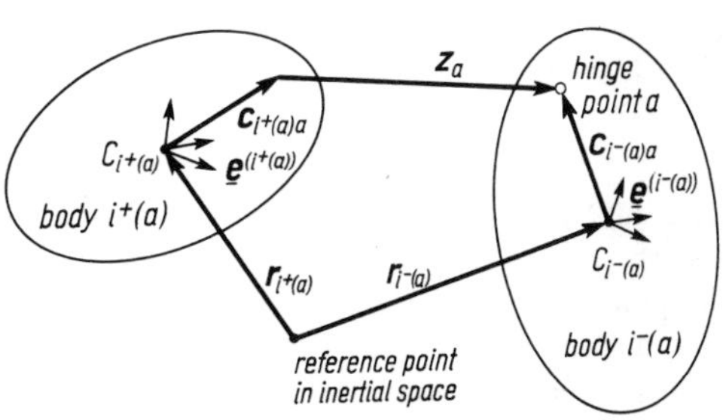

Fig. 5.36
Vectors describing the location of contiguous bodies relative to inertial space and relative to each other

The convention is adopted that the position and motion of body $i^-(a)$ relative to body $i^+(a)$ are described. For the description of the position it is sufficient to specify two quantities as functions of the generalized coordinates (and of time if the constraints in the hinge are rheonomic). One quantity is the radius vector in the base $\underline{e}^{(i^+(a))}$ of a single point fixed on body $i^-(a)$ and the other is the transformation matrix $\underline{G}_a$ between the bases $\underline{e}^{(i^+(a))}$ and $\underline{e}^{(i^-(a))}$. This matrix is defined by the equation

$$\underline{e}^{(i^-(a))} = \underline{G}_a \underline{e}^{(i^+(a))} \qquad a=1\dots n. \tag{5.110}$$

The point fixed on body $i^-(a)$ is called the hinge point. In principle, any point can be chosen. In practice, a point should be selected whose coordinates in the base $\underline{e}^{(i^+(a))}$ can be expressed as functions of the generalized coordinates and of time in a particularly simple way. The location of the hinge point on body $i^-(a)$ is specified by the body-fixed vector $\boldsymbol{c}_{i^-(a)a}$ which originates from the point $C_{i^-(a)}$. The variable radius vector of the hinge point in the base $\underline{e}^{(i^+(a))}$ is split into two parts. One part is a vector $\boldsymbol{c}_{i^+(a)a}$ which is fixed on body $i^+(a)$. The other part is called the hinge vector $\boldsymbol{z}_a$ (see Fig. 5.36). In the base $\underline{e}^{(i^+(a))}$ the coordinates of $\boldsymbol{z}_a$ only are functions of $q_{a1}\dots q_{an_a}$ and of time. How the body-fixed starting and terminating points of $\boldsymbol{z}_a$ are chosen in practice is illustrated by two examples. In the first example a ball-and-socket joint is considered. In this case the geometric center of the joint which is fixed on both bodies should be chosen as end point for both vectors $\boldsymbol{c}_{i^-(a)a}$ and $\boldsymbol{c}_{i^+(a)a}$. The hinge vector $\boldsymbol{z}_a$ is then identically zero. Fig. 5.36 becomes identical with Fig. 5.12. This example illustrates the similarity between the mathematical formulations in this section and in earlier sections. In the second example the hinge allows a translation of body $i^-(a)$ along an axis fixed on body $i^+(a)$ and a rotation about the same axis (Fig. 5.37). As generalized coordinates a cartesian coordinate q_{a1} along the axis and a rotation angle q_{a2} about the axis are chosen. As hinge point a point on the axis should be selected because its coordinates in the base $\underline{e}^{(i^+(a))}$ are linear functions of q_{a1} and independent of q_{a2}. For all other points the coordinates are linear in q_{a1} and, in addition, circular functions of q_{a2}. The location of the point fixed on body $i^+(a)$ has little influence on the mathematical expression for the coordinates of $\boldsymbol{z}_a$ since these coordinates are linear functions of q_{a1} anyway. If a point on the axis is chosen as is shown in Fig. 5.37

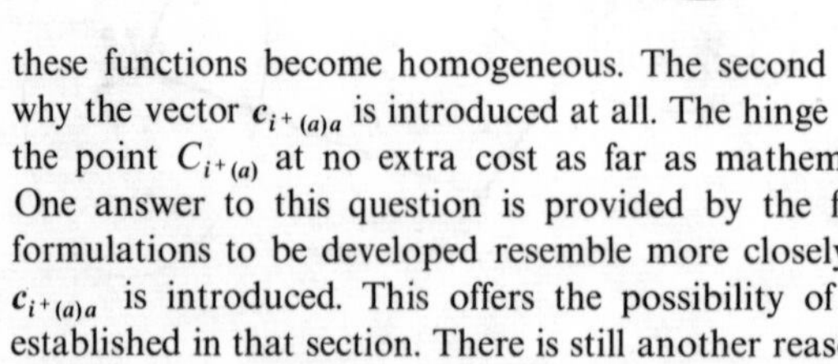

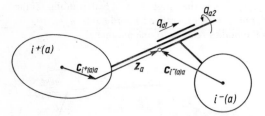

Fig. 5.37
One way of defining the hinge vector $\boldsymbol{z}_a$ for a particular hinge

these functions become homogeneous. The second example stimulates the question why the vector $\boldsymbol{c}_{i^+(a)a}$ is introduced at all. The hinge vector $\boldsymbol{z}_a$ could always start from the point $C_{i^+(a)}$ at no extra cost as far as mathematical difficulties are concerned. One answer to this question is provided by the first example. The mathematical formulations to be developed resemble more closely those of Sec. 5.2.6 if the vector $\boldsymbol{c}_{i^+(a)a}$ is introduced. This offers the possibility of using relationships which were established in that section. There is still another reason for the definition of the vector $\boldsymbol{c}_{i^+(a)a}$. The vectors $\boldsymbol{c}_{i^+(a)a}$ and $\boldsymbol{c}_{i^-(a)a}$ appear together in expressions of such general form as $\boldsymbol{C}_{ia} = S_{ia}\boldsymbol{c}_{ia}$ $(i,a=1\dots n)$, for instance. As long as both vectors are arbitrary body-fixed vectors no difficulties arise in interpreting such expressions. But difficulties would arise if it were necessary to distinguish between the two because one of them is zero and the other is not. Note that in Fig. 5.36 in the case $a=1$ one of the two bodies is body 0 and that also on this body such a vector is defined, namely the vector $\boldsymbol{c}_{01}$. As in Sec. 5.2.2 the following definition will be used. The vector $\boldsymbol{c}_{ia}$ ($i=0 \dots n$, $a=1 \dots n$) is zero if i equals neither $i^+(a)$ nor $i^-(a)$, i.e. if hinge a is not located on body i.

A final remark is required concerning the matrix $\underline{G}_a$ of Eq. (5.110). Its expression as a function of $q_{a1} \dots q_{an_a}$ and of t depends, first, on the choice of generalized coordinates and, second, on the orientation of the bases $\underline{e}^{(i^+(a))}$ and $\underline{e}^{(i^-(a))}$ on the bodies. This orientation should be chosen so as to render the expression for $\underline{G}_a$ as simple as possible. That this requirement can, in general, be satisfied for only one hinge on each body was shown in Problem 5.13. Let us summarize what has been said so far. For each of the hinges $a=1 \dots n$ the user of the formalism is free to choose

(i) the nature of the generalized coordinates $q_{a1} \dots q_{an_a}$,

(ii) the location of the hinge point on body $i^-(a)$ and of the point on body $i^+(a)$ from which $\boldsymbol{z}_a$ starts,

(iii) the orientation of the body-fixed vector bases $\underline{e}^{(i^+(a))}$ and $\underline{e}^{(i^-(a))}$.

These decisions should be made so as to render the functions

$$\begin{aligned} \boldsymbol{z}_a &= \boldsymbol{z}_a(q_{a1} \dots q_{an_a}, t) \qquad a=1 \dots n \\ \underline{G}_a &= \underline{G}_a(q_{a1} \dots q_{an_a}, t) \qquad a=1 \dots n \end{aligned} \tag{5.111}$$

as simple as possible. In principle, the user is also free in choosing the directions of the arcs in the directed system graph. However, for reasons explained earlier, the arcs should be directed either all toward s_0 or all away from s_0. The programming instructions given in Sec. 5.2.7 which will be useful also for the present case require all arcs to be directed away from s_0. For most types of hinges the difficulties in formulating the coordinates of $\boldsymbol{z}_a$ in the base $\underline{e}^{(i^+(a))}$ are independent of which body is body $i^+(a)$ and which $i^-(a)$. There are exceptions, however. One such case is the hinge shown in Fig. 5.34d (see Problem 5.15).

Problems

5.14 For each of the hinges of Fig. 5.34 select one body as body $i^+(a)$. Then, choose generalized coordinates and locations for the starting and terminating points of the hinge vector $\boldsymbol{z}_a$ and formulate as functions of the generalized coordinates the matrix $\underline{G}_a$ and the coordinates of $\boldsymbol{z}_a$ in a base fixed on body $i^+(a)$. For the sake of simplicity assume that the guides in Fig. 5.34d and e are either straight lines or circular.

5.15 For the hinge of Fig. 5.34d compare the amount of labor necessary to formulate the coordinates of $\boldsymbol{z}_a$ in the base $\underline{e}^{(i^+(a))}$ if (i) the left and (ii) the right body is chosen to be body $i^+(a)$.

Having described the position and angular orientation of body $i^-(a)$ in the base $\underline{e}^{(i^+(a))}$ we now concern ourselves with its motion relative to the same base. The velocity distribution of the body is uniquely defined by two vectors, namely by the velocity of the hinge point a and by the angular velocity of body $i^-(a)$, both relative to $\underline{e}^{(i^+(a))}$. The first vector is the time derivative of $\boldsymbol{z}_a$ in the base $\underline{e}^{(i^+(a))}$. It is called $\mathring{\boldsymbol{z}}_a$ with an open circle instead of a dot in order to distinguish it from the time derivative in inertial space. The relative angular velocity is called $\boldsymbol{\Omega}_a$. Differentiating $\mathring{\boldsymbol{z}}_a$ and $\boldsymbol{\Omega}_a$ with respect to time in the base $\underline{e}^{(i^+(a))}$ one obtains the acceleration $\overset{\circ\circ}{z}_a$ of the hinge point a and the angular acceleration $\mathring{\boldsymbol{\Omega}}_a$ of body $i^-(a)$, both relative to $\underline{e}^{(i^+(a))}$. The vector $\mathring{\boldsymbol{\Omega}}_a$ is also the time derivative of $\boldsymbol{\Omega}_a$ in the base $\underline{e}^{(i^-(a))}$ since the two derivatives differ only by the term $\boldsymbol{\Omega}_a \times \boldsymbol{\Omega}_a = \mathbf{0}$. It is advisable to interpret $\mathring{\boldsymbol{\Omega}}_a$ as derivative of $\boldsymbol{\Omega}_a$ in $\underline{e}^{(i^-(a))}$ since the scalar coordinates of $\boldsymbol{\Omega}_a$ will be available more easily in this base than in the reference base $\underline{e}^{(i^+(a))}$. For $\mathring{z}_a$ and $\overset{\circ\circ}{z}_a$ explizit expressions are obtained from Eq. (5.111):

$$\mathring{z}_a = \sum_{i=1}^{n_a} \frac{\partial z_a}{\partial q_{ai}} \dot{q}_{ai} + \frac{\partial z_a}{\partial t}$$

$$a = 1 \dots n\,.$$

$$\overset{\circ\circ}{z}_a = \sum_{i=1}^{n_a} \left[\frac{\partial z_a}{\partial q_{ai}} \ddot{q}_{ai} + \sum_{j=1}^{n_a} \frac{\partial^2 z_a}{\partial q_{ai} \partial q_{aj}} \dot{q}_{ai} \dot{q}_{aj} + 2 \frac{\partial^2 z_a}{\partial q_{ai} \partial t} \dot{q}_{ai} \right] + \frac{\partial^2 z_a}{\partial t^2}$$

In actual calculations the partial differentiations have to be applied to the scalar coordinates of z_a in the base $\underline{e}^{(i^+(a))}$ (see Problem 5.16). With the abbreviations

$$\boldsymbol{k}_{ai} = \frac{\partial z_a}{\partial q_{ai}}, \qquad \boldsymbol{k}_{a0} = \frac{\partial z_a}{\partial t} \qquad a = 1 \dots n,\ i = 1 \dots n_a \tag{5.112}$$

and

$$\boldsymbol{s}_a = \sum_{i=1}^{n_a} \left[\sum_{j=1}^{n_a} \frac{\partial^2 z_a}{\partial q_{ai} \partial q_{aj}} \dot{q}_{ai} \dot{q}_{aj} + 2 \frac{\partial^2 z_a}{\partial q_{ai} \partial t} \dot{q}_{ai} \right] + \frac{\partial^2 z_a}{\partial t^2} \qquad a = 1 \dots n \tag{5.113}$$

the equations take the form

$$\mathring{z}_a = \sum_{i=1}^{n_a} \boldsymbol{k}_{ai} \dot{q}_{ai} + \boldsymbol{k}_{a0} \tag{5.114}$$

$$a = 1 \dots n\,.$$

$$\overset{\circ\circ}{z}_a = \sum_{i=1}^{n_a} \boldsymbol{k}_{ai} \ddot{q}_{ai} + \boldsymbol{s}_a \tag{5.115}$$

The relative angular velocity $\boldsymbol{\Omega}_a$ has the general form

$$\boldsymbol{\Omega}_a = \sum_{i=1}^{n_a} \boldsymbol{p}_{ai}(q_{a1} \dots q_{an_a}, t) \dot{q}_{ai} + \boldsymbol{p}_{a0}(q_{a1} \dots q_{an_a}, t) \qquad a = 1 \dots n\,. \tag{5.116}$$

The explicit dependency on t exists only in the presence of rheonomic constraints. If all constraints are scleronomic then $\boldsymbol{p}_{a0}$ is identically zero and $\boldsymbol{p}_{ai}$ $(i = 1 \dots n_a)$ depends on $q_{a1} \dots q_{an_a}$ only. Two examples illustrate the rheonomic case. In the first example a hinge with n_a generalized coordinates is considered in which, first, all constraints are scleronomic so that Eq. (5.116) has the special form $\boldsymbol{\Omega}_a = \sum_{i=1}^{n_a} \boldsymbol{p}_{ai}(q_{a1} \dots q_{an_a}) \dot{q}_{ai}$. Now, a rheonomic constraint is introduced by the condition that the variable q_{an_a} is prescribed as a function of time. Also $\dot{q}_{an_a}$ is then a known function of time. The hinge loses one degree of freedom, and $\boldsymbol{\Omega}_a$ becomes

$$\boldsymbol{\Omega}_a = \sum_{i=1}^{n_a - 1} \boldsymbol{p}_{ai}(q_{a1} \dots q_{a,n_a-1}, q_{an_a}(t))\, \dot{q}_{ai} + \boldsymbol{p}_{an_a}(q_{a1} \dots q_{a,n_a-1}, q_{an_a}(t))\, \dot{q}_{an_a}(t)\,.$$

This has the general form of Eq. (5.116). The second example is of a different nature. The body $i^-(a)$ is moving on a curved surface of body $i^+(a)$. It is in touch with this surface in three support points (imagine a three-legged table moving on the surface). It has three degrees of freedom of relative motion (translation along the surface and one degree of freedom of rotation). A rheonomic constraint is introduced by the condition that the shape of the surface changes with time in a prescribed manner. Obviously, body $i^-(a)$ has an angular velocity relative to the base $\underline{e}^{(i^+(a))}$ even in the case where all three generalized coordinates are kept constant. This is the term called $\boldsymbol{p}_{a0}$ in Eq. (5.116). The sum in front of it accounts for changes in the generalized coordinates. That it is, indeed, a linear combination of the generalized velocities $\dot{q}_{ai}$ $(i=1\ldots n_a)$ follows from Poisson's Eq. (2.27) which in the present notation reads $\underline{\tilde{\Omega}}_a = -\underline{\dot{G}}_a \underline{G}_a^{\mathrm{T}}$.

The relative angular acceleration $\mathring{\boldsymbol{\Omega}}_a$ is derived from Eq. (5.116) to be

$$\mathring{\boldsymbol{\Omega}}_a = \sum_{i=1}^{n_a} \boldsymbol{p}_{ai}\ddot{q}_{ai} + \boldsymbol{w}_a \qquad a=1\ldots n \tag{5.117}$$

with

$$\boldsymbol{w}_a = \sum_{i=1}^{n_a} \left[\sum_{j=1}^{n_a} \frac{\partial \boldsymbol{p}_{ai}}{\partial q_{aj}} \dot{q}_{aj} + \frac{\partial \boldsymbol{p}_{ai}}{\partial t} + \frac{\partial \boldsymbol{p}_{a0}}{\partial q_{ai}} \right] \dot{q}_{ai} + \frac{\partial \boldsymbol{p}_{a0}}{\partial t} \qquad a=1\ldots n\,. \tag{5.118}$$

In Eq. (5.104) for d'Alembert's principle occur vectors $\delta \boldsymbol{r}_i$ and $\boldsymbol{\delta\pi}_i$ $(i=1\ldots n)$ which describe variations of the position and angular orientation of the bodies relative to inertial space. With the quantities introduced so far only variations of the position and angular orientation of body $i^-(a)$ relative to the base $\underline{e}^{(i^+(a))}$ can be described. To do so two quantities $\mathring{\delta} \boldsymbol{z}_a$ and $\boldsymbol{\delta\varkappa}_a$ are introduced. The first is the variation of the location of hinge point a in the base $\underline{e}^{(i^+(a))}$. The open circle on top of the delta points to the fact that it is not the variation $\delta \boldsymbol{z}_a$ of $\boldsymbol{z}_a$ in inertial space. The two variations are obviously different because the base $\underline{e}^{(i^+(a))}$ itself undergoes variations of orientation in inertial space. The second quantity $\boldsymbol{\delta\varkappa}_a$ describes the variation of the angular orientation of body $i^-(a)$ in the base $\underline{e}^{(i^+(a))}$. Like $\boldsymbol{\delta\pi}_i$ it is the product of a unit vector with an infinitesimal rotation angle. Both quantities, $\mathring{\delta} \boldsymbol{z}_a$ and $\boldsymbol{\delta\varkappa}_a$ can easily be expressed in terms of the generalized coordinates $q_{a1}\ldots q_{an_a}$, of their variations and of time. The expression for $\mathring{\delta} \boldsymbol{z}_a$ is obtained from Eq. (5.111). It must be remembered that in d'Alembert's principle the time t is kept constant during any variation. Consequently,

$$\mathring{\delta} \boldsymbol{z}_a = \sum_{i=1}^{n_a} \boldsymbol{k}_{ai} \delta q_{ai} \qquad a=1\ldots n\,. \tag{5.119}$$

The vectors $\boldsymbol{k}_{ai}$ are the partial derivatives defined by Eq. (5.112). For $\boldsymbol{\delta\varkappa}_a$ Eq. (5.116) yields

$$\boldsymbol{\delta\varkappa}_a = \sum_{i=1}^{n_a} \boldsymbol{p}_{ai} \delta q_{ai} \qquad a=1\ldots n\,. \tag{5.120}$$

The equations (5.111) to (5.120) give a complete description of the position, motion and variation of position of contiguous bodies relative to each other.

Problem

5.16 Continue Problem 5.14 by calculating for each of the hinges in Fig. 5.34 the coordinates of $\mathring{\boldsymbol{z}}_a$ and $\mathring{\mathring{\boldsymbol{z}}}_a$ in $\underline{e}^{(i^+(a))}$ and of $\boldsymbol{\Omega}_a$ and $\mathring{\boldsymbol{\Omega}}_a$ in $\underline{e}^{(i^-(a))}$.

5.2.8.3 Kinematics of motion of bodies relative to inertial space We now turn to the kinematics of motion of bodies relative to inertial space. First, the absolute angular velocities $\boldsymbol{\omega}_i$ $(i=1\ldots n)$ will be considered[1]. They are related to the relative angular velocities $\boldsymbol{\Omega}_a$ through the equation

$$\boldsymbol{\Omega}_a=\boldsymbol{\omega}_{i^-(a)}-\boldsymbol{\omega}_{i^+(a)} \qquad a=1\ldots n. \tag{5.121}$$

This can be written in the form

$$\boldsymbol{\Omega}_a=-\sum_{i=0}^{n} S_{ia}\boldsymbol{\omega}_i=-S_{0a}\boldsymbol{\omega}_0-\sum_{i=1}^{n} S_{ia}\boldsymbol{\omega}_i \qquad a=1\ldots n$$

or as a matrix equation

$$\underline{\boldsymbol{\Omega}}=-\boldsymbol{\omega}_0\underline{S}_0^{\mathrm{T}}-\underline{S}^{\mathrm{T}}\underline{\boldsymbol{\omega}}$$

with the column matrices $\underline{\boldsymbol{\Omega}}=[\boldsymbol{\Omega}_1\ldots\boldsymbol{\Omega}_n]^{\mathrm{T}}$ and $\underline{\boldsymbol{\omega}}=[\boldsymbol{\omega}_1\ldots\boldsymbol{\omega}_n]^{\mathrm{T}}$. Premultiplication with $\underline{T}^{\mathrm{T}}$ leads to

$$\underline{\boldsymbol{\omega}}=-\underline{T}^{\mathrm{T}}\underline{\boldsymbol{\Omega}}+\boldsymbol{\omega}_0\underline{1}_n. \tag{5.122}$$

This yields for the absolute angular velocity of body i the formula

$$\boldsymbol{\omega}_i=-\sum_{a=1}^{n} T_{ai}\boldsymbol{\Omega}_a+\boldsymbol{\omega}_0 \qquad i=1\ldots n \tag{5.123}$$

and, hence, for the absolute angular acceleration

$$\dot{\boldsymbol{\omega}}_i=-\sum_{a=1}^{n} T_{ai}(\mathring{\boldsymbol{\Omega}}_a+\boldsymbol{w}_a^*)+\dot{\boldsymbol{\omega}}_0 \qquad i=1\ldots n$$

with $\quad \boldsymbol{w}_a^*=\boldsymbol{\omega}_{i^-(a)}\times\boldsymbol{\Omega}_a \qquad a=1\ldots n.$

These n equations are, again, combined in the single matrix equation

$$\dot{\underline{\boldsymbol{\omega}}}=-\underline{T}^{\mathrm{T}}(\mathring{\underline{\boldsymbol{\Omega}}}+\underline{\boldsymbol{w}}^*)+\dot{\boldsymbol{\omega}}_0\underline{1}_n \tag{5.124}$$

with the column matrices $\mathring{\underline{\boldsymbol{\Omega}}}=[\mathring{\boldsymbol{\Omega}}_1\ldots\mathring{\boldsymbol{\Omega}}_n]^{\mathrm{T}}$ and $\underline{\boldsymbol{w}}^*=[\boldsymbol{w}_1^*\ldots\boldsymbol{w}_n^*]^{\mathrm{T}}$. According to Eq. (5.117) a single element $\mathring{\boldsymbol{\Omega}}_a$ of $\mathring{\underline{\boldsymbol{\Omega}}}$ is a linear combination of the second time derivatives of generalized coordinates. The column matrix $\mathring{\underline{\boldsymbol{\Omega}}}$ can, therefore, be expressed in the form

$$\mathring{\underline{\boldsymbol{\Omega}}}=\underline{\boldsymbol{p}}^{\mathrm{T}}\ddot{\underline{q}}+\underline{\boldsymbol{w}} \tag{5.125}$$

if the matrices on the right hand side are defined as follows. The symbol $\ddot{\underline{q}}$ stands for the column matrix

$$\ddot{\underline{q}}=[\ddot{q}_{11}\cdots\ddot{q}_{1n_1}\quad \ddot{q}_{21}\cdots\ddot{q}_{2n_2}\cdots\cdots\ddot{q}_{n1}\cdots\ddot{q}_{nn_n}]^{\mathrm{T}}.$$

The number of elements in it equals the total number of degrees of freedom of the entire system. This number will be called N in what follows. It is calculated as

[1] The development up to Eq. (5.128) is identical in form with the one leading from Eq. (5.86) to Eq. (5.93). It is more general, however, because the hinges can have more than three degrees of freedom.

$N = \sum_{a=1}^{n} n_a$. The rectangular matrix $\underline{p}$ has N rows and n columns. Each column is associated with one hinge. The matrix has a quasi-diagonal structure. Its transpose is

$$\underline{p}^{\mathrm{T}} = \begin{bmatrix} \boldsymbol{p}_{11} \cdots \boldsymbol{p}_{1n_1} & & & \underline{0} \\ & \boldsymbol{p}_{21} \cdots \boldsymbol{p}_{2n_2} & & \\ & & \ddots & \\ \underline{0} & & & \boldsymbol{p}_{n1} \cdots \boldsymbol{p}_{nn_n} \end{bmatrix}. \tag{5.126}$$

Finally, the column matrix $\underline{\boldsymbol{w}} = [\boldsymbol{w}_1 \ldots \boldsymbol{w}_n]^{\mathrm{T}}$ has as elements the terms defined in Eq. (5.118). With Eq. (5.125) the column matrix of the absolute angular accelerations in Eq. (5.124) becomes

$$\underline{\dot{\boldsymbol{\omega}}} = -\underline{T}^{\mathrm{T}}(\underline{p}^{\mathrm{T}}\underline{\ddot{q}} + \underline{\boldsymbol{f}}) + \dot{\boldsymbol{\omega}}_0 \underline{1}_n \tag{5.127}$$

where $\quad \underline{\boldsymbol{f}} = \underline{\boldsymbol{w}} + \underline{\boldsymbol{w}}^*.$ (5.128)

This expression for $\underline{\dot{\boldsymbol{\omega}}}$ is the first which is ready for substitution into Eq. (5.104) for d'Alembert's principle. It should be clear that all vectors which build up the matrices $\underline{p}$, $\underline{\boldsymbol{f}}$ and $\dot{\boldsymbol{\omega}}_0\underline{1}_n$ are known functions of the generalized coordinates q_{ai} $(a=1\ldots n, i=1\ldots n_a)$, of their first derivatives $\dot{q}_{ai}$ and of time t. Consider, for example, the vector $\boldsymbol{f}_a$. The term $\boldsymbol{w}_a$ is given by Eq. (5.118), $\boldsymbol{\Omega}_a$ in $\boldsymbol{w}_a^*$ by Eq. (5.116), and $\boldsymbol{\omega}_{i^-(a)}$ is with Eq. (5.123) a sum of relative angular velocities and of the absolute angular velocity $\boldsymbol{\omega}_0$ which is a known function of time. The description of angular velocities and accelerations is concluded with two formulas for the column matrices $\underline{\boldsymbol{\omega}}$ and $\underline{\boldsymbol{\Omega}} = [\boldsymbol{\Omega}_1 \ldots \boldsymbol{\Omega}_n]^{\mathrm{T}}$. Eqs. (5.116) and (5.122) yield with $\underline{p}_0 = [\boldsymbol{p}_{10} \cdots \boldsymbol{p}_{n0}]^{\mathrm{T}}$ the expressions

$$\underline{\boldsymbol{\Omega}} = \underline{p}^{\mathrm{T}}\underline{\dot{q}} + \underline{p}_0 \tag{5.129}$$

and $$\underline{\boldsymbol{\omega}} = -\underline{T}^{\mathrm{T}}(\underline{p}^{\mathrm{T}}\underline{\dot{q}} + \underline{p}_0) + \boldsymbol{\omega}_0 \underline{1}_n. \tag{5.130}$$

It is now a simple matter to find an explicit expression for the column matrix $\underline{\boldsymbol{\delta\pi}}$ occuring in Eq. (5.104). Consider, first, the difference $\boldsymbol{\delta\pi}_{i^-(a)} - \boldsymbol{\delta\pi}_{i^+(a)}$ of the infinitesimal rotations of two contiguous bodies $i^-(a)$ and $i^+(a)$, both relative to the same base fixed in inertial space. Since infinitesimal rotations can be added like vectors this difference is interpreted as the infinitesimal rotation of body $i^-(a)$ relative to body $i^+(a)$. This was, on the other hand, the definition of the vector $\boldsymbol{\delta\varkappa}_a$. Hence,

$$\boldsymbol{\delta\varkappa}_a = \boldsymbol{\delta\pi}_{i^-(a)} - \boldsymbol{\delta\pi}_{i^+(a)} \qquad a = 1 \ldots n.$$

This has the same form as Eq. (5.121) for the absolute and relative angular velocities. The further treatment of the equation is also the same. First, it is rewritten in the form

$$\boldsymbol{\delta\varkappa}_a = -S_{0a}\boldsymbol{\delta\pi}_0 - \sum_{i=1}^{n} S_{ia}\boldsymbol{\delta\pi}_i \qquad a = 1 \ldots n.$$

On the right hand side, $\boldsymbol{\delta\pi}_0$ is the variation of the orientation of the reference base $\underline{e}^{(0)}$

relative to inertial space. It is zero because the orientation is prescribed as a function of time. The equation reduces, therefore, to

$$\delta\varkappa_a = -\sum_{i=1}^{n} S_{ia}\,\delta\pi_i \qquad a=1\dots n.$$

In matrix form all n equations are combined in the single equation

$$\underline{\delta\varkappa} = -\underline{S}^{\mathrm{T}}\,\underline{\delta\pi} \tag{5.131}$$

with the column matrices $\underline{\delta\pi}$ known from Eq. (5.104) and $\underline{\delta\varkappa}=[\delta\varkappa_1\dots\delta\varkappa_n]^{\mathrm{T}}$. Premultiplication with $\underline{T}^{\mathrm{T}}$ yields

$$\underline{\delta\pi} = -\underline{T}^{\mathrm{T}}\,\underline{\delta\varkappa}\,.$$

The desired relationship between $\underline{\delta\pi}$ and the variations of the generalized coordinates is obtained when Eq. (5.120) is used for the elements of $\underline{\delta\varkappa}$. This leads to

$$\underline{\delta\pi} = -\underline{T}^{\mathrm{T}}\,\underline{\boldsymbol{p}}^{\mathrm{T}}\,\delta\underline{q} \tag{5.132}$$

where $\underline{\boldsymbol{p}}$ is defined by Eq. (5.126) and $\delta\underline{q}$ is the column matrix

$$\delta\underline{q}=[\delta q_{11}\dots\delta q_{1n_1}\quad \delta q_{21}\dots\delta q_{2n_2}\dots\dots\delta q_{n1}\dots\delta q_{nn_n}]^{\mathrm{T}}. \tag{5.133}$$

With this equation for $\underline{\delta\pi}$ the second term required for Eq. (5.104) is ready for substitution.

The angular orientation of body i ($i=1\dots n$) in inertial space is known if it is known relative to the base $\underline{e}^{(0)}$. In order to specify the latter the transformation matrix $\underline{A}_i$ defined by the equation

$$\underline{e}^{(0)}=\underline{A}_i\,\underline{e}^{(i)} \qquad i=1\dots n$$

is introduced. The matrices $\underline{A}_1\dots\underline{A}_n$ can be calculated recursively as functions of the generalized coordinates and of time with the help of the alternative relationships

$$\underline{A}_{i^-(a)}=\underline{A}_{i^+(a)}\,\underline{G}_a^{\mathrm{T}}, \qquad \underline{A}_{i^+(a)}=\underline{A}_{i^-(a)}\,\underline{G}_a \qquad a=1\dots n. \tag{5.134}$$

An example for the practical application of these relationships was given following Eq. (5.95).

We now turn to the terms $\ddot{\underline{r}}$ and $\delta\underline{r}$ of Eq. (5.104). As a basis for both, explicit expressions are needed for the radius vectors $\boldsymbol{r}_i$ ($i=1\dots n$) as functions of the generalized coordinates and of time. These are found from Fig. 5.36 where $\boldsymbol{r}_{i^+(a)}$ and $\boldsymbol{r}_{i^-(a)}$ are the radius vectors in inertial space of the points $C_{i^+(a)}$ and $C_{i^-(a)}$, respectively. The figure yields the relationship

$$\boldsymbol{z}_a=(\boldsymbol{r}_{i^-(a)}+\boldsymbol{c}_{i^-(a)a})-(\boldsymbol{r}_{i^+(a)}+\boldsymbol{c}_{i^+(a)a}) \qquad a=1\dots n$$

which can be written in the form

$$\boldsymbol{z}_a = -\sum_{i=0}^{n} S_{ia}(\boldsymbol{r}_i+\boldsymbol{c}_{ia}) \qquad a=1\dots n$$

or with the vectors

$$C_{ia}=S_{ia}c_{ia} \qquad i=0\dots n,\ a=1\dots n$$

in the form

$$z_a=-S_{0a}r_0-C_{0a}-\sum_{i=1}^{n}(S_{ia}r_i+C_{ia}) \qquad a=1\dots n\,.$$

All n equations are combined in the matrix equation

$$\underline{z}=-r_0\underline{S}_0^{\mathrm{T}}-\underline{C}_0^{\mathrm{T}}-\underline{S}^{\mathrm{T}}\underline{r}-\underline{C}^{\mathrm{T}}\underline{1}_n \tag{5.135}$$

with the column matrices $\underline{r}=[r_1\dots r_n]^{\mathrm{T}}$ and $\underline{z}=[z_1\dots z_n]^{\mathrm{T}}$, the row matrix

$$\underline{C}_0=[C_{01}\dots C_{0n}]$$

and the $(n\times n)$ matrix

$$\underline{C}=\begin{bmatrix}C_{11}\dots C_{1n}\\ \vdots\\ C_{n1}\dots C_{nn}\end{bmatrix}.$$

$\underline{C}$ is identical with the matrix defined in Eq. (5.11). In $\underline{C}_0$ only the first element $C_{01}=S_{01}c_{01}$ is different from zero since hinge 1 is the only hinge on body 0. Premultiplication of Eq. (5.135) with $\underline{T}^{\mathrm{T}}$ yields for $\underline{r}$ the explicit result

$$\underline{r}=-(\underline{C}\,\underline{T})^{\mathrm{T}}\underline{1}_n-\underline{T}^{\mathrm{T}}\underline{z}+r_0\underline{1}_n-(\underline{C}_0\underline{T})^{\mathrm{T}}\,. \tag{5.136}$$

From this equation expressions can now be deduced for $\ddot{\underline{r}}$ and $\delta\underline{r}$. The second time derivative is

$$\ddot{\underline{r}}=-(\ddot{\underline{C}}\,\underline{T})^{\mathrm{T}}\underline{1}_n-\underline{T}^{\mathrm{T}}\ddot{\underline{z}}+\ddot{r}_0\underline{1}_n-(\ddot{\underline{C}}_0\underline{T})^{\mathrm{T}}\,. \tag{5.137}$$

The elements of the matrix $\underline{C}\,\underline{T}$ are the vectors d_{ij} $(i,j=1\dots n)$ which were discussed following Eq. (5.17) and illustrated in Fig. 5.14. Each vector d_{ij} is fixed on the body which is denoted by the first index i. The definition of the vectors is now extended to include

$$d_{0j}=(\underline{C}_0\underline{T})_j=\sum_{a=1}^{n}T_{aj}S_{0a}c_{0a} \qquad j=1\dots n\,. \tag{5.138}$$

The product $T_{aj}S_{0a}$ is different from zero (and equal to -1) only for hinges on body 0, i.e. only for hinge 1, whence follows

$$d_{0j}=-c_{01} \qquad j=1\dots n\,. \tag{5.139}$$

The second time derivative of d_{ij} is

$$\ddot{d}_{ij}=-d_{ij}\times\dot{\omega}_i+\omega_i\times(\omega_i\times d_{ij}) \qquad i=0\dots n,\ j=1\dots n\,.$$

The contribution of the first term $-d_{ij}\times\dot{\omega}_i$ to the column matrix $(\ddot{\underline{C}}\,\underline{T})^{\mathrm{T}}\underline{1}_n$ in Eq. (5.137) is

$$-\begin{bmatrix}d_{11}\times\dot{\omega}_1\dots d_{n1}\times\dot{\omega}_n\\ \vdots\\ d_{1n}\times\dot{\omega}_1 \quad d_{nn}\times\dot{\omega}_n\end{bmatrix}\underline{1}_n=-\begin{bmatrix}d_{11}\dots d_{n1}\\ \vdots\\ d_{1n} \quad d_{nn}\end{bmatrix}\times\begin{bmatrix}\dot{\omega}_1\\ \vdots\\ \dot{\omega}_n\end{bmatrix}=-(\underline{C}\,\underline{T})^{\mathrm{T}}\times\dot{\underline{\omega}}\,. \tag{5.140}$$

The contribution of the second term $\boldsymbol{\omega}_i \times (\boldsymbol{\omega}_i \times \boldsymbol{d}_{ij})$ to $(\underline{\dot{C}}\,\underline{T})^{\mathrm{T}} \underline{1}_n$ is a column matrix $\underline{g} = [\boldsymbol{g}_1 \dots \boldsymbol{g}_n]^{\mathrm{T}}$ with the elements

$$\boldsymbol{g}_i = \sum_{j=1}^{n} \boldsymbol{\omega}_j \times (\boldsymbol{\omega}_j \times \boldsymbol{d}_{ji}) \qquad i=1 \dots n. \tag{5.141}$$

Similarly, the last term in Eq. (5.137) is the column matrix

$$(\underline{\ddot{C}}_0 \underline{T})^{\mathrm{T}} = \begin{bmatrix} \dot{\boldsymbol{\omega}}_0 \times \boldsymbol{d}_{01} + \boldsymbol{\omega}_0 \times (\boldsymbol{\omega}_0 \times \boldsymbol{d}_{01}) \\ \vdots \\ \dot{\boldsymbol{\omega}}_0 \times \boldsymbol{d}_{0n} + \boldsymbol{\omega}_0 \times (\boldsymbol{\omega}_0 \times \boldsymbol{d}_{0n}) \end{bmatrix} \tag{5.142}$$

or, in view of the identity (5.139),

$$(\underline{\ddot{C}}_0 \underline{T})^{\mathrm{T}} = -[\dot{\boldsymbol{\omega}}_0 \times \boldsymbol{c}_{01} + \boldsymbol{\omega}_0 \times (\boldsymbol{\omega}_0 \times \boldsymbol{c}_{01})]\, \underline{1}_n. \tag{5.143}$$

The expression for $\underline{\ddot{\boldsymbol{r}}}$ is now

$$\underline{\ddot{\boldsymbol{r}}} = (\underline{C}\,\underline{T})^{\mathrm{T}} \times \underline{\dot{\boldsymbol{\omega}}} - \underline{T}^{\mathrm{T}} \underline{\ddot{\boldsymbol{z}}} - \underline{\boldsymbol{g}} + \ddot{\boldsymbol{r}}_0 \underline{1}_n - (\underline{\ddot{C}}_0 \underline{T})^{\mathrm{T}}. \tag{5.144}$$

Next, the column matrix $\underline{\ddot{\boldsymbol{z}}}$ is reformulated. Its individual elements are

$$\ddot{\boldsymbol{z}}_a = \overset{\circ\circ}{\boldsymbol{z}}_a + \dot{\boldsymbol{\omega}}_{i^+(a)} \times \boldsymbol{z}_a + 2\boldsymbol{\omega}_{i^+(a)} \times \overset{\circ}{\boldsymbol{z}}_a + \boldsymbol{\omega}_{i^+(a)} \times (\boldsymbol{\omega}_{i^+(a)} \times \boldsymbol{z}_a) \qquad a=1 \dots n. \tag{5.145}$$

This follows from the definition of $\overset{\circ}{\boldsymbol{z}}_a$ and $\overset{\circ\circ}{\boldsymbol{z}}_a$ as time derivatives of $\boldsymbol{z}_a$ in the base $\underline{e}^{(i^+(a))}$. In order to give the elements involving $\dot{\boldsymbol{\omega}}_{i^+(a)}$ and $\boldsymbol{\omega}_{i^+(a)}$ a convenient form, new dimensionless scalar quantities S^+_{ia} $(i=0 \dots n,\ a=1 \dots n)$ are introduced which satisfy the identities

$$\dot{\boldsymbol{\omega}}_{i^+(a)} = \sum_{j=0}^{n} S^+_{ja} \dot{\boldsymbol{\omega}}_j \qquad a=1 \dots n. \tag{5.146}$$

From this follows for S^+_{ia} the definition

$$S^+_{ia} = \begin{cases} +1 & \text{if } i=i^+(a) \\ 0 & \text{otherwise} \end{cases} \qquad i=0 \dots n,\ a=1 \dots n. \tag{5.147}$$

These scalars are used to construct the row matrix

$$\underline{S}^+_0 = [S^+_{01} \dots S^+_{0n}]$$

and the $(n \times n)$ matrix

$$\underline{S}^+ = \begin{bmatrix} S^+_{11} & \dots & S^+_{1n} \\ \vdots & & \\ S^+_{n1} & & S^+_{nn} \end{bmatrix}. \tag{5.148}$$

Comparison with Eq. (5.1) reveals that $\underline{S}^+_0$ and $\underline{S}^+$ are obtained from $\underline{S}_0$ and $\underline{S}$, respectively, if in the latter matrices each element -1 is replaced by zero. With Eq. (5.115) for $\overset{\circ\circ}{\boldsymbol{z}}_a$ and Eq. (5.146) for $\dot{\boldsymbol{\omega}}_{i^+(a)}$ Eq. (5.145) can be rewritten in the form

$$\ddot{\boldsymbol{z}}_a = \sum_{i=1}^{n_a} \boldsymbol{k}_{ai} \ddot{q}_{ai} + \boldsymbol{s}_a + \sum_{j=1}^{n} [-S^+_{ja} \boldsymbol{z}_a \times \dot{\boldsymbol{\omega}}_j + 2\boldsymbol{\omega}_j \times S^+_{ja} \overset{\circ}{\boldsymbol{z}}_a + \boldsymbol{\omega}_j \times (\boldsymbol{\omega}_j \times S^+_{ja} \boldsymbol{z}_a)] +$$
$$+ \dot{\boldsymbol{\omega}}_0 \times S^+_{0a} \boldsymbol{z}_a + 2\boldsymbol{\omega}_0 \times S^+_{0a} \overset{\circ}{\boldsymbol{z}}_a + \boldsymbol{\omega}_0 \times (\boldsymbol{\omega}_0 \times S^+_{0a} \boldsymbol{z}_a).$$

At this point all n expressions $\ddot{z}_1 \dots \ddot{z}_n$ are combined, again, in the column matrix $\ddot{\underline{z}}$. This yields

$$\ddot{\underline{z}} = \underline{k}^{\mathrm{T}} \ddot{\underline{q}} + \underline{s} - \underline{Z}^{\mathrm{T}} \times \dot{\underline{\omega}} + 2\underline{h} + \underline{g}^* + \underline{u} \,. \tag{5.149}$$

The matrices on the right hand side are defined as follows. The $(N \times n)$ matrix $\underline{k}$ whose transpose is

$$\underline{k}^{\mathrm{T}} = \begin{bmatrix} \boldsymbol{k}_{11} \dots \boldsymbol{k}_{1n_1} & & & \underline{0} \\ & \boldsymbol{k}_{21} \dots \boldsymbol{k}_{2n_2} & & \\ & & \ddots & \\ \underline{0} & & & \boldsymbol{k}_{n1} \dots \boldsymbol{k}_{nn_n} \end{bmatrix} \tag{5.150}$$

has the same structure as the matrix $\underline{p}$ in Eq. (5.126). The column matrix $\underline{s} = [s_1 \dots s_n]^{\mathrm{T}}$ is composed of the expressions defined by Eq. (5.113). The $(n \times n)$ matrix $\underline{Z}$ is

$$\underline{Z} = \begin{bmatrix} \boldsymbol{Z}_{11} \dots \boldsymbol{Z}_{1n} \\ \vdots \\ \boldsymbol{Z}_{n1} \quad \boldsymbol{Z}_{nn} \end{bmatrix} \tag{5.151}$$

with the elements

$$\boldsymbol{Z}_{ia} = S_{ia}^{+} \boldsymbol{z}_a \qquad i,a = 1 \dots n \,. \tag{5.152}$$

Finally, $\underline{h}$, $\underline{g}^*$ and $\underline{u}$ are column matrices with the elements

$$\boldsymbol{h}_a = \begin{cases} \boldsymbol{\omega}_{i^+(a)} \times \mathring{\boldsymbol{z}}_a & \text{for } i^+(a) \neq 0 \\ \boldsymbol{0} & \text{for } i^+(a) = 0 \end{cases} \qquad a = 1 \dots n$$

$$\boldsymbol{g}_a^* = \sum_{j=1}^{n} \boldsymbol{\omega}_j \times (\boldsymbol{\omega}_j \times \boldsymbol{Z}_{ja}) \qquad a = 1 \dots n \tag{5.153}$$

$$\boldsymbol{u}_a = \dot{\boldsymbol{\omega}}_0 \times \boldsymbol{Z}_{0a} + 2\boldsymbol{\omega}_0 \times \mathring{\boldsymbol{Z}}_{0a} + \boldsymbol{\omega}_0 \times (\boldsymbol{\omega}_0 \times \boldsymbol{Z}_{0a}) \qquad a = 1 \dots n \,, \tag{5.154}$$

respectively. The vector $\boldsymbol{Z}_{0a} = S_{0a}^{+} \boldsymbol{z}_a$ in $\boldsymbol{u}_a$ is zero for $a = 2 \dots n$ since only hinge 1 is located on body 0. Therefore, $\boldsymbol{u}_a$ can be simplified to

$$\boldsymbol{u}_a = [\dot{\boldsymbol{\omega}}_0 \times \boldsymbol{z}_1 + 2\boldsymbol{\omega}_0 \times \mathring{\boldsymbol{z}}_1 + \boldsymbol{\omega}_0 \times (\boldsymbol{\omega}_0 \times \boldsymbol{z}_1)] S_{0a}^{+} \qquad a = 1 \dots n$$

and, hence, the column matrix $\underline{u}$ to

$$\underline{u} = [\dot{\boldsymbol{\omega}}_0 \times \boldsymbol{z}_1 + 2\boldsymbol{\omega}_0 \times \mathring{\boldsymbol{z}}_1 + \boldsymbol{\omega}_0 \times (\boldsymbol{\omega}_0 \times \boldsymbol{z}_1)] \underline{S}_0^{+\mathrm{T}} \,. \tag{5.155}$$

The expression for $\ddot{\underline{z}}$ is now substituted into Eq. (5.144) for $\ddot{\underline{r}}$:

$$\ddot{\underline{r}} = [(\underline{C} + \underline{Z})\underline{T}]^{\mathrm{T}} \times \dot{\underline{\omega}} - \underline{T}^{\mathrm{T}} (\underline{k}^{\mathrm{T}} \ddot{\underline{q}} + \underline{s} + 2\underline{h} + \underline{g}^*) - \underline{g} + \ddot{\boldsymbol{r}}_0 \underline{1}_n - (\ddot{\boldsymbol{C}}_0 \underline{T})^{\mathrm{T}} - \underline{T}^{\mathrm{T}} \underline{u} \,. \tag{5.156}$$

For $\dot{\underline{\omega}}$ Eq. (5.127) is substituted. This leads to the equation

$$\ddot{\underline{r}} = -\{[\underline{T}(\underline{C} + \underline{Z})\underline{T}]^{\mathrm{T}} \times \underline{p}^{\mathrm{T}} + \underline{T}^{\mathrm{T}} \underline{k}^{\mathrm{T}}\} \ddot{\underline{q}} + \underline{U}$$

in which $\underline{U}$ is an abbreviation for the sum of all terms which do not depend on second time derivatives of generalized coordinates. It is

$$\underline{U} = -[(\underline{C}+\underline{Z})\underline{T}]^{\mathrm{T}} \times (\underline{T}^{\mathrm{T}}\underline{f} - \dot{\omega}_0 \underline{1}_n) - \underline{T}^{\mathrm{T}}(\underline{s} + 2\underline{h} + \underline{g}^*) - \underline{g} + \ddot{r}_0 \underline{1}_n - (\underline{\ddot{C}}_0 \underline{T})^{\mathrm{T}} - \underline{T}^{\mathrm{T}}\underline{u}. \tag{5.157}$$

The matrix in curled brackets in front of $\ddot{q}$ in the equation for $\ddot{\underline{r}}$ is the transpose of the matrix $-\underline{p} \times \underline{T}(\underline{C}+\underline{Z})\underline{T} + \underline{k}\,\underline{T}$. The minus sign in front of the first term is caused by the interchange of the factors in the vector-cross product. Thus, the result

$$\ddot{\underline{r}} = [\underline{p} \times \underline{T}(\underline{C}+\underline{Z})\underline{T} - \underline{k}\,\underline{T}]^{\mathrm{T}} \ddot{\underline{q}} + \underline{U} \tag{5.158}$$

is obtained. This expression will later be substituted into Eq. (5.104).

Before the matrix $\delta\underline{r}$ is considered the expression for $\underline{U}$ in Eq. (5.157) will be further simplified. First, the elements of the column matrix $\underline{T}^{\mathrm{T}}\underline{g}^* + \underline{g}$ can be rewritten with the help of Eqs. (5.153) and (5.141) in the form

$$(\underline{T}^{\mathrm{T}}\underline{g}^* + \underline{g})_i = \sum_{a=1}^{n} \sum_{j=1}^{n} T_{ai}\boldsymbol{\omega}_j \times (\boldsymbol{\omega}_j \times S_{ja}^+ \boldsymbol{z}_a) + \sum_{j=1}^{n} \boldsymbol{\omega}_j \times (\boldsymbol{\omega}_j \times \boldsymbol{d}_{ji})$$

or using Eq. (5.17) for $\boldsymbol{d}_{ji}$

$$(\underline{T}^{\mathrm{T}}\underline{g}^* + \underline{g})_i = \sum_{a=1}^{n} \sum_{j=1}^{n} \boldsymbol{\omega}_j \times [\boldsymbol{\omega}_j \times T_{ai}(S_{ja}\boldsymbol{c}_{ja} + S_{ja}^+ \boldsymbol{z}_a)].$$

From the definition of S_{ja}^+ follows the identity $S_{ja}^+ = S_{ja} S_{ja}^+$ $(j, a = 1 \dots n)$. Hence,

$$(\underline{T}^{\mathrm{T}}\underline{g}^* + \underline{g})_i = \sum_{a=1}^{n} \sum_{j=1}^{n} \boldsymbol{\omega}_j \times [\boldsymbol{\omega}_j \times T_{ai} S_{ja}(\boldsymbol{c}_{ja} + S_{ja}^+ \boldsymbol{z}_a)]. \tag{5.159}$$

The vectors

$$\boldsymbol{c}_{ia}^* = \boldsymbol{c}_{ia} + S_{ia}^+ \boldsymbol{z}_a = \boldsymbol{c}_{ia} + \boldsymbol{Z}_{ia} \qquad i = 0 \dots n,\ a = 1 \dots n \tag{5.160}$$

in this expression have a simple physical interpretation. Obviously

$$\boldsymbol{c}_{ia}^* = \begin{cases} \boldsymbol{c}_{ia} + \boldsymbol{z}_a & \text{for } i = i^+(a) \\ \boldsymbol{c}_{ia} & \text{for } i = i^-(a) \\ \boldsymbol{0} & \text{otherwise} \end{cases} \qquad i = 0 \dots n,\ a = 1 \dots n.$$

Fig. 5.38 is a repetition of Fig. 5.36. On the two bodies the vectors $\boldsymbol{c}_{i^+(a)a}^*$ and $\boldsymbol{c}_{i^-(a)a}^*$ are shown. They connect the body centers of mass with the hinge point just as the vectors $\boldsymbol{c}_{i^+(a)a}$ and $\boldsymbol{c}_{i^-(a)a}$ do in the absence of hinge vectors $\boldsymbol{z}_a$ (cf. Fig. 5.12). This suggests to introduce the vectors

$$\boldsymbol{d}_{ji}^* = \sum_{a=1}^{n} T_{ai} S_{ja} \boldsymbol{c}_{ja}^* \qquad j = 0 \dots n,\ i = 1 \dots n \tag{5.161}$$

as generalizations of the vectors

$$\boldsymbol{d}_{ji} = \sum_{a=1}^{n} T_{ai} S_{ja} \boldsymbol{c}_{ja} \qquad j = 0,\ i = 1 \dots n.$$

For a particular value of i the vectors $\boldsymbol{d}_{ji}$ were illustrated in Fig. 5.14. For systems with hinge vectors $z_1 \dots z_n$ this figure takes the generalized form shown in Fig. 5.39. Just as Fig. 5.14 led to the relationship $\boldsymbol{r}_i = \boldsymbol{r}_0 - \sum_{j=1}^{n} \boldsymbol{d}_{ji}$ (Eq. (5.16)) we now have

$$\boldsymbol{r}_i = \boldsymbol{r}_0 - \sum_{j=0}^{n} \boldsymbol{d}^*_{ji} \qquad i = 1 \dots n .$$

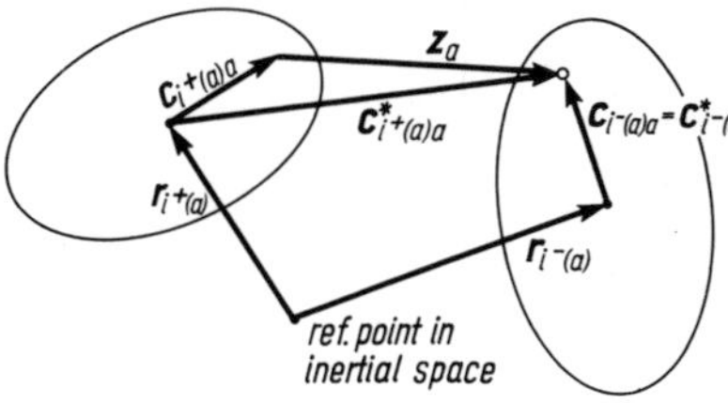

Fig. 5.38
Illustration of the vectors $\boldsymbol{c}^*_{ia}$

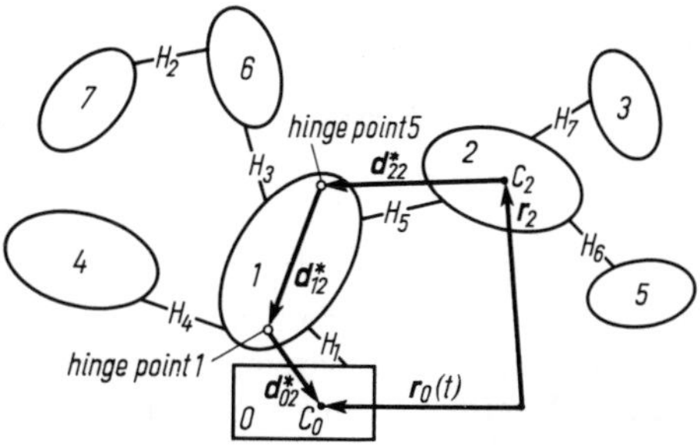

Fig. 5.39 Illustration of the vectors $\boldsymbol{d}^*_{j2}$ ($j = 0 \dots n$). The hinge points 1 and 5 are, according to the directed system graph of Fig. 5.8c, fixed on body 1. Compare with Fig. 5.14

The different summation ranges in the two formulas are explained by the fact that in Sec. 5.2.2 in contrast to the present section the origin C_0 of the base $\underline{e}^{(0)}$ was located such that $\boldsymbol{d}_{01}$ was zero. In terms of the vectors $\boldsymbol{d}^*_{ji}$ Eq. (5.159) reduces to

$$(\underline{T}^{\mathrm{T}} \underline{g}^* + \underline{g})_i = \sum_{j=1}^{n} \boldsymbol{\omega}_j \times (\boldsymbol{\omega}_j \times \boldsymbol{d}^*_{ji}) \qquad i = 1 \dots n .$$

The vectors $\boldsymbol{d}^*_{ij}$ play an important role in Eqs. (5.157) and (5.158) also in that they are the elements of the matrix $(\underline{C} + \underline{Z})\underline{T}$:

$$[(\underline{C} + \underline{Z})\underline{T}]_{ij} = \boldsymbol{d}^*_{ij} \qquad i,j = 1 \dots n . \tag{5.162}$$

This formula represents a generalization of the equation $(\underline{C}\,\underline{T})_{ij} = \boldsymbol{d}_{ij}$ through which the vectors $\boldsymbol{d}_{ij}$ were originally introduced (cf. Eq. (5.15)).

Consider now the last term $\underline{T}^{\mathrm{T}} \underline{u}$ in Eq. (5.157). In Eq. (5.155) for $\underline{u}$ the first element S^+_{01} in the column matrix $\underline{S}_0^{+\mathrm{T}}$ is either zero or $+1$, and all other elements are zero. Hence, $\underline{T}^{\mathrm{T}} \underline{S}_0^{+\mathrm{T}}$ is identical with $-S^+_{01} \underline{1}_n$. With this and with Eqs. (5.143) and (5.160) the last three terms of $\underline{U}$ can be combined in the form

$$\ddot{\boldsymbol{r}}_0 \underline{1}_n - (\underline{\ddot{C}}_0 \underline{T})^{\mathrm{T}} - \underline{T}^{\mathrm{T}} \underline{u} = [\ddot{\boldsymbol{r}}_0 + \dot{\boldsymbol{\omega}}_0 \times \boldsymbol{c}^*_{01} + 2\, S^+_{01} \boldsymbol{\omega}_0 \times \mathring{\boldsymbol{z}}_1 + \boldsymbol{\omega}_0 \times (\boldsymbol{\omega}_0 \times \boldsymbol{c}^*_{01})] \underline{1}_n . \tag{5.163}$$

The expression in square brackets represents the absolute acceleration of the hinge point number 1 minus the relative acceleration $S^+_{01} \mathring{\mathring{\boldsymbol{z}}}_1$ which is part of $\underline{\boldsymbol{k}}^{\mathrm{T}} \ddot{\underline{q}} + \underline{\boldsymbol{s}}$ in Eq. (5.149).

The last term in Eq. (5.104) for d'Alembert's principle which has to be developed is $\delta \underline{\boldsymbol{r}}$. Like $\ddot{\underline{\boldsymbol{r}}}$ it is derived from Eq. (5.136) for $\underline{\boldsymbol{r}}$. The variation of given functions of time is zero so

$$\delta \underline{r} = -\underline{T}^{\mathrm{T}}(\delta \underline{C}^{\mathrm{T}} \underline{1}_n + \delta \underline{z}).$$

First, the product $\delta \underline{C}^{\mathrm{T}} \underline{1}_n$ is treated. A single element of $\underline{C}$ is the vector $\boldsymbol{C}_{ia}$ which is either fixed on body i or zero. Its variation is, therefore, $\delta \boldsymbol{C}_{ia} = -\boldsymbol{C}_{ia} \times \boldsymbol{\delta\pi}_i$. With this the product $\delta \underline{C}^{\mathrm{T}} \underline{1}_n$ becomes

$$-\begin{bmatrix} \boldsymbol{C}_{11} \times \boldsymbol{\delta\pi}_1 \ldots \boldsymbol{C}_{n1} \times \boldsymbol{\delta\pi}_n \\ \vdots \\ \boldsymbol{C}_{1n} \times \boldsymbol{\delta\pi}_1 \ldots \boldsymbol{C}_{nn} \times \boldsymbol{\delta\pi}_n \end{bmatrix} \underline{1}_n = -\begin{bmatrix} \boldsymbol{C}_{11} \ldots \boldsymbol{C}_{n1} \\ \vdots \\ \boldsymbol{C}_{1n} \quad \boldsymbol{C}_{nn} \end{bmatrix} \times \begin{bmatrix} \boldsymbol{\delta\pi}_1 \\ \vdots \\ \boldsymbol{\delta\pi}_n \end{bmatrix} = -\underline{C}^{\mathrm{T}} \times \underline{\boldsymbol{\delta\pi}}. \tag{5.164}$$

A single element $\delta \boldsymbol{z}_a$ of the column matrix $\delta \underline{z}$ is

$$\delta \boldsymbol{z}_a = \mathring{\delta} \boldsymbol{z}_a + \boldsymbol{\delta\pi}_{i^+(a)} \times \boldsymbol{z}_a \qquad a = 1 \ldots n. \tag{5.165}$$

This follows from the definition that $\mathring{\delta} \boldsymbol{z}_a$ is the variation of $\boldsymbol{z}_a$ in the base $\underline{e}^{(i^+(a))}$. When the terms $\delta \boldsymbol{z}_1 \ldots \delta \boldsymbol{z}_n$ are, again, combined in a column matrix, each of the two terms on the right hand side produces a column matrix. The first one, in view of Eq. (5.119), is

$$[\mathring{\delta} \boldsymbol{z}_1 \ldots \mathring{\delta} \boldsymbol{z}_n]^{\mathrm{T}} = \underline{k}^{\mathrm{T}} \delta \underline{q}. \tag{5.166}$$

The matrices $\underline{k}$ and $\delta \underline{q}$ were defined by Eqs. (5.150) and (5.133), respectively. In the second term in Eq. (5.165) $\boldsymbol{\delta\pi}_{i^+(a)}$ is rewritten as $\sum_{j=0}^{n} S_{ja}^{+} \boldsymbol{\delta\pi}_j$. The variation of the position of body 0 is zero. The summation range is, therefore only $j = 1 \ldots n$. This yields

$$\boldsymbol{\delta\pi}_{i^+(a)} \times \boldsymbol{z}_a = -\sum_{j=1}^{n} S_{ja}^{+} \boldsymbol{z}_a \times \boldsymbol{\delta\pi}_j = -\sum_{j=1}^{n} \boldsymbol{Z}_{ja} \times \boldsymbol{\delta\pi}_j.$$

With this and with Eqs. (5.166) and (5.165) the column matrix $\delta \underline{z}$ is $\underline{k}^{\mathrm{T}} \delta \underline{q} - \underline{Z}^{\mathrm{T}} \times \underline{\boldsymbol{\delta\pi}}$. Substituting this and Eq. (5.164) into the equation for $\delta \underline{r}$ one gets

$$\delta \underline{r} = \underline{T}^{\mathrm{T}} [(\underline{C} + \underline{Z})^{\mathrm{T}} \times \underline{\boldsymbol{\delta\pi}} - \underline{k}^{\mathrm{T}} \delta \underline{q}] \tag{5.167}$$

or with Eq. (5.132) for $\underline{\boldsymbol{\delta\pi}}$

$$\delta \underline{r} = -\underline{T}^{\mathrm{T}} [(\underline{C} + \underline{Z})^{\mathrm{T}} \underline{T}^{\mathrm{T}} \times \underline{p}^{\mathrm{T}} + \underline{k}^{\mathrm{T}}] \delta \underline{q} = [\underline{p} \times \underline{T} (\underline{C} + \underline{Z}) \underline{T} - \underline{k}\, \underline{T}]^{\mathrm{T}} \delta \underline{q}. \tag{5.168}$$

This equation concludes the study on system kinematics.

5.2.8.4 Virtual work done in the hinges Before the results obtained for $\underline{\dot{\omega}}$, $\underline{\boldsymbol{\delta\pi}}$, $\underline{\ddot{r}}$ and $\delta \underline{r}$ are substituted into Eq. (5.104) for d'Alembert's principle an expression must be developed for the total virtual work δW done in the hinges. This is easy to do. Let δW_a be the virtual work done in a single hinge a $(a = 1 \ldots n)$. It can be expressed in terms of the quantities $\boldsymbol{z}_a$, $\mathring{\boldsymbol{z}}_a$, $\mathring{\delta} \boldsymbol{z}_a$, $\underline{G}_a$, $\boldsymbol{\Omega}_a$ and $\boldsymbol{\delta\varkappa}_a$ for this hinge a and in terms of physical parameters such as spring constants, for example. The expression for δW_a will always be linear in $\mathring{\delta} \boldsymbol{z}_a$ and $\boldsymbol{\delta\varkappa}_a$ so that it can be written in the form

$$\delta W_a = -\mathring{\delta} \boldsymbol{z}_a \cdot \boldsymbol{X}_a - \boldsymbol{\delta\varkappa}_a \cdot \boldsymbol{Y}_a \qquad a = 1 \ldots n. \tag{5.169}$$

Through this equation a force $-\boldsymbol{X}_a$ and a torque $-\boldsymbol{Y}_a$ are defined which together do the same virtual work as the actually existing hinge forces. The term $-\mathring{\delta} z_a \cdot X_a$ shows that the line of action of the force $-X_a$ is passing through the hinge point a. Both quantities, $-X_a$ and $-Y_a$, are applied to the body whose position is subject to a variation, i.e. to body $i^-(a)$. It follows that $+X_a$ and $+Y_a$ are acting on body $i^+(a)$. The vectors $\mathring{\delta} z_a$ and $\delta \varkappa_a$ are related to variations of the generalized coordinates through Eqs. (5.119) and (5.120) so that the virtual work in hinge a becomes

$$\delta W_a = -\sum_{i=1}^{n_a} \delta q_{ai}(k_{ai} \cdot X_a + p_{ai} \cdot Y_a) \qquad a=1 \dots n .$$

The total virtual work δW is the sum over $\delta W_1 \dots \delta W_n$. This can be written in the form

$$\delta W = -\delta \underline{q}^{\mathrm{T}}(\underline{k} \cdot \underline{X} + \underline{p} \cdot \underline{Y}) \tag{5.170}$$

with the matrices $\underline{k}$ and $\underline{p}$ from Eqs. (5.126) and (5.150) and with the column matrices $\underline{X} = [X_1 \dots X_n]^{\mathrm{T}}$ and $\underline{Y} = [Y_1 \dots Y_n]^{\mathrm{T}}$. It should be noted that in every particular case of application not the forces X_a and torques Y_a $(a=1 \dots n)$ are calculated but directly the scalar products $k_{ai} \cdot X_a$ and $p_{ai} \cdot Y_a$.

5.2.8.5 Equations of motion Now the expressions for $\underline{\dot{\omega}}$, $\underline{\delta\pi}$, $\underline{\ddot{r}}$, $\delta \underline{r}$ and δW from Eqs. (5.127), (5.132), (5.158), (5.168) and (5.170) are substituted into Eq. (5.104). This yields

$$\delta \underline{q}^{\mathrm{T}} \{[\underline{p} \times \underline{T}(\underline{C}+\underline{Z})\underline{T} - \underline{k}\,\underline{T}] \cdot \{\underline{F} - \underline{m}[\underline{p} \times \underline{T}(\underline{C}+\underline{Z})\underline{T} - \underline{k}\,\underline{T}]^{\mathrm{T}} \underline{\ddot{q}} - \underline{m}\,\underline{U}\} -$$
$$- \underline{p}\,\underline{T} \cdot \{\underline{M} - \underline{J} \cdot [-\underline{T}^{\mathrm{T}}(\underline{p}^{\mathrm{T}} \underline{\ddot{q}} + \underline{f}) + \dot{\omega}_0 \underline{1}_n] - \underline{V}\} - \underline{k} \cdot \underline{X} - \underline{p} \cdot \underline{Y}\} = 0$$

or in abbreviated form

$$\delta \underline{q}^{\mathrm{T}}(-\underline{A}\,\underline{\ddot{q}} + \underline{B}) = 0$$

with

$$\underline{A} = [\underline{p} \times \underline{T}(\underline{C}+\underline{Z})\underline{T} - \underline{k}\,\underline{T}] \cdot \underline{m}[\underline{p} \times \underline{T}(\underline{C}+\underline{Z})\underline{T} - \underline{k}\,\underline{T}]^{\mathrm{T}} + (\underline{p}\,\underline{T}) \cdot \underline{J} \cdot (\underline{p}\,\underline{T})^{\mathrm{T}} \tag{5.171}$$

$$\underline{B} = [\underline{p} \times \underline{T}(\underline{C}+\underline{Z})\underline{T} - \underline{k}\,\underline{T}] \cdot (\underline{F} - \underline{m}\,\underline{U}) - \underline{p}\,\underline{T} \cdot [\underline{M} + \underline{J} \cdot (\underline{T}^{\mathrm{T}} \underline{f} - \dot{\omega}_0 \underline{1}_n) - \underline{V}] -$$
$$- \underline{k} \cdot \underline{X} - \underline{p} \cdot \underline{Y}. \tag{5.172}$$

Since the variations of the generalized coordinates are independent one has

$$\underline{A}\,\underline{\ddot{q}} = \underline{B}. \tag{5.173}$$

These are the equations of motion of the system in their final form. Their number equals the number N of generalized coordinates. The coefficient matrix $\underline{A}$ is symmetric. It is also positive definite. This follows from the fact that a coefficient matrix of $\underline{\ddot{q}}$ found from d'Alembert's principle can always be interpreted as coefficient matrix of a kinetic energy expression.

Special Case D'Alembert's principle for the special case of systems with a fictitious hinge 1 was formulated in Eqs. (5.108) and (5.109). Only the latter equation needs be considered here. The column matrices $\underline{\dot{\omega}}$, $\underline{\delta\pi}$, $\underline{\ddot{R}}$ and $\delta \underline{R}$ have to be expressed in terms of generalized coordinates, of their derivatives and variations and of time. To do this no new considerations of system kinematics are necessary! The matrices $\underline{\dot{\omega}}$ and $\underline{\delta\pi}$

are the same as in Eq. (5.104) for d'Alembert's principle describing the general case. They are given by Eqs. (5.127) and (5.132), respectively. The matrices $\ddot{\underline{R}}$ and $\delta\underline{R}$ can be obtained from the corresponding Eqs. (5.158) and (5.168) for $\ddot{\underline{r}}$ and $\delta\underline{r}$, respectively, by a method which was used in Sec. 5.2.4 already. The vectors $\boldsymbol{R}_i$ and $\boldsymbol{r}_i$ $(i=1\dots n)$ are related by the equation (cf. Eq. (5.106)) $\boldsymbol{R}_i=\boldsymbol{r}_i-\boldsymbol{r}_C$. All n equations are combined in matrix form as $\underline{R}=\underline{r}-\boldsymbol{r}_C\underline{1}_n$. Using Eq. (5.136) for $\underline{r}$ this becomes

$$\underline{R}=-(\underline{C}\,\underline{T})^{\mathrm{T}}\underline{1}_n-\underline{T}^{\mathrm{T}}\underline{z}+(\boldsymbol{r}_0+\boldsymbol{c}_{01}-\boldsymbol{r}_C)\underline{1}_n$$

(the last term is explained by the comment on the matrix $\underline{C}_0$ prior to Eq. (5.136)). This equation is now premultiplied by the transpose of the matrix $\underline{\mu}$ whose elements are defined in Eq. (5.43). The left hand side yields $\underline{R}$, again, because of Eq. (5.46). On the right hand side the term $\underline{\mu}^{\mathrm{T}}\underline{1}_n$ is zero according to Eq. (5.49). This leaves

$$\underline{R}=-(\underline{C}\,\underline{T}\,\underline{\mu})^{\mathrm{T}}\underline{1}_n-(\underline{T}\,\underline{\mu})^{\mathrm{T}}\underline{z}.$$

From this, expressions for $\ddot{\underline{R}}$ and $\delta\underline{R}$ could be developed. However, it is unnecessary to do this because these expressions can be obtained more directly by premultiplying Eqs. (5.158) and (5.168) for $\ddot{\underline{r}}$ and $\delta\underline{r}$, respectively, by $\underline{\mu}^{\mathrm{T}}$:

$$\ddot{\underline{R}}=[\underline{p}\times\underline{T}(\underline{C}+\underline{Z})\underline{T}\underline{\mu}-\underline{k}\,\underline{T}\underline{\mu}]^{\mathrm{T}}\ddot{\underline{q}}+\underline{\mu}^{\mathrm{T}}\underline{U}\qquad \delta\underline{R}=[\underline{p}\times\underline{T}(\underline{C}+\underline{Z})\underline{T}\underline{\mu}-\underline{k}\,\underline{T}\underline{\mu}]^{\mathrm{T}}\delta\underline{q}.$$

The elements in the first row of $\underline{T}$ are either all $+1$ or all -1. The first row of the product $\underline{T}\,\underline{\mu}$ is, therefore, either $+\underline{1}_n^{\mathrm{T}}\underline{\mu}$ or $-\underline{1}_n^{\mathrm{T}}\underline{\mu}$ and, hence, according to Eq. (5.49) zero. This has the consequence that in the expressions for $\ddot{\underline{R}}$ and $\delta\underline{R}$ the first column of each of the matrices $\underline{C}$, $\underline{Z}$ and $\underline{k}$ is multiplied by zero. In these first columns only quantities associated with translational motions of the base $\underline{e}^{(0)}$ and of body 1 relative to one another appear, namely $\boldsymbol{c}_{11}$, $\boldsymbol{z}_1$ and the partial derivatives $\partial\boldsymbol{z}_1/\partial q_{1i}$ $(i=1\dots n_1)$ where $n_1=6$ is the number of degrees of freedom in the fictitious hinge number 1. For these relative translational motions then, no differential equations can be obtained from Eq. (5.109) into which $\ddot{\underline{R}}$ and $\delta\underline{R}$ will be substituted. This had to be expected since such equations have been replaced by Eq. (5.108) for the motion of the composite system center of mass. It is, therefore, not necessary to define six generalized coordinates for hinge 1. Only three are needed which describe the angular orientation of the base $\underline{e}^{(0)}$ and of body 1 relative to one another. Summarizing we can state the following. In the column matrices $\ddot{\underline{q}}$ and $\delta\underline{q}$ hinge 1 is represented by only three quantities $\ddot{q}_{1i}$ and δq_{1i} $(i=1,2,3)$, respectively. Accordingly, the first columns of the matrices $\underline{p}$ and $\underline{k}$ also contain three elements only. In the matrix $\underline{k}$, in particular, these three elements $\boldsymbol{k}_{11}$, $\boldsymbol{k}_{12}$ and $\boldsymbol{k}_{13}$ are zero. This is not important, however, since the first column of $\underline{k}$ is multiplied by zero anyway. In the matrix $\underline{p}$ the elements $\boldsymbol{p}_{11}$, $\boldsymbol{p}_{12}$ and $\boldsymbol{p}_{13}$ are nonzero, and they do effect $\ddot{\underline{R}}$ and $\delta\underline{R}$.

The virtual work δW in Eq. (5.109) has the same form as in the general case. It is given by Eq. (5.170). The force $\boldsymbol{X}_1$ and the torque $\boldsymbol{Y}_1$ in hinge 1 are equal to zero. The expressions for δW, $\ddot{\underline{R}}$, $\delta\underline{R}$, $\dot{\underline{\omega}}$ and $\underline{\delta\pi}$ can now be substituted into Eq. (5.109). This results in an equation of the form $\delta\underline{q}^{\mathrm{T}}(-\underline{A}\,\ddot{\underline{q}}+\underline{B})=0$. As a consequence of the independence of all variations δq_{ai} $(a=1\dots n,\ i=1\dots n_a)$ equations of motion in the form

$$\underline{A}\,\ddot{\underline{q}}=\underline{B} \tag{5.174}$$

are obtained, again. The matrices $\underline{A}$ and $\underline{B}$ are now

$$\underline{A}=[\underline{p}\times\underline{T}(\underline{C}+\underline{Z})\underline{T}\,\underline{\mu}-\underline{k}\,\underline{T}\,\underline{\mu}]\cdot\underline{m}[\underline{p}\times\underline{T}(\underline{C}+\underline{Z})\underline{T}\,\underline{\mu}-\underline{k}\,\underline{T}\,\underline{\mu}]^{\mathrm{T}}+(\underline{p}\,\underline{T})\cdot\underline{J}\cdot(\underline{p}\,\underline{T})^{\mathrm{T}}. \tag{5.175}$$

$$\underline{B}=[\underline{p}\times\underline{T}(\underline{C}+\underline{Z})\underline{T}\,\underline{\mu}-\underline{k}\,\underline{T}\,\underline{\mu}]\cdot(\underline{F}-\underline{m}\underline{\mu}^{\mathrm{T}}\underline{U})-\underline{p}\,\underline{T}\cdot[\underline{M}+\underline{J}\cdot(\underline{T}^{\mathrm{T}}\underline{f}-\dot{\omega}_0\underline{1}_n)-\underline{V}]- \\ -\underline{k}\cdot\underline{X}-\underline{p}\cdot\underline{Y}. \tag{5.176}$$

These equations in combination with Eq. (5.108) fully describe the behavior of the system.
The product $\underline{\mu}^{\mathrm{T}}\underline{U}$ in the matrix $\underline{B}$ is with Eq. (5.157)

$$\underline{\mu}^{\mathrm{T}}\underline{U}=-[(\underline{C}+\underline{Z})\underline{T}\,\underline{\mu}]^{\mathrm{T}}\times(\underline{T}^{\mathrm{T}}\underline{f}-\dot{\omega}_0\underline{1}_n)-\underline{\mu}^{\mathrm{T}}[\underline{T}^{\mathrm{T}}(\underline{s}+2\underline{h}+\underline{g}^*)+\underline{g}].$$

In verifying this it must be recognized that because of the identity $\underline{\mu}^{\mathrm{T}}\underline{1}_n=\underline{0}$ the last three terms of $\underline{U}$ which have the form shown in Eq. (5.163) do not give a contribution. Here and in other places in the expressions for $\underline{A}$ and $\underline{B}$ appears the matrix product $(\underline{C}+\underline{Z})\underline{T}\,\underline{\mu}$. Just as the matrices $\underline{C}+\underline{Z}$ and $(\underline{C}+\underline{Z})\underline{T}$ were interpreted in terms of vectors $\boldsymbol{c}_{ia}^*$ and $\boldsymbol{d}_{ij}^*$, respectively, we can now give a physical interpretation also to the product $(\underline{C}+\underline{Z})\underline{T}\,\underline{\mu}$. This is done by generalizing the concept of the vectors $\boldsymbol{b}_{ij}$ which were defined on the augmented bodies (Fig. 5.15) and shown to be the elements of the matrix $-\underline{C}\,\underline{T}\,\underline{\mu}$ (Eq. (5.53)). In analogy to $\boldsymbol{b}_{ij}$ new vectors $\boldsymbol{b}_{ij}^*=-[(\underline{C}+\underline{Z})\underline{T}\,\underline{\mu}]_{ij}$ for $i,j=1\dots n$ are now defined. They are to be interpreted as follows. For each body of the multi-body system, again, an augmented body is constructed in the usual fashion. This time, however, the point masses are not attached to the end points of the vectors $\boldsymbol{c}_{ia}$ but to the end points of the vectors $\boldsymbol{c}_{ia}^*$ defined by Eq. (5.160), i.e. to the hinge points. Since each hinge point is moving relative to one of the two respective contiguous bodies the augmented bodies thus constructed are, in general, not rigid. The vectors $\boldsymbol{b}_{ij}^*$ are illustrated in Fig. 5.40 which should be compared with Fig. 5.15.

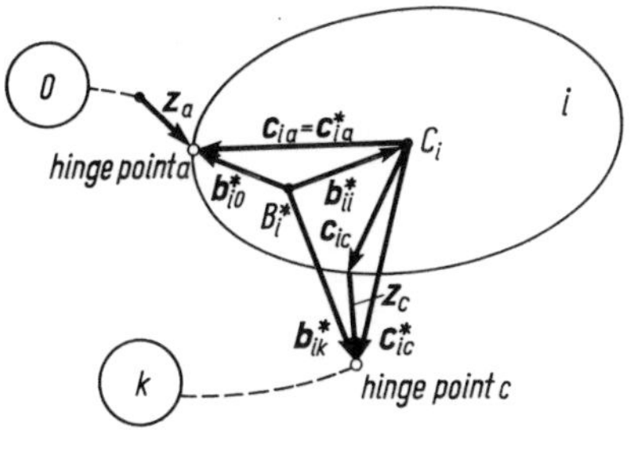

Fig. 5.40 Body i with hinge points 1 and 2 which are located by $\boldsymbol{c}_{ia}^*$ (fixed on body i) and $\boldsymbol{c}_{ic}^*=\boldsymbol{c}_{ic}+\boldsymbol{z}_c$ (not fixed on body i), with center of mass C_i, generalized barycenter B_i^* (not fixed on body i) and vectors $\boldsymbol{b}_{ii}^*$, $\boldsymbol{b}_{i0}^*$ and $\boldsymbol{b}_{ik}^*$. Compare with Fig. 5.15

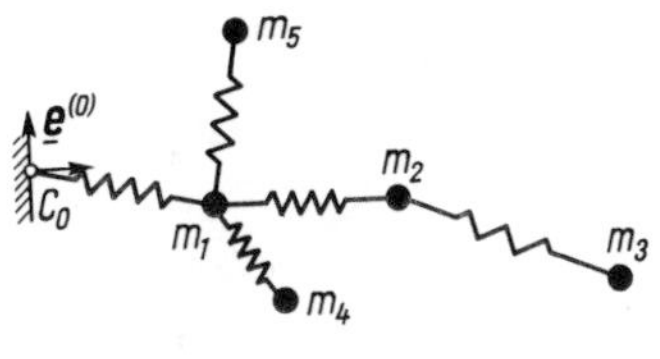

Fig. 5.41
A system of spring-connected point masses in spatial motion

Problems

5.17 Verify that the matrices $\underline{A}$ and $\underline{B}$ in Eq. (5.173) are identical with those in Eq. (5.97) if the system has ball-and-socket, universal and pin joints only.

5.18 Develop Eq. (5.34) of Sec. 5.2.2 from d'Alembert's principle. In what respect is the development different from that leading to Eq. (5.173)?

5.19 Formulate the matrices $\underline{A}$ and $\underline{B}$ in Eq. (5.173) for a system in which the hinges allow only purely translational motions of contiguous bodies relative to one another and in which, furthermore, body 0 does not rotate in inertial space.

5.20 Indicate which terms in Eq. (5.173) are zero if the system under consideration is composed of spring-connected point masses as shown in Fig. 5.41. The suspension point C_0 is fixed in inertial space.

5.2.9 Internal forces and torques in the hinges of a system with arbitrary holonomic constraints

During motions of a multi-body system, contiguous bodies act upon each other with forces and torques which are transmitted by the hinges. The engineer needs to know these forces and torques in order to be able to design sufficiently strong hinge mechanisms or to judge whether existing hinge mechanisms can sustain a given motion. Internal forces and torques have two sources, first, constraints in the hinges and, second, springs, dampers and similar devices. In this section no distinction will be made between these two. Instead, the resultant of both will be determined for given motions of a system. For this purpose the bodies of the system are isolated by cutting all hinges. The internal forces and torques in hinge a ($a=1 \dots n$) are then replaced by an equivalent pair of a single force $\boldsymbol{X}_a$ and a single torque $\boldsymbol{Y}_a$. The line of action of $\boldsymbol{X}_a$ is chosen to pass through the hinge point of hinge a. This point is defined as explained in Fig. 5.36. In Fig. 5.42 a representative body i ($i=1 \dots n$) is shown on which two hinges labeled b and c are located. In this figure, it is assumed that in the directed

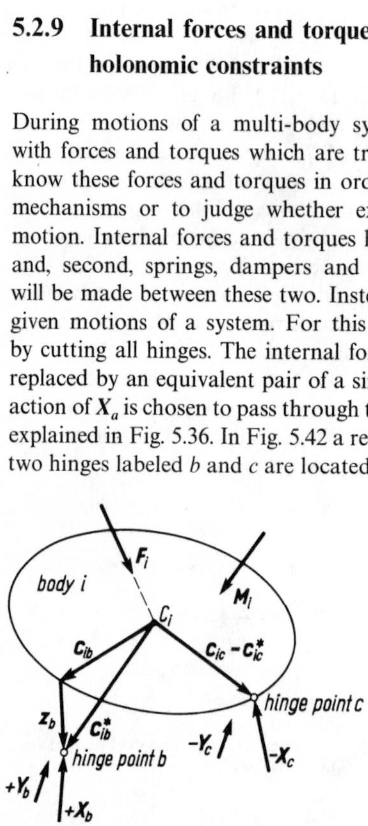

Fig. 5.42
Free-body diagram for body i. The hinge points b and c are fixed on a contiguous body and on body i, respectively

system graph the arc u_b is pointing away from the vertex s_i and the arc u_c is pointing towards it. Thus, the hinge point b is located at the end point of the vector $\boldsymbol{c}_{ib}+\boldsymbol{z}_b=\boldsymbol{c}_{ib}^*$ and the hinge point c at the end point of $\boldsymbol{c}_{ic}=\boldsymbol{c}_{ic}^*$. For the internal forces and torques the sign convention of Sec. 5.2.2 is used which says that $+\boldsymbol{X}_a$ and $+\boldsymbol{Y}_a$ are applied to body $i^+(a)$ and $-\boldsymbol{X}_a$ and $-\boldsymbol{Y}_a$ to body $i^-(a)$. This explains why $+\boldsymbol{X}_b$, $+\boldsymbol{Y}_b$, $-\boldsymbol{X}_c$ and $-\boldsymbol{Y}_c$ are acting on body i in Fig. 5.42. The figure is identical with Fig. 5.11 except for the existence of the hinge vector $\boldsymbol{z}_b$. To body i are also applied a resultant external force $\boldsymbol{F}_i$ (with its line of action passing through the body center of mass) and a resultant external torque $\boldsymbol{M}_i$. Newton's law and the law of moment of momentum for the body read

$$\left.\begin{aligned} m_i \ddot{\boldsymbol{r}}_i &= \boldsymbol{F}_i + \sum_{a=1}^{n} S_{ia} \boldsymbol{X}_a \\ \dot{\boldsymbol{L}}_i &= \boldsymbol{M}_i + \sum_{a=1}^{n} S_{ia} [(\boldsymbol{c}_{ia} + S_{ia}^{+} \boldsymbol{z}_a) \times \boldsymbol{X}_a + \boldsymbol{Y}_a] \end{aligned}\right\} \quad i = 1 \dots n . \tag{5.177}$$

This is identical with Eqs. (5.7) and (5.8) except for the term $S_{ia}^{+} \boldsymbol{z}_a$ which accounts for the additional lever $\boldsymbol{z}_a$ in the case where i equals $i^{+}(a)$. The expressions $S_{ia} \boldsymbol{c}_{ia}$ and $S_{ia} S_{ia}^{+} \boldsymbol{z}_a = S_{ia}^{+} \boldsymbol{z}_a$ are abbreviated $\boldsymbol{C}_{ia}$ and $\boldsymbol{Z}_{ia}$, respectively, as before. The equations of motion can then be rewritten in matrix form as

$$\underline{m} \ddot{\underline{\boldsymbol{r}}} = \underline{\boldsymbol{F}} + \underline{S} \underline{\boldsymbol{X}} \tag{5.178}$$

$$\dot{\underline{\boldsymbol{L}}} = \underline{\boldsymbol{M}} + (\underline{\boldsymbol{C}} + \underline{\boldsymbol{Z}}) \times \underline{\boldsymbol{X}} + \underline{S} \underline{\boldsymbol{Y}} . \tag{5.179}$$

All matrices used in these equations are known from previous sections. The internal hinge forces and torques can now be obtained in explicit form by premultiplying both equations by $\underline{T}$:

$$\underline{\boldsymbol{X}} = \underline{T} (\underline{m} \ddot{\underline{\boldsymbol{r}}} - \underline{\boldsymbol{F}}) \qquad \underline{\boldsymbol{Y}} = \underline{T} [\dot{\underline{\boldsymbol{L}}} - \underline{\boldsymbol{M}} - (\underline{\boldsymbol{C}} + \underline{\boldsymbol{Z}}) \times \underline{\boldsymbol{X}}] . \tag{5.180}$$

All terms on the right hand side of the first equation and all terms except $\underline{\boldsymbol{X}}$ on the right hand side of the second equation are known quantities if the motion of the system is given. The terms $\ddot{\underline{\boldsymbol{r}}}$ and $\dot{\underline{\boldsymbol{L}}}$, for instance, are related to generalized coordinates, velocities and accelerations through Eqs. (5.158) and (5.104), respectively:

$$\ddot{\underline{\boldsymbol{r}}} = [\underline{\boldsymbol{p}} \times \underline{T} (\underline{\boldsymbol{C}} + \underline{\boldsymbol{Z}}) \underline{T} - \underline{\boldsymbol{k}} \underline{T}]^{\mathrm{T}} \ddot{\underline{q}} + \underline{\boldsymbol{U}} \qquad \dot{\underline{\boldsymbol{L}}} = \underline{\boldsymbol{J}} \cdot \dot{\underline{\boldsymbol{\omega}}} + \underline{\boldsymbol{V}} .$$

For $\dot{\underline{\boldsymbol{\omega}}}$ Eq. (5.127) has to be substituted. Numerical calculations of $\underline{\boldsymbol{X}}$ and $\underline{\boldsymbol{Y}}$ are preferably executed in parallel with numerical integrations of the equations of motion (5.173) because all terms needed for one set of equations have to be evaluated also for the other.

5.3 Multi-body systems with closed chains and with arbitrary constraints

Most multi-body systems found in engineering practice do not have tree structure but an interconnection structure with closed chains. An example is illustrated in Fig. 5.43a without specification of the kinematic properties of the hinges. In general, the system will contain, both, closed kinematic chains and closed non-kinematic chains. As was explained in Sec. 5.1 all hinges in a closed kinematic chain have kinematic constraints whereas in a closed non-kinematic chain there is at least one hinge without constraints. In this section not only holonomic but also nonholonomic constraints will be taken into consideration. In setting up equations of motion for such systems the results of previous investigations can be used. It is always possible to cut hinges in such a way that a system with tree structure is produced. Obviously, this can be done in different ways. Criteria for selecting a particular system with tree structure will become evident later. For the moment the choice is made arbitrarily. Fig. 5.43 b depicts one possible system with tree structure which is obtained from Fig. 5.43a. If this system with tree structure has only holonomic

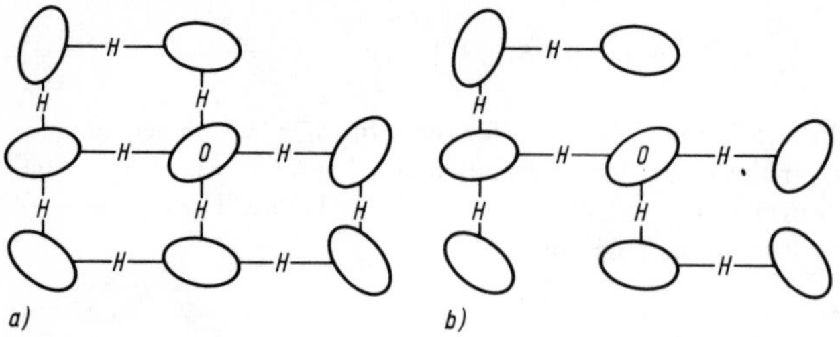

Fig. 5.43 A system with closed chains (a)) and a corresponding reduced system with tree structure (b)). The motion of body 0 is prescribed. Hinges with arbitrary constraints are indicated by the letter "H"

constraints then it is called the reduced system. Otherwise another system with tree structure is constructed which differs from the given one in that all nonholonomic constraints are removed. This new system is then called the reduced system. First, only the reduced system will be considered. Later, all constraints and all internal hinge forces other than constraint forces which have been eliminated in the process of generating the reduced system will be re-introduced. In this way, the original system with closed chains will be recovered, again.

For the reduced system equations of motion can be easily set up with the help of the results obtained in Sec. 5.2.8. It should be clear that Eqs. (5.171) to (5.173) are not, in general, immediately applicable. The reason is that these equations are valid only for such systems with tree structure in which the external body whose motion is prescribed as a function of time is coupled with one single body of the system. In the present case this cannot be assumed. Suppose, for example, that in Fig. 5.43a the motion of the body labeled body 0 is prescribed. If Fig. 5.43b represents the reduced system then this system is divided by body 0 into three dynamically independent subsystems each of which is coupled with body 0 by a single hinge. Each subsystem is governed by equations of motion which can be written in the form $\underline{A}^k \ddot{\underline{q}}^k = \underline{B}^k$ where k is an index identifying the subsystem, and $\underline{A}^k$ and $\underline{B}^k$ are given by Eqs. (5.171) and (5.172), respectively. From this follows that the equations of motion for the reduced system as a whole are

$$\begin{bmatrix} \underline{A}^1 & & & \underline{0} \\ & \underline{A}^2 & & \\ & & \ddots & \\ \underline{0} & & & \underline{A}^s \end{bmatrix} \begin{bmatrix} \ddot{\underline{q}}^1 \\ \ddot{\underline{q}}^2 \\ \vdots \\ \ddot{\underline{q}}^s \end{bmatrix} = \begin{bmatrix} \underline{B}^1 \\ \underline{B}^2 \\ \vdots \\ \underline{B}^s \end{bmatrix} \tag{5.181}$$

where s is the total number of independent subsystems (three in the present case). What makes the formulation unwieldy is the indexing of bodies, hinges and generalized coordinates. Bodies and hinges must be identified by two indices, namely the index of the particular subsystem and an index within the subsystem. The range of the second index is different for different subsystems. Generalized coordinates require a third index. When the effects of the cut hinges on the dynamics of the complete system are investigated this difficulty proves particularly disadvantageous. For this reason a new labeling of bodies and hinges is introduced in which the system is treated as a whole. It leads in a very natural way to generalizations of the graph-theoretical concepts of Sec. 5.2.1.

5.3.1 The mathematical description of the interconnection structure. A generalization of Section 5.2.1

Let $n+1$ be the number of bodies including body 0. Then, the reduced system has n hinges. The number of cut hinges is called n^*. For the system of Fig. 5.43, for example, n and n^* are seven and three, respectively. The bodies and hinges are labeled as follows. Body 0 is already identified. The remaining bodies and the hinges of the reduced system are labeled in an arbitrary order from 1 to n each. The cut hinges are labeled in an arbitrary order from $n+1$ to $n+n^*$. Fig. 5.44a illustrates one possible labeling. For the complete system of $n+1$ bodies and $n+n^*$ hinges a directed graph is drawn whose

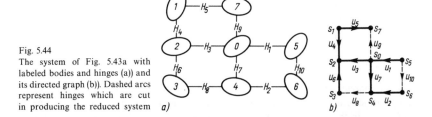

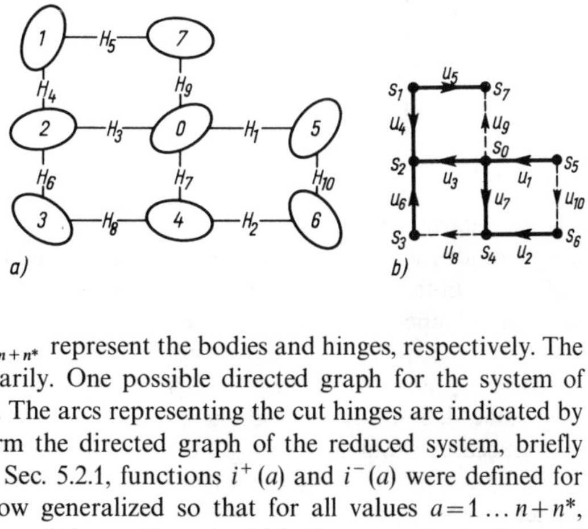

Fig. 5.44
The system of Fig. 5.43a with labeled bodies and hinges (a)) and its directed graph (b)). Dashed arcs represent hinges which are cut in producing the reduced system

vertices $s_0 \dots s_n$ and arcs $u_1 \dots u_{n+n^*}$ represent the bodies and hinges, respectively. The arc directions are chosen arbitrarily. One possible directed graph for the system of Fig. 5.44a is shown in Fig. 5.44b. The arcs representing the cut hinges are indicated by dashed lines. The solid lines form the directed graph of the reduced system, briefly called the r e d u c e d g r a p h. In Sec. 5.2.1, functions $i^+(a)$ and $i^-(a)$ were defined for $a=1 \dots n$. Their definition is now generalized so that for all values $a=1 \dots n+n^*$, $i^+(a)$ and $i^-(a)$ denote the indices of the vertices at which the arc u_a is starting and terminating, respectively. For the directed graph of Fig. 5.44b the functions read

a	1	2	3	4	5	6	7	8	9	10
$i^+(a)$	5	6	0	1	1	3	0	4	0	5
$i^-(a)$	0	4	2	2	7	2	4	3	7	6

In Sec. 5.2.1, the functions $i^+(a)$ and $i^-(a)$ were used to define the incidence matrix of a directed graph with tree structure (see Eq. (5.1)). The generalized equation

$$S_{ia}=\begin{cases}+1 & \text{for } i=i^{+}(a)\\ -1 & \text{for } i=i^{-}(a)\\ 0 & \text{otherwise}\end{cases}\qquad i=0\ldots n,\quad a=1\ldots n+n^{*}$$

defines a rectangular incidence matrix for the graph without tree structure. As before, each vertex is represented by one row and each arc by one column. The first n rows represent the incidence matrix of the reduced graph. The complete matrix is partitioned into the following four submatrices:

$$\underline{S}_0=[S_{01}\ldots S_{0n}]\qquad \underline{S}_0^{*}=[S_{0,n+1}\ldots S_{0,n+n^{*}}]$$

$$\underline{S}=\begin{bmatrix}S_{11}\ldots S_{1n}\\ \vdots\\ S_{n1}\quad S_{nn}\end{bmatrix}\qquad \underline{S}^{*}=\begin{bmatrix}S_{1,n+1}\ldots S_{1,n+n^{*}}\\ \vdots\\ S_{n,n+1}\quad S_{n,n+n^{*}}\end{bmatrix}.$$

For the directed graph of Fig. 5.44b, for example, the incidence matrix and its submatrices are

$$\begin{array}{c} \underline{S}_0\rightarrow\\ \\ \\ \\ \underline{S}\rightarrow\\ \\ \\ \\ \end{array}\left[\begin{array}{ccccccc|ccc} -1 & 0 & +1 & 0 & 0 & 0 & +1 & 0 & +1 & 0\\ \hline 0 & 0 & 0 & +1 & +1 & 0 & 0 & 0 & 0 & 0\\ 0 & 0 & -1 & -1 & 0 & -1 & 0 & 0 & 0 & 0\\ 0 & 0 & 0 & 0 & 0 & +1 & 0 & -1 & 0 & 0\\ 0 & -1 & 0 & 0 & 0 & 0 & -1 & +1 & 0 & 0\\ +1 & 0 & 0 & 0 & 0 & 0 & 0 & 0 & 0 & +1\\ 0 & +1 & 0 & 0 & 0 & 0 & 0 & 0 & 0 & -1\\ 0 & 0 & 0 & 0 & -1 & 0 & 0 & 0 & -1 & 0 \end{array}\right]\begin{array}{c} \leftarrow\underline{S}_0^{*}\\ \\ \\ \\ \leftarrow\underline{S}^{*}\\ \\ \\ \\ \end{array}$$

The submatrices $\underline{S}_0$ and $\underline{S}$ are determined by the reduced graph with tree structure. They are identical with the matrices of equal name in Sec. 5.2.1 except for one minor difference. In that section only the first element of $\underline{S}_0$ was different from zero, whereas now any number of elements can be different from zero depending on the number of hinges on body 0. In the equations of motion for systems with tree structure the matrices $\underline{S}_0^{+}$ and $\underline{S}^{+}$ defined by Eqs. (5.147) and (5.148) played an important role. The same definitions are used here, again. In addition, matrices $\underline{S}_0^{*+}$ and $\underline{S}^{*+}$ are introduced. All four matrices $\underline{S}_0^{+}$, $\underline{S}^{+}$, $\underline{S}_0^{*+}$ and $\underline{S}^{*+}$ are obtained from the matrices $\underline{S}_0$, $\underline{S}$, $\underline{S}_0^{*}$ and $\underline{S}^{*}$, respectively, when in the latter ones every element -1 is replaced by zero.

The definition of the matrix $\underline{T}$ for graphs with tree structure (see Eq. (5.4)) is directly applicable to the reduced graph. For the reduced graph of Fig. 5.44b, for example, $\underline{T}$ reads

$$\underline{T}=\begin{bmatrix} 0 & 0 & 0 & 0 & +1 & 0 & 0\\ 0 & 0 & 0 & 0 & 0 & +1 & 0\\ -1 & -1 & -1 & 0 & 0 & 0 & -1\\ +1 & 0 & 0 & 0 & 0 & 0 & +1\\ 0 & 0 & 0 & 0 & 0 & 0 & -1\\ 0 & 0 & +1 & 0 & 0 & 0 & 0\\ 0 & 0 & 0 & -1 & 0 & -1 & 0 \end{bmatrix}.$$

The definition of $\underline{T}$ cannot be generalized for graphs without tree structure in the sense that the arcs $u_{n+1} \ldots u_{n+n^*}$ are represented by additional rows. The reason is that in a graph with closed chains the path between the vertices s_0 and s_i is not uniquely defined for all vertices s_i $(i = 1 \ldots n)$. It should be noted that not all elements in the first row of $\underline{T}$ are different from zero as was the case for graphs with tree structure and with only one single arc incident with s_0.

The essential relationships

$$\underline{T}^{\mathrm{T}} \underline{S}_0^{\mathrm{T}} = -\underline{1}_n \quad \text{and} \quad \underline{T}\,\underline{S} = \underline{S}\,\underline{T} = \underline{E} \tag{5.182}$$

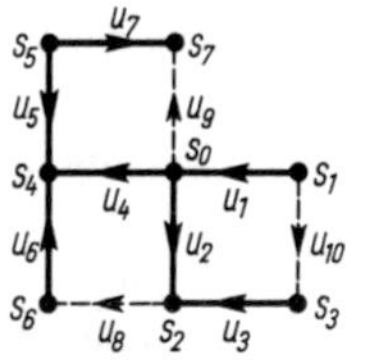

Fig. 5.45
Directed system graph with regular labeling
for the system of Fig. 5.44a

are still valid. The proof for the latter is exactly the same as in Sec. 5.2.1. The i-th element of $\underline{T}^{\mathrm{T}} \underline{S}_0^{\mathrm{T}}$ $(i = 1 \ldots n)$ is $\displaystyle\sum_{a=1}^{n} T_{ai} S_{0a}$. Only one single arc u_a contributes to the sum, namely the one which belongs to the path between s_0 and s_i and which is incident with s_0. For this arc T_{ai} and S_{0a} have opposite signs. This proves the first equation.

In practice it is not advisable to label the bodies and hinges of the reduced system as arbitrarily as has been done so far. In Sec. 5.2.1 the notion of regular labeling was defined for a system with tree structure in which there is only one hinge on body 0. The notion is now generalized for the case where, in a system with tree structure like the reduced system, the number of hinges on body 0 is arbitrary. It was stated previously that the reduced system consists of as many independent subsystems as there are hinges on body 0. A regular labeling has the following properties. The indices of, both, bodies and hinges of each subsystem belong to one and the same uninterrupted sequence of numbers. In addition, the labeling of each subsystem is regular in the sense defined in Sec. 5.2.1, i.e. the inboard arc of each vertex carries the same label as that vertex and the sequence of labels along the path from vertex s_0 to any other vertex is monotonically increasing. To give an example Fig. 5.45 shows a directed graph for the system of Fig. 5.44b in which the subgraph representing the reduced system has a regular labeling. The reduced system consists of three independent subsystems with one, two and four bodies, respectively (not counting body 0). The uninterrupted sequences of numbers were chosen to be (1), (2, 3) and (4, 5, 6, 7). The incidence matrix and the matrix $\underline{T}$ for this graph have a particularly simple structure. They are

$$
\underline{S}_0 \rightarrow
\begin{array}{c}
\begin{array}{cccccccccc}
-1 & +1 & 0 & +1 & 0 & 0 & 0 & 0 & +1 & 0
\end{array} \leftarrow \underline{S}_0^* \\[4pt]
\hline
\end{array}
$$

$$
\underline{S} \rightarrow
\begin{bmatrix}
+1 & 0 & 0 & 0 & 0 & 0 & 0 & 0 & 0 & +1 \\
0 & -1 & -1 & 0 & 0 & 0 & 0 & +1 & 0 & 0 \\
0 & 0 & +1 & 0 & 0 & 0 & 0 & 0 & 0 & -1 \\
0 & 0 & 0 & -1 & -1 & -1 & 0 & 0 & 0 & 0 \\
0 & 0 & 0 & 0 & +1 & 0 & +1 & 0 & 0 & 0 \\
0 & 0 & 0 & 0 & 0 & +1 & 0 & -1 & 0 & 0 \\
0 & 0 & 0 & 0 & 0 & 0 & -1 & 0 & -1 & 0
\end{bmatrix} \leftarrow \underline{S}_0^*
$$

$$\underline{T} = \begin{bmatrix} +1 & 0 & 0 & 0 & 0 & 0 & 0 \\ 0 & -1 & -1 & 0 & 0 & 0 & 0 \\ 0 & 0 & +1 & 0 & 0 & 0 & 0 \\ 0 & 0 & 0 & -1 & -1 & -1 & -1 \\ 0 & 0 & 0 & 0 & +1 & 0 & +1 \\ 0 & 0 & 0 & 0 & 0 & +1 & 0 \\ 0 & 0 & 0 & 0 & 0 & 0 & -1 \end{bmatrix}.$$

5.3.2 Equations of motion

The generalization of the graph theoretical concepts allows the formulation of equations of motion for the reduced system in a form which is more convenient than the one of Eq. (5.181). This requires no new development! The reader may verify for himself that the entire Section 5.2.8 (with one minor modification to be explained at once) remains valid if in all equations the matrices $\underline{S}_0$, $\underline{S}$ and $\underline{T}$ are understood to be the new matrices just defined. Hence, the equations of motion in the formulation of Eqs. (5.171) to (5.173) remain valid. Only the following modification is necessary. The identity (5.139) for the vectors $\boldsymbol{d}_{0j}$ defined in Eq. (5.138) is no longer valid since there may be more than one hinge on body 0. This has the consequence that the expression for the matrix $(\ddot{\underline{C}}_0\underline{T})^{\mathrm{T}}$ given in Eq. (5.142) can no longer be reduced to the form of Eq. (5.143). For the same reason the matrix $\underline{\boldsymbol{u}}$ is no longer given by Eq. (5.155). Instead, its elements are defined by Eq. (5.154). These are the only modifications. As is seen in Eq. (5.157) only the matrix $\underline{U}$—and only the last two terms in this matrix—are affected. The necessary change can be summarized as follows. The expression $\ddot{\boldsymbol{r}}_0\underline{1}_n-(\ddot{\underline{C}}_0\underline{T})^{\mathrm{T}}-\underline{T}^{\mathrm{T}}\underline{\boldsymbol{u}}$ in Eq. (5.157) is no longer given by Eq. (5.163). Instead, it is a column matrix with the elements

$$[\ddot{\boldsymbol{r}}_0\underline{1}_n-(\ddot{\underline{C}}_0\underline{T})^{\mathrm{T}}-\underline{T}^{\mathrm{T}}\underline{\boldsymbol{u}}]_i$$
$$=\ddot{\boldsymbol{r}}_0+\dot{\boldsymbol{\omega}}_0\times\boldsymbol{c}^*_{0\alpha(i)}+2S^+_{0\alpha(i)}\boldsymbol{\omega}_0\times\mathring{\boldsymbol{z}}_{\alpha(i)}+\boldsymbol{\omega}_0\times(\boldsymbol{\omega}_0\times\boldsymbol{c}^*_{0\alpha(i)}) \qquad i=1\dots n.$$

The index $\alpha(i)$ of the vectors $\boldsymbol{c}^*_{0\alpha(i)}$ and $\boldsymbol{z}_{\alpha(i)}$ is explained as follows. To each body i $(i=1\dots n)$ of the reduced system belongs a particular hinge on body 0 by means of which body i is connected either directly or indirectly with body 0. The index of this hinge is called $\alpha(i)$. In other words: $\alpha(i)$ is the index of the arc in the reduced graph which is incident with s_0 and which, furthermore, belongs to the path between s_0 and s_i. For the directed graph of Fig. 5.44b $\alpha(i)$ is given by the list

i	1	2	3	4	5	6	7
$\alpha(i)$	3	3	3	7	1	7	3

In the Special Case in which a multi-body system is not physically connected with a body whose motion is prescribed as a function of time equations of motion were obtained in the form of Eqs. (5.108) and (5.174) to (5.176). In these equations the matrix $\underline{U}$ need not be modified! This follows from the fact that the moving base $\underline{e}^{(0)}$ is connected by a fictitious hinge with only one body of the system.

5.3.2.1 Incorporation of constraint equations After these comments on the reduced system, d'Alembert's principle can now be formulated for the complete system with closed chains. It was stated earlier that this formulation must take into consideration all constraints and all internal hinge forces other than constraint forces which were removed in setting up the reduced system. In the system of Fig. 5.44a, for example, constraint forces as well as spring forces may have been removed in producing the reduced system of Fig. 5.44b. As a result of the re-introduction of kinematic constraints the variations of the generalized coordinates of the reduced system are no longer independent. Constraints can be either holonomic or nonholonomic. The total number of removed constraints is called ν_2. Out of these, ν_1 ($\nu_1 \leqslant \nu_2$) are assumed to be holonomic. The two sets are formulated as follows:

Holonomic constraints

$$f_i(\underline{q},t)=0 \qquad i=1\ldots\nu_1. \tag{5.183}$$

Nonholonomic constraints

$$g_i(\underline{q},\underline{\dot{q}},t)=0 \qquad i=\nu_1+1\ldots\nu_2.$$

These equations are, by definition, not integrable with respect to time since otherwise they would represent holonomic constraints. In practice, g_i is a linear function of the generalized velocities so that the equations can be written in the form

$$g_i=a_{i1}\dot{q}_{11}+\cdots+a_{iN}\dot{q}_{nn_n}+a_i=0 \qquad i=\nu_1+1\ldots\nu_2 \tag{5.184}$$

where $N=\sum_{a=1}^{n} n_a$ is the total number of generalized coordinates.

The functions f_i, $a_{i1}\ldots a_{iN}$ and a_i are, in general, nonlinear functions of either all or part of the generalized coordinates as well as of time t. This is indicated by the notation $f_i(\underline{q},t)$, for example. How these functions are actually formulated in practical cases will be demonstrated in Illustrative Examples 5.3 to 5.5.

The variation of f_i ($i=1\ldots\nu_1$) caused by a variation of the generalized coordinates (with time t held fixed) is zero, whence follows

$$\delta f_i=\delta\underline{q}^{\mathrm{T}}\underline{f}_i'^{\mathrm{T}}=0 \qquad i=1\ldots\nu_1$$

with the row matrices

$$\underline{f}_i'=\left[\frac{\partial f_i}{\partial q_{11}}\cdots\frac{\partial f_i}{\partial q_{nn_n}}\right] \qquad i=1\ldots\nu_1. \tag{5.185}$$

From Eq. (5.184) follows that the variations of the generalized coordinates also satisfy the relationships

$$a_{i1}\delta q_{11}+\cdots+a_{iN}\delta q_{nn_n}=0 \qquad i=\nu_1+1\ldots\nu_2$$

or

$$\delta\underline{q}^{\mathrm{T}}\underline{a}_i^{\mathrm{T}}=0 \qquad i=\nu_1+1\ldots\nu_2$$

with the row matrices

$$\underline{a}_i=[a_{i1}\ldots a_{iN}] \qquad i=\nu_1+1\ldots\nu_2. \tag{5.186}$$

D'Alembert's principle for the complete system with closed chains can now be written in the form

$$\delta \underline{q}^{\mathrm{T}}\left[-\underline{A}\ddot{\underline{q}}+\underline{B}+\sum_{i=1}^{\nu_1}\lambda_i \underline{f}_i'^{\mathrm{T}}+\sum_{i=\nu_1+1}^{\nu_2}\lambda_i \underline{a}_i^{\mathrm{T}}\right]+\delta W^*=0. \tag{5.187}$$

Each constraint equation is associated with a Lagrangian multiplier $\lambda_i\,(i=1\ldots\nu_2)$. The total virtual work done in all cut hinges together is called δW^*. It can be written in the form $\delta W^*=\delta\underline{q}^{\mathrm{T}}\underline{B}^*$ with a column matrix $\underline{B}^*$ which is composed of generalized forces that act in the cut hinges. An explicit formulation for $\underline{B}^*$ will be developed later. D'Alembert's principle then reads

$$\delta \underline{q}^{\mathrm{T}}\left[-\underline{A}\ddot{\underline{q}}+\underline{B}+\underline{B}^*+\sum_{i=1}^{\nu_1}\lambda_i \underline{f}_i'^{\mathrm{T}}+\sum_{i=\nu_1+1}^{\nu_2}\lambda_i \underline{a}_i^{\mathrm{T}}\right]=0. \tag{5.188}$$

The terms involving $\lambda_1\ldots\lambda_{\nu_2}$ can be combined in the matrix product $\underline{H}^{\mathrm{T}}\underline{\lambda}$ where $\underline{\lambda}$ is the column matrix $[\lambda_1\ldots\lambda_{\nu_2}]^{\mathrm{T}}$ and $\underline{H}$ the rectangular matrix

$$\underline{H}=\begin{bmatrix}\underline{f}_1'\\ \vdots\\ \underline{f}_{\nu_1}'\\ \underline{a}_{\nu_1+1}\\ \vdots\\ \underline{a}_{\nu_2}\end{bmatrix} \tag{5.189}$$

with ν_2 rows and N columns. The submatrix formed by the first ν_1 rows represents the Jacobian for the holonomic constraints. Eq. (5.188) now reads

$$\delta\underline{q}^{\mathrm{T}}(-\underline{A}\ddot{\underline{q}}+\underline{B}+\underline{B}^*+\underline{H}^{\mathrm{T}}\underline{\lambda})=0. \tag{5.190}$$

It is assumed that all ν_2 constraint equations are independent of one another. Thus, there exists a set of ν_2 variations of generalized coordinates (elements of $\delta\underline{q}$) which are linearly dependent. The remaining $N-\nu_2$ variations are independent. The multipliers $\lambda_1\ldots.\lambda_{\nu_2}$ are chosen in such a way that in Eq. (5.190) the coefficient of each variation belonging to the first set equals zero. The coefficient of each variation belonging to the second set must be zero because of the independence of these variations. Hence, one obtains

$$\underline{A}\ddot{\underline{q}}=\underline{B}+\underline{B}^*+\underline{H}^{\mathrm{T}}\underline{\lambda}. \tag{5.191}$$

These equations together with the constraint equations (5.183) and (5.184) describe the motions of the complete system. For practical applications this formulation is not yet satisfactory, however, since it does not permit the application of numerical integration algorithms. In order to arrive at equations which can be integrated numerically the unknowns $\lambda_1\ldots\lambda_{\nu_2}$ must be eliminated. For this purpose, the holonomic constraint equations (5.183) are differentiated with respect to time:

$$\dot{f}_i=\frac{\partial f_i}{\partial q_{11}}\dot{q}_{11}+\cdots+\frac{\partial f_i}{\partial q_{nn_n}}\dot{q}_{nn_n}+\frac{\partial f_i}{\partial t}=\underline{f}_i'\dot{\underline{q}}+\frac{\partial f_i}{\partial t}=0 \qquad i=1\ldots\nu_1 \tag{5.192}$$

with the row matrix $\underline{f}'_i$ from Eq. (5.185) and the column matrix $\dot{\underline{q}}=[\dot{q}_{11} \dots \dot{q}_{nn_n}]^T$. One more differentiation with respect to time yields

$$\ddot{f}_i=\underline{f}'_i\ddot{\underline{q}}+\varphi_i(\underline{q},\dot{\underline{q}},t)=0 \qquad i=1\dots\nu_1\,. \tag{5.193}$$

The actual calculation of the functions φ_i is a straightforward procedure although in practice it may lead to unwieldy expressions (see Illustrative Examples 5.3 and 5.5). Next, the holonomic constraint equations (5.184) are differentiated once with respect to time:

$$\begin{aligned}\dot{g}_i&=a_{i1}\,\ddot{q}_{11}+\dots+a_{iN}\ddot{q}_{nn_n}+\varphi_i(\underline{q},\dot{\underline{q}},t)\\ &=\underline{a}_i\ddot{\underline{q}}+\varphi_i(\underline{q},\dot{\underline{q}},t)=0 \qquad i=\nu_1+1\dots\nu_2\end{aligned} \tag{5.194}$$

with the row matrix $\underline{a}_i$ from Eq. (5.186). What was said about $\varphi_1\dots\varphi_{\nu_1}$ applies also to the functions $\varphi_{\nu_1+1}\dots\varphi_{\nu_2}$ (see Illustrative Example 5.4). Eqs. (5.192) and (5.184) are combined in the equation

$$\underline{H}\dot{\underline{q}}=-\underline{\psi} \tag{5.195}$$

where $\underline{H}$ is the matrix defined by Eq. (5.189) and $\underline{\psi}$ is the column matrix

$$\underline{\psi}(\underline{q},\dot{\underline{q}},t)=\left[\frac{\partial f_1}{\partial t}\dots\frac{\partial f\nu_1}{\partial t}\quad a_{\nu_1+1}\dots a_{\nu_2}\right]^T. \tag{5.196}$$

The latter is zero if all constraints are scleronomic. Similarly Eqs. (5.193) and (5.194) are combined in the matrix equation

$$\underline{H}\ddot{\underline{q}}=-\underline{\varphi} \tag{5.197}$$

with $\qquad \underline{\varphi}(\underline{q},\dot{\underline{q}},t)=[\varphi_1\dots\varphi_{\nu_2}]^T.$

Suppose that numerical integration starts at time $t=t_0$. First, initial values must be determined for the generalized coordinates and velocities. Those for the generalized coordinates must satisfy Eq. (5.183). Once a set of initial values $q_{11}\dots q_{nn_n}$ has been determined Eq. (5.195) is used to find a set of initial generalized velocities $\dot{\underline{q}}$. Starting from initial conditions thus determined it is possible, in principle, to proceed as follows. The two sets of second-order differential equations (5.191) and (5.197) are used to eliminate $\underline{\lambda}$ (by solving Eq. (5.191) for $\ddot{\underline{q}}$, substituting this into Eq. (5.197), solving the latter for $\underline{\lambda}$ and substituting $\underline{\lambda}$ back into Eq. (5.191)). The resulting differential equations for the generalized coordinates can then be integrated numerically by some standard algorithm. In practice, however, this method is bound to fail for the following reason. Eq. (5.197) is identical with

$$\ddot{f}_i=0 \quad i=1\dots\nu_1; \qquad \dot{g}_i=0 \quad i=\nu_1+1\dots\nu_2.$$

Provided proper initial conditions are chosen these equations have the exact solutions $f_i\equiv 0$ $(i=1\dots\nu_1)$ and $g_i\equiv 0$ $(i=\nu_1+1\dots\nu_2)$. In the course of numerical integrations, however, unavoidable numerical errors will cause an unbound increase of $|f_i|$ and $|g_i|$. Thus, the constraint equations will be violated.

It was Baumgarte's idea [18] to replace the differential equations (5.197) by others which for admissable initial conditions (satisfying Eqs. (5.183) and (5.195)) have the same solutions $f_i\equiv 0$ $(i=1\dots\nu_1)$ and $g_i\equiv 0$ $(i=\nu_1+1\dots\nu_2)$ and which are, furthermore, asymptotically stable in the sense of Ljapunov. There is an infinite variety of such equations. The simplest ones are

$$\ddot{f}_i + 2\alpha\dot{f}_i + \beta^2 f_i = 0 \qquad \alpha = \text{const} > 0 \text{ and } \beta = \text{const} \quad i = 1 \dots \nu_1$$

and

$$\dot{g}_i + \sigma g_i = 0 \qquad \sigma = \text{const} > 0 \qquad i = \nu_1 + 1 \dots \nu_2.$$

Any numerical error which causes f_i, $\dot{f}_i$ or g_i to deviate from the nominal value zero will now be damped out automatically. For efficient stabilization of $f_i = 0$ it is useful to choose $\alpha = \beta$ (critical damping). It is not true that the stabilizing effect is the better the larger α, β and σ are chosen! It must be avoided that the stabilizing terms $2\alpha\dot{f}_i + \beta^2 f_i$ and σg_i (which differ from zero only because of numerical errors) become dominant terms in the numerical integration of the complete set of differential equations. Attention must also be paid to the fact that through the stabilization of the holonomic constraints additional eigenfrequencies are introduced into the system[1]. In view of Eqs. (5.193) and (5.194) the newly constructed differential equations can be written in the form

$$\underline{f}'_i \ddot{\underline{q}} + \varphi_i^*(\underline{q}, \dot{\underline{q}}, t) = 0 \qquad i = 1 \dots \nu_1$$

with

$$\varphi_i^* = \varphi_i + 2\alpha\dot{f}_i + \beta^2 f_i \qquad i = 1 \dots \nu_1 \tag{5.198}$$

and

$$\underline{a}_i \ddot{\underline{q}} + \varphi_i^*(\underline{q}, \dot{\underline{q}}, t) = 0 \qquad i = \nu_1 + 1 \dots \nu_2$$

with

$$\varphi_i^* = \varphi_i + \sigma g_i \qquad i = \nu_1 + 1 \dots \nu_2. \tag{5.199}$$

The equations are combined in matrix form as

$$\underline{H}\ddot{\underline{q}} = -\underline{\varphi}^* \tag{5.200}$$

where $\underline{\varphi}^*(\underline{q}, \dot{\underline{q}}, t)$ is the column matrix

$$\underline{\varphi}^* = [\varphi_1^* \dots \varphi_{\nu_2}^*]^{\mathrm{T}}. \tag{5.201}$$

Except for a different right hand side Eq. (5.200) is identical with Eq. (5.197). Now the Lagrangian multipliers are eliminated. Eq. (5.191) yields $\ddot{\underline{q}} = \underline{A}^{-1}(\underline{B} + \underline{B}^* + \underline{H}^{\mathrm{T}}\underline{\lambda})$. Substitution into Eq. (5.200) leads to

$$\underline{H}\,\underline{A}^{-1}\underline{H}^{\mathrm{T}}\underline{\lambda} = -[\underline{H}\,\underline{A}^{-1}(\underline{B} + \underline{B}^*) + \underline{\varphi}^*]. \tag{5.202}$$

If all constraint equations are independent of each other the matrix $\underline{H}$ has the full row rank ν_2. There is a theorem stating that $\underline{H}\,\underline{A}^{-1}\underline{H}^{\mathrm{T}}$ has the same rank ν_2 if $\underline{A}^{-1}$ is a definite square matrix (Zurmühl [19]). This is the case. Consequently, the inverse exists so that

$$\underline{\lambda} = -(\underline{H}\,\underline{A}^{-1}\underline{H}^{\mathrm{T}})^{-1}[\underline{H}\,\underline{A}^{-1}(\underline{B} + \underline{B}^*) + \underline{\varphi}^*]. \tag{5.203}$$

Substituting this back into Eq. (5.191) one obtains

$$\ddot{\underline{q}} = \underline{A}^{-1}\{\underline{B} + \underline{B}^* - \underline{H}^{\mathrm{T}}(\underline{H}\,\underline{A}^{-1}\underline{H}^{\mathrm{T}})^{-1}[\underline{H}\,\underline{A}^{-1}(\underline{B} + \underline{B}^*) + \underline{\varphi}^*]\}. \tag{5.204}$$

This is the final form of the equations of motion for a system with closed chains.

---

[1] For further developments of this method see Baumgarte [40].

The number of equations is equal to the number of generalized coordinates in the reduced system. The constraint equations are taken care of by the terms $\underline{H}$ and $\underline{\varphi}^*$. The matrix $\underline{\varphi}^*$, in particular, guarantees the numerical stabilization of the constraints. Numerical integrations start from initial conditions for t, $\underline{q}$ and $\dot{\underline{q}}$ which satisfy Eqs. (5.183) and (5.195).

In two places it was assumed that the constraint equations are independent of each other. These places are the transition from Eq. (5.190) to Eq. (5.191) and the solution of Eq. (5.202) for $\underline{\lambda}$. In practice, it can easily happen that constraint equations are formulated which are not independent of each other. One reason for this is that even for relatively simple systems the functions f_i and g_i in Eqs. (5.183) and (5.184) can be so complicated that the recognition of dependencies is difficult. Such a case is treated in Illustrative Example 5.3. Furthermore, it is possible to have constraint equations which are independent of each other except for a special set of values of the generalized coordinates. When these values are accidentally met in the course of numerical integrations the program will fail because the matrix $\underline{H}\,\underline{A}^{-1}\underline{H}^{\mathrm{T}}$ in Eq. (5.202) becomes singular (in practice, the inversion of the matrix will fail not only at the critical point but in a certain neighborhood of it). This problem is avoided if Eq. (5.202) for $\underline{\lambda}$ is solved not by inversion of the coefficient matrix but by some algorithm for the solution of linear equations. Suppose that the rank of the coefficient matrix and, thus, also the row rank of $\underline{H}$ is smaller than ν_2, say $\nu<\nu_2$. Then, only ν multipliers can be determined from Eq. (5.202). Those $\nu_2-\nu$ columns of $\underline{H}$ in Eq. (5.190) which correspond to the undeterminate multipliers are linear combinations of the remaining ν columns. In other words, $\nu_2-\nu$ constraint equations are a consequence of the remaining ν constraint equations. The undeterminate multipliers which correspond to the dependent constraints are superfluous. They can, therefore, be set equal to zero. The solution thus found for $\underline{\lambda}$ replaces Eq. (5.203). The column matrix $\ddot{\underline{q}}$ is calculated not from Eq. (5.204) but through direct substitution of the numerical result for $\underline{\lambda}$ into Eq. (5.191).

5.3.2.2 Virtual work done in cut hinges In the equations of motion (5.204) all quantities except $\underline{B}^*$ have been specified in detail. In the remainder of this section an explicit expression will be developed for this matrix. The investigation starts from d'Alembert's principle in the form of Eq. (5.187) where $\delta W^*=\delta\underline{q}^{\mathrm{T}}\underline{B}^*$ was defined to be the virtual work done in all cut hinges together as a result of variations $\delta\underline{q}$ of the generalized coordinates of the reduced system. Let δW_a $(a=n+1\dots n+n^*)$ be the virtual work in the cut hinge number a, so that

$$\delta\underline{q}^{\mathrm{T}}\underline{B}^* = \sum_{a=n+1}^{n+n^*} \delta W_a\,. \tag{5.205}$$

The virtual work δW_a can most easily be expressed in terms of quantities which describe the position and angular orientation relative to one another of the two bodies that are coupled by hinge a. It is, therefore, natural to introduce for the cut hinges the same kinematic quantities which are used also for the hinges of the reduced system. These quantities are hinge vectors $\boldsymbol{z}_a$ together with first time derivatives $\overset{\circ}{\boldsymbol{z}}_a$ and variations $\overset{\circ}{\delta}\boldsymbol{z}_a$ (both relative to the respective bases $\underline{e}^{(i^+(a))}$), transformation matrices $\underline{G}_a$, relative angular velocities $\boldsymbol{\Omega}_a$ and vectors $\boldsymbol{\delta\varkappa}_a$ which denote variations of relative angular orientations. Along with the vectors $\boldsymbol{z}_a$ new body-fixed vectors $\boldsymbol{c}_{i^+(a)a}$ and $\boldsymbol{c}_{i^-(a)a}$

must be introduced for the cut hinges number $a=n+1 \dots n+n^*$. Fig. 5.36 then applies also to these hinges. In Sec. 5.2.2 it was defined that the vector $\boldsymbol{c}_{ia}$ $(i, a=1 \dots n)$ is zero if hinge a is not located on body i. This definition is now extended to all hinges $a=1 \dots n+n^*$.

It is clear that all kinematic quantities associated with the cut hinges, i.e. $\boldsymbol{z}_a$, $\mathring{\boldsymbol{z}}_a$, $\delta \mathring{\boldsymbol{z}}_a$, $\underline{G}_a$, $\boldsymbol{\Omega}_a$ and $\boldsymbol{\delta\varkappa}_a$ are determined by the generalized coordinates for the reduced system and by the first time derivatives and variations of these coordinates. In what follows, explizit formulations will be developed for these relationships. The first quantity to be examined is $\boldsymbol{z}_a$ $(a=n+1 \dots n+n^*)$. Fig. 5.36 yields

$$\boldsymbol{z}_a = (\boldsymbol{r}_{i^-(a)} + \boldsymbol{c}_{i^-(a)a}) - (\boldsymbol{r}_{i^+(a)} + \boldsymbol{c}_{i^+(a)a}) \qquad a=n+1 \dots n+n^* .$$

Using the elements S_{ia} of the newly introduced matrices $\underline{S}_0^*$ and $\underline{S}^*$ this becomes

$$\boldsymbol{z}_a = -\sum_{i=0}^{n} S_{ia}(\boldsymbol{r}_i + \boldsymbol{c}_{ia}) = -S_{0a}(\boldsymbol{r}_0 + \boldsymbol{c}_{0a}) - \sum_{i=1}^{n} S_{ia}(\boldsymbol{r}_i + \boldsymbol{c}_{ia}) \qquad a=n+1 \dots n+n^* .$$

All n^* equations can be combined in the matrix equation

$$\underline{z}^* = -\boldsymbol{r}_0 \underline{S}_0^{*\mathrm{T}} - \underline{C}_0^{*\mathrm{T}} - \underline{S}^{*\mathrm{T}} \underline{r} - \underline{C}^{*\mathrm{T}} \underline{1}_n \tag{5.206}$$

with the column matrix $\underline{z}^* = [\boldsymbol{z}_{n+1} \dots \boldsymbol{z}_{n+n^*}]^{\mathrm{T}}$, the row matrix

$$\underline{C}_0^* = [S_{0,n+1} \boldsymbol{c}_{0,n+1} \dots S_{0,n+n^*} \boldsymbol{c}_{0,n+n^*}]$$

and a matrix $\underline{C}^*$ whose elements are the body-fixed vectors

$$\boldsymbol{C}_{ia} = S_{ia} \boldsymbol{c}_{ia} \qquad i=1 \dots n,\ a=n+1 \dots n+n^* .$$

The matrices $\underline{C}_0^*$ and $\underline{C}^*$ correspond to the matrices $\underline{C}_0$ and $\underline{C}$, respectively, which are associated with the reduced system. Note that quantities like δW^*, $\underline{z}^*$, $\underline{S}^*$, $\underline{C}^*$ etc. which are associated with the set of all cut hinges are designated by an asterisk, whereas the asterisk is not used for quantities associated with a single cut hinge like δW_a, $\boldsymbol{z}_a$, S_{ia}, $\boldsymbol{c}_{ia}$ etc. which are sufficiently designated by their indices[1]. Substitution of Eq. (5.136) for $\underline{r}$ into Eq. (5.206) yields

$$\begin{aligned}\underline{z}^* &= -\boldsymbol{r}_0 \underline{S}_0^{*\mathrm{T}} - \underline{C}_0^{*\mathrm{T}} - \underline{S}^{*\mathrm{T}}[-(\underline{C}\,\underline{T})^{\mathrm{T}} \underline{1}_n - \underline{T}^{\mathrm{T}} \underline{z} + \boldsymbol{r}_0 \underline{1}_n - \underline{T}^{\mathrm{T}} \underline{C}_0^{\mathrm{T}}] - \underline{C}^{*\mathrm{T}} \underline{1}_n \\ &= -\boldsymbol{r}_0(\underline{S}_0^{*\mathrm{T}} + \underline{S}^{*\mathrm{T}} \underline{1}_n) - (\underline{C}_0^* - \underline{C}_0 \underline{T}\,\underline{S}^*)^{\mathrm{T}} - (\underline{C}^* - \underline{C}\,\underline{T}\,\underline{S}^*)^{\mathrm{T}} \underline{1}_n + (\underline{T}\,\underline{S}^*)^{\mathrm{T}} \underline{z} .\end{aligned}$$

The term $\underline{S}_0^{*\mathrm{T}} + \underline{S}^{*\mathrm{T}} \underline{1}_n$ is a column matrix whose a-th element $(a=n+1 \dots n+n^*)$ is the sum of all elements in the a-th column of the incidence matrix. Since each column contains exactly one element $+1$ and one element -1 this sum is zero. This leaves as final result

$$\underline{z}^* = -(\underline{C}_0^* - \underline{C}_0 \underline{T}\,\underline{S}^*)^{\mathrm{T}} - (\underline{C}^* - \underline{C}\,\underline{T}\,\underline{S}^*)^{\mathrm{T}} \underline{1}_n + (\underline{T}\,\underline{S}^*)^{\mathrm{T}} \underline{z} . \tag{5.207}$$

All quantities on the right hand side are known functions of the generalized coordinates.

Next, an expression is developed for the column matrix $\mathring{\underline{z}}^* = [\mathring{\boldsymbol{z}}_{n+1} \dots \mathring{\boldsymbol{z}}_{n+n^*}]^{\mathrm{T}}$. Since each of its elements is the velocity of one point in the system relative to another point

[1] Through Eq. (5.160) vectors $\boldsymbol{c}_{ia}^* = \boldsymbol{c}_{ia} + S_{ia}^+ \boldsymbol{z}_a$ $(i=0 \dots n,\ a=1 \dots n)$ were defined. The asterisk in this symbol does not refer to cut hinges, of course.

in the system it can be predicted that $\underline{\mathring{z}}^*$ is independent of $\dot{\boldsymbol{r}}_0$ and of $\boldsymbol{\omega}_0$. These quantities will, therefore, merely be indicated by dots in what follows. The column matrix $\underline{\dot{z}}^* = [\dot{z}_{n+1} \dots \dot{z}_{n+n^*}]^{\mathrm{T}}$ of the absolute time derivatives is according to Eq. (5.206)

$$\underline{\dot{z}}^* = -\underline{S}^{*\mathrm{T}} \underline{\dot{r}} - \underline{\dot{C}}^{*\mathrm{T}} \underline{1}_n + \cdots .$$

The derivatives $\mathring{z}_a$ and $\dot{z}_a$ are related by the equation

$$\mathring{z}_a = \dot{z}_a - \boldsymbol{\omega}_{i^+(a)} \times z_a = \dot{z}_a + \sum_{i=1}^{n} S_{ia}^+ z_a \times \boldsymbol{\omega}_i + \cdots \qquad a = n+1 \dots n+n^* .$$

In analogy to the matrix $\underline{Z}$ (cf. Eq. (5.151)) a matrix $\underline{Z}^*$ is defined. Its elements are

$$\boldsymbol{Z}_{ia} = S_{ia}^+ z_a \qquad i = 1 \dots n,\ a = n+1 \dots n+n^* .$$

With it the equations for $\mathring{z}_a$ can be combined in the matrix equation

$$\underline{\mathring{z}}^* = \underline{\dot{z}}^* + \underline{Z}^{*\mathrm{T}} \times \underline{\omega} + \cdots$$

or with the above expression for $\underline{\dot{z}}^*$

$$\underline{\mathring{z}}^* = -\underline{S}^{*\mathrm{T}} \underline{\dot{r}} - \underline{\dot{C}}^{*\mathrm{T}} \underline{1}_n + \underline{Z}^{*\mathrm{T}} \times \underline{\omega} + \cdots .$$

The matrix $\underline{\dot{r}}$ is obtained by differentiating Eq. (5.136). The last two terms contain only $\boldsymbol{r}_0$ and vectors which are fixed on body 0 so that their time derivatives can be omitted. This leaves

$$\underline{\dot{r}} = -(\underline{\dot{C}}\, \underline{T})^{\mathrm{T}} \underline{1}_n - \underline{T}^{\mathrm{T}} \underline{\dot{z}} + \cdots$$

whence follows

$$\underline{\mathring{z}}^* = -(\underline{\dot{C}}^* - \underline{\dot{C}}\, \underline{T}\, \underline{S}^*)^{\mathrm{T}} \underline{1}_n + (\underline{T}\, \underline{S}^*)^{\mathrm{T}} \underline{\dot{z}} + \underline{Z}^{*\mathrm{T}} \times \underline{\omega} + \cdots .$$

The first term is

$$-(\underline{\dot{C}}^* - \underline{\dot{C}}\, \underline{T}\, \underline{S}^*)^{\mathrm{T}} \underline{1}_n = (\underline{C}^* - \underline{C}\, \underline{T}\, \underline{S}^*)^{\mathrm{T}} \times \underline{\omega}$$

for reasons explained by Eq. (5.140). For a single element of $\underline{\dot{z}}$ the relationship holds

$$\dot{z}_a = \mathring{z}_a + \boldsymbol{\omega}_{i^+(a)} \times z_a = \mathring{z}_a - \sum_{i=1}^{n} S_{ia}^+ z_a \times \boldsymbol{\omega}_i + \cdots \qquad a = 1 \dots n .$$

In matrix form this reads

$$\underline{\dot{z}} = \underline{\mathring{z}} - \underline{Z}^{\mathrm{T}} \times \underline{\omega} + \cdots$$

so that $\underline{\mathring{z}}^*$ is the expression

$$\underline{\mathring{z}}^* = [\underline{C}^* + \underline{Z}^* - (\underline{C} + \underline{Z})\, \underline{T}\, \underline{S}^*]^{\mathrm{T}} \times \underline{\omega} + (\underline{T}\, \underline{S}^*)^{\mathrm{T}} \underline{\mathring{z}} + \cdots .$$

For $\underline{\omega}$ Eq. (5.130) is substituted. Finally, with $\underline{k}_0 = [\boldsymbol{k}_{10} \dots \boldsymbol{k}_{n0}]^{\mathrm{T}}$, $\underline{\mathring{z}}$ equals $\underline{k}^{\mathrm{T}} \underline{\dot{q}} + \underline{k}_0$ as follows from Eq. (5.114). This yields

$$\begin{aligned} \underline{\mathring{z}}^* = {} & \{\underline{p}\, \underline{T} \times [\underline{C}^* + \underline{Z}^* - (\underline{C} + \underline{Z})\, \underline{T}\, \underline{S}^*] + \underline{k}\, \underline{T}\, \underline{S}^*\}^{\mathrm{T}} \underline{\dot{q}} - \\ & - \{\underline{T}\, [\underline{C}^* + \underline{Z}^* - (\underline{C} + \underline{Z})\, \underline{T}\, \underline{S}^*]\}^{\mathrm{T}} \times \underline{p}_0 + (\underline{T}\, \underline{S}^*)^{\mathrm{T}} \underline{k}_0 . \end{aligned} \qquad (5.208)$$

No dots indicating omissions are needed any more since it is known that in the final result all omitted terms cancel each other[1]. All quantities of the expression on the right hand side are known functions of the generalized coordinates. From the equation for $\mathring{\underline{z}}^*$ follows at once that the column matrix $\mathring{\delta}\underline{z}^* = [\mathring{\delta}z_{n+1} \ldots \mathring{\delta}z_{n+n^*}]^{\mathrm{T}}$ is

$$\mathring{\delta}\underline{z}^* = \{\underline{p}\,\underline{T} \times [\underline{C}^* + \underline{Z}^* - (\underline{C} + \underline{Z})\,\underline{T}\,\underline{S}^*] + \underline{k}\,\underline{T}\,\underline{S}^*\}^{\mathrm{T}}\delta\underline{q}. \tag{5.209}$$

The next quantities to be related to the generalized coordinates are the transformation matrices $\underline{G}_a$ for $a = n+1 \ldots n+n^*$. The transformation matrices $\underline{A}_i$ $(i = 1 \ldots n)$ defined by the equation $\underline{e}^{(0)} = \underline{A}_i\underline{e}^{(i)}$ are known as functions of the generalized coordinates (cf. Eq. (5.134)). Hence,

$$\underline{G}_a = \underline{A}^{\mathrm{T}}_{i^-(a)}\underline{A}_{i^+(a)} \qquad a = n+1 \ldots n+n^* \tag{5.210}$$

is already the desired relationship.
Next, the relative angular velocities $\boldsymbol{\Omega}_a$ $(a = n+1 \ldots n+n^*)$ are examined. By definition

$$\boldsymbol{\Omega}_a = \boldsymbol{\omega}_{i^-(a)} - \boldsymbol{\omega}_{i^+(a)} \qquad a = n+1 \ldots n+n^*$$

or

$$\boldsymbol{\Omega}_a = -S_{0a}\boldsymbol{\omega}_0 - \sum_{i=1}^{n} S_{ia}\boldsymbol{\omega}_i \qquad a = n+1 \ldots n+n^*.$$

In matrix form this reads

$$\underline{\boldsymbol{\Omega}}^* = -\boldsymbol{\omega}_0\underline{S}_0^{*\mathrm{T}} - \underline{S}^{*\mathrm{T}}\underline{\boldsymbol{\omega}}$$

with the column matrix $\underline{\boldsymbol{\Omega}}^* = [\boldsymbol{\Omega}_{n+1} \ldots \boldsymbol{\Omega}_{n+n^*}]^{\mathrm{T}}$. Substituting Eq. (5.122) for $\underline{\boldsymbol{\omega}}$ one obtains

$$\underline{\boldsymbol{\Omega}}^* = -\boldsymbol{\omega}_0(\underline{S}_0^{*\mathrm{T}} + \underline{S}^{*\mathrm{T}}\underline{1}_n) + (\underline{T}\,\underline{S}^*)^{\mathrm{T}}\underline{\boldsymbol{\Omega}}.$$

The factor of $\boldsymbol{\omega}_0$ is zero so that the final result becomes with Eq. (5.129)

$$\underline{\boldsymbol{\Omega}}^* = (\underline{T}\,\underline{S}^*)^{\mathrm{T}}\underline{\boldsymbol{\Omega}} = (\underline{p}\,\underline{T}\,\underline{S}^*)^{\mathrm{T}}\dot{\underline{q}} + (\underline{T}\,\underline{S}^*)^{\mathrm{T}}\underline{p}_0. \tag{5.211}$$

For the column matrix $\underline{\boldsymbol{\delta\varkappa}}^* = [\boldsymbol{\delta\varkappa}_{n+1} \ldots \boldsymbol{\delta\varkappa}_{n+n^*}]^{\mathrm{T}}$ this yields

$$\underline{\boldsymbol{\delta\varkappa}}^* = (\underline{p}\,\underline{T}\,\underline{S}^*)^{\mathrm{T}}\delta\underline{q}. \tag{5.212}$$

In Eqs. (5.207) to (5.212) all kinematic quantities associated with the cut hinges have been expressed explicitly as functions of the generalized coordinates for the reduced system and of the first time derivatives and variations of these coordinates.

We are now in a position to develop from Eq. (5.205) an expression for the matrix $\underline{B}^*$. The expression for the virtual work δW_a in the cut hinge number a must be linear in

---

[1] It is an exercise for the reader to show that the omitted term in Eq. (5.208) has the form

$$\boldsymbol{\omega}_0 \times \{[\underline{C}_0^* + \underline{Z}_0^* - (\underline{C}_0 + \underline{Z}_0)\,\underline{T}\,\underline{S}^*]^{\mathrm{T}} + [\underline{C}^* + \underline{Z}^* - (\underline{C} + \underline{Z})\,\underline{T}\,\underline{S}^*]^{\mathrm{T}}\underline{1}_n\}$$

where $\underline{Z}_0$ and $\underline{Z}_0^*$ are row matrices with elements $Z_{0a} = S_{0a}z_a$ $(a = 1 \ldots n)$ and $Z_{0a} = S_{0a}z_a$ $(a = n+1 \ldots n+n^*)$, respectively. Write down in detail all matrices in this expression for the system whose directed system graph is illustrated in Fig. 5.44b and interpret the elements of the column matrix in curled brackets.

$\overset{\circ}{\delta} z_a$ and $\boldsymbol{\delta\varkappa}_a$ so that it can be written in the form

$$\delta W_a = -\overset{\circ}{\delta} z_a \cdot \boldsymbol{X}_a - \boldsymbol{\delta\varkappa}_a \cdot \boldsymbol{Y}_a \qquad a = n+1 \dots n+n^* .$$

Through this equation an equivalent internal hinge force $\boldsymbol{X}_a$ and torque $\boldsymbol{Y}_a$ are introduced. The equation is the same as for the hinges of the reduced system (Eq. (5.169)). The total virtual work done in all cut hinges together becomes

$$\begin{aligned} \delta \underline{q}^{\mathrm{T}} \underline{B}^* &= -\sum_{a=n+1}^{n+n^*} (\overset{\circ}{\delta} z_a \cdot \boldsymbol{X}_a + \boldsymbol{\delta\varkappa}_a \cdot \boldsymbol{Y}_a) \\ &= -\overset{\circ}{\delta} \underline{z}^{*\mathrm{T}} \cdot \underline{\boldsymbol{X}}^* - \boldsymbol{\delta} \underline{\boldsymbol{\varkappa}}^{*\mathrm{T}} \cdot \underline{\boldsymbol{Y}}^* \end{aligned}$$

where $\underline{\boldsymbol{X}}^*$ and $\underline{\boldsymbol{Y}}^*$ are the column matrices $[\boldsymbol{X}_{n+1} \dots \boldsymbol{X}_{n+n^*}]^{\mathrm{T}}$ and $[\boldsymbol{Y}_{n+1} \dots \boldsymbol{Y}_{n+n^*}]^{\mathrm{T}}$, respectively. When now $\overset{\circ}{\delta} \underline{z}^*$ and $\boldsymbol{\delta}\underline{\boldsymbol{\varkappa}}^*$ are replaced by the expressions in Eqs. (5.209) and (5.212), respectively, it is seen that $\underline{B}^*$ is the matrix

$$\underline{B}^* = -\{\underline{\boldsymbol{p}}\, \underline{T} \times [\underline{\boldsymbol{C}}^* + \underline{\boldsymbol{Z}}^* - (\underline{\boldsymbol{C}} + \underline{\boldsymbol{Z}})\, \underline{T}\, \underline{S}^*] + \underline{\boldsymbol{k}}\, \underline{T}\, \underline{S}^*\} \cdot \underline{\boldsymbol{X}}^* - \underline{\boldsymbol{p}}\, \underline{T}\, \underline{S}^* \cdot \underline{\boldsymbol{Y}}^* . \qquad (5.213)$$

The elements of the matrices on the right hand side are either constant scalars ($\underline{T}$, $\underline{S}^*$) or vectors with constant coordinates in body-fixed bases ($\underline{\boldsymbol{C}}$, $\underline{\boldsymbol{C}}^*$) or variable vectors whose coordinates in body-fixed bases are known functions of the generalized co-ordinates ($\underline{\boldsymbol{p}}$, $\underline{\boldsymbol{k}}$, $\underline{\boldsymbol{Z}}$) or variable vectors which are functions of quantities which, in turn, are known functions of the generalized coordinates ($\underline{\boldsymbol{Z}}^*$). The forces and torques of which $\underline{\boldsymbol{X}}^*$ and $\underline{\boldsymbol{Y}}^*$ are composed can, in every particular case of application, be expressed as functions of z_a, $\dot{z}_a$, $\underline{G}_a$ and $\boldsymbol{\Omega}_a$ ($a = n+1 \dots n+n^*$) and, thus, through Eqs. (5.207) to (5.211) as functions of the generalized coordinates. An example is treated in Illustrative Example 5.3. With the expression for $\underline{B}^*$ the equations of motion (5.204) are now in a form which is applicable to any system of rigid bodies and which requires from the user a minimum amount of preparatory work.

5.3.2.3 Criteria for selecting a reduced system In this subsection we return to the very beginning of Chap. 5.3. It has not yet been said by which criterion a reduced system should be produced from a given system with closed kinematic and non-kinematic chains. In other words, how should the hinges which are to be cut be chosen? The number n^* of hinges to be cut does not depend on the selection of a particular reduced system since it is the difference between two given numbers. One is the total number of hinges in the given system, and the other is the number of hinges in the reduced system which is equal to the number of bodies (not counting body 0). It is obvious that the formulation of constraint equations and the subsequent construction of the matrices $\underline{H}$, $\underline{\varphi}^*$ and $\underline{\psi}$ requires much work which has to be done by the user of the formalism for each particular system separately. Normally, each constraint equation involves a large number of generalized coordinates so that the derivatives $\dot{f}_i$, $\ddot{f}_i$ and $\dot{g}_i$ in Eqs. (5.192) to (5.194) are unwieldy expressions. Compared with this task it is straightforward to develop expressions for the forces and torques of which the matrices $\underline{\boldsymbol{X}}^*$ and $\underline{\boldsymbol{Y}}^*$ in Eq. (5.213) are composed because each force and torque is related to a single cut hinge. Hence, it can be concluded that, on principle, among the n^* cut hinges as many as possible should be without constraints. In other words, in closed non-kinematic chains hinges without constraints should be cut. In closed kinematic chains hinges with constraints must be cut. As regards the selection of these hinges the following can be said. The total number of degrees of freedom of the system with closed kinematic chains is a given number. It is the difference $N-\nu_2$ between the total number N of degrees of freedom of the reduced system and the total number ν_2 of constraints in all cut hinges together. Both, N and ν_2 depend on the choice of hinges to be cut. From the point of view of preparatory work required for the explicit formulation of the matrices $\underline{H}$, φ^* and $\underline{\psi}$ in Eqs. (5.189), (5.198), (5.199) and (5.196) it is desirable to keep ν_2 as small as possible. From the point of view of calculation time required for the evaluation and inversion of the matrices $\underline{A}$ and $\underline{H}\,\underline{A}^{-1}\underline{H}^{\mathrm{T}}$ it is desirable to keep, both, N and ν_2 as small as possible. Since $N-\nu_2$ is a fixed number N can be minimized together with ν_2. From this it is concluded that the hinges to be cut should be selected such that the total number of constraints in these hinges is a minimum. This criterion still leaves a choice, in general, which may be welcome from some other point of view. The criterion is useless if, for instance, all hinges available for cutting have the same number of constraints.

Illustrative Example 5.3 In Fig. 5.46 a linkage consisting of six bodies labeled 0 to 5 is shown. Bodies 0, 2 and 4 are identical, and bodies 1, 3 and 5 are identical. Furthermore, body 1 is a mirror image of body 0. The bodies are coupled by six pin joints labeled 1 to 6 to form a closed kinematic chain. On each hinge axis a torsional spring

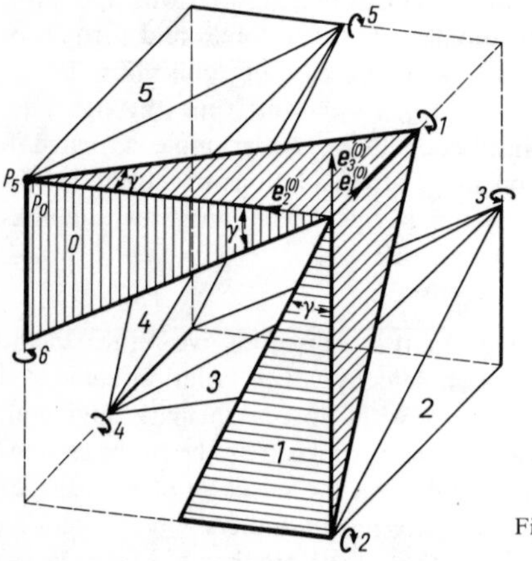

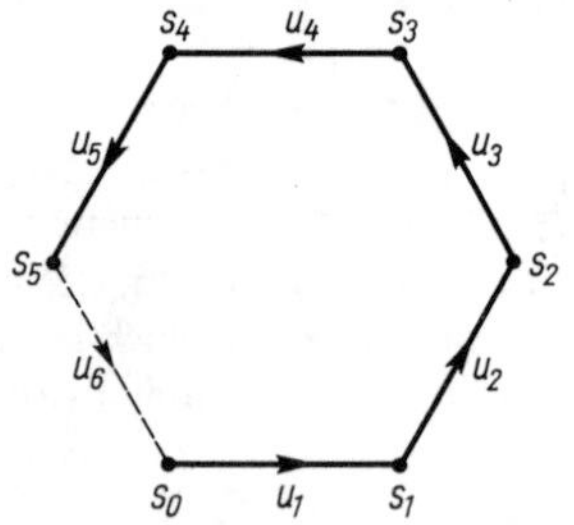

Fig. 5.47
The directed system graph for Fig. 5.46. In the reduced system hinge 6 is cut

Fig. 5.46 Closed kinematic chain with six pin joints. Body 0 is fixed in inertial space

and a damper are mounted whose constant coefficients are denoted k_a and d_a, respectively ($a=1\ldots6$). All springs are unstretched in the position shown in Fig. 5.46 in which the linkage fits into a cube of unit side length. Body 0 is fixed in inertial space. Determine the total number of degrees of freedom of the linkage and formulate the equations of motion.

Solution. From mere inspection it is not clear whether the system has any degree of freedom at all. In order to solve this part of the problem and as a preparation for the formulation of eventually necessary equations of motion hinge 6 is cut. The result is a system with tree structure (the reduced system) with five degrees of freedom, one in each of the hinges 1 to 5. On the bodies vector bases $\underline{e}^{(0)} \ldots \underline{e}^{(5)}$ are fixed. In the configuration shown in the figure (briefly called the cube configuration) all vector bases are parallel to each other, and all base vectors are parallel to edges of the cube. Next, generalized coordinates for the uncut hinges are defined as follows. The arc directions in the graph are chosen as shown in Fig. 5.47. For $a=1 \ldots 5$ q_a is defined as the angle through which body $i^-(a)$ is rotated relative to body $i^+(a)$ in the positive sense about the body-fixed base vector in the hinge axis a. All angles $q_1 \ldots q_5$ are zero in the cube configuration. For the transformation matrices $\underline{G}_a$ $(a=1 \ldots 5)$ defined by Eq. (5.85) Fig. 5.46 yields the expressions (with the abbreviations $c_a = \cos q_a$, $s_a = \sin q_a$)

$$\underset{\text{for } a=1,4}{\begin{bmatrix} 1 & 0 & 0 \\ 0 & c_a & s_a \\ 0 & -s_a & c_a \end{bmatrix}}, \quad \underset{\text{for } a=2,5}{\begin{bmatrix} c_a & 0 & -s_a \\ 0 & 1 & 0 \\ s_a & 0 & c_a \end{bmatrix}}, \quad \underset{\text{for } a=3}{\begin{bmatrix} c_a & s_a & 0 \\ -s_a & c_a & 0 \\ 0 & 0 & 1 \end{bmatrix}}. \tag{5.214}$$

The cut hinge has one degree of freedom. Hence, five constraint equations must be formulated. This equals the number of degrees of freedom of the reduced system so that the closed linkage has zero degrees of freedom unless the five constraints are dependent of each other. Three constraints are derived from the condition that the point P_5 on body 5 coincide with the point P_0 on body 0 (see Fig. 5.46). This is expressed by the equation $-\boldsymbol{e}_2^{(0)} - \boldsymbol{e}_3^{(1)} + \boldsymbol{e}_1^{(2)} + \boldsymbol{e}_2^{(3)} + \boldsymbol{e}_3^{(4)} - \boldsymbol{e}_1^{(5)} = \mathbf{0}$. The scalar form of this equation is simplest if the vectors are resolved either in $\underline{e}^{(2)}$ or in $\underline{e}^{(3)}$. Choosing $\underline{e}^{(3)}$ we get

$$\begin{aligned} &\underline{G}_3 \underline{G}_2 \underline{G}_1 [0 \ \ -1 \ \ 0]^{\mathrm{T}} + \underline{G}_3 \underline{G}_2 [0 \ \ 0 \ \ -1]^{\mathrm{T}} + \underline{G}_3 [\ \ 1 \ \ 0 \ \ 0]^{\mathrm{T}} \\ &\quad + [0 \ \ \ 1 \ \ 0]^{\mathrm{T}} + \underline{G}_4^{\mathrm{T}} [0 \ \ 0 \ \ \ 1]^{\mathrm{T}} + \underline{G}_4^{\mathrm{T}} \underline{G}_5^{\mathrm{T}} [-1 \ \ 0 \ \ 0]^{\mathrm{T}} = [0 \ \ 0 \ \ 0]^{\mathrm{T}} \end{aligned}$$

or explicitly

$$-s_3 c_1 - c_3 s_2 s_1 + c_3 s_2 + c_3 - c_5 = 0 \tag{5.215}$$

$$-c_3 c_1 + s_3 s_2 s_1 - s_3 s_2 - s_3 + 1 - s_4 - s_4 s_5 = 0 \tag{5.216}$$

$$c_2 s_1 - c_2 + c_4 + c_4 s_5 = 0. \tag{5.217}$$

The remaining two constraints require that the base vectors $\boldsymbol{e}_1^{(5)}$ and $\boldsymbol{e}_2^{(5)}$ be orthogonal to $\boldsymbol{e}_3^{(0)}$. This means that the elements with indices (1, 3) and (2, 3) of the matrix $\underline{G}_5 \underline{G}_4 \underline{G}_3 \underline{G}_2 \underline{G}_1$ must be zero or explicitly

$$c_5(-c_3 s_2 c_1 + s_3 s_1) + s_5 [s_4 (s_3 s_2 c_1 + c_3 s_1) - c_4 c_2 c_1] = 0 \tag{5.218}$$

$$c_4 (s_3 s_2 c_1 + c_3 s_1) + s_4 c_2 c_1 = 0. \tag{5.219}$$

These five constraint equations can be simplified substantially if it is recognized that they must be compatible with the following three constraint equations which are valid for reasons of symmetry

$$f_1 = q_3 - q_1 = 0, \qquad f_2 = q_5 + q_1 = 0, \qquad f_3 = q_4 + q_2 = 0. \tag{5.220}$$

When this is substituted Eqs. (5.215) to (5.219) can be reduced to (in this order)

$$\begin{aligned}
(s_1+s_1 s_2-s_2)c_1 &= 0\\
(s_1+s_1 s_2-s_2)(s_1-1)s_2 &= 0\\
0 &= 0\\
(s_1+s_1 s_2-s_2)(1+s_1+s_1 s_2)c_1 &= 0\\
(s_1+s_1 s_2-s_2)c_1 c_2 &= 0.
\end{aligned}$$

All equations are satisfied by the constraint

$$f_4=\sin q_1+\sin q_1 \sin q_2-\sin q_2=0. \tag{5.221}$$

Together with Eq. (5.220) this represents four independent constraints. Hence, the closed kinematic chain has one degree of freedom. From Eq. (5.221) some important information can be extracted. The relationship $\sin q_1=\sin q_2/(1+\sin q_2)$ is illustrated in Fig. 5.48. The conditions $|\sin q_1|\leqslant 1$ and $|\sin q_2|\leqslant 1$ imply $-210°\leqslant q_1\leqslant +30°$ and $-30°\leqslant q_2\leqslant +210°$. From this it can be concluded that the linkage moves without collisions of contiguous bodies provided the angle γ in Fig. 5.46 does not exceed 30°.

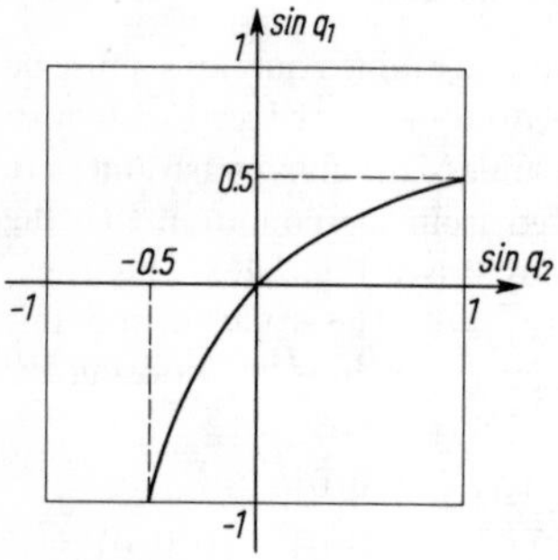

Fig. 5.48
The relationship between q_1 and q_2 as a result of the constraints in the closed kinematic chain

Next, the equations of motion (5.204) are formulated. The matrices $\underline{A}$ and $\underline{B}$ for the reduced system are given by Eqs. (5.171) and (5.172). Since all hinges are pin joints these expressions are identical with the ones given by Eqs. (5.98) and (5.99). All quantities which are necessary for their evaluation have been defined. The matrix $\underline{A}$ has five rows and columns. It remains to construct the matrices $\underline{H}$, φ^*, ψ and $\underline{B}^*$. $\underline{H}$ is the Jacobian for the constraint equations $f_i=0$ $(i=1,2,3,4)$. It has the form

$$\underline{H}=\begin{bmatrix} -1 & 0 & +1 & 0 & 0\\ +1 & 0 & 0 & 0 & +1\\ 0 & +1 & 0 & +1 & 0\\ c_1(1+s_2) & c_2(s_1-1) & 0 & 0 & 0 \end{bmatrix}.$$

It is easy to verify that $\underline{H}$ has the full row rank $\nu_2=4$ for all values of q_1. Hence, the inverse in Eq. (5.203) exists. The elements of φ^* are calculated from Eq. (5.198) in which the terms $\varphi_1\ldots\varphi_4$ are defined by Eq. (5.193) as the parts of the second derivatives $\ddot{f}_i$ not containing $\ddot{q}_1\ldots\ddot{q}_5$. The results are

$$\varphi_1^*=2\alpha(\dot{q}_3-\dot{q}_1)+\beta^2(q_3-q_1) \qquad \varphi_2^*=2\alpha(\dot{q}_5+\dot{q}_1)+\beta^2(q_5+q_1)$$

$$\varphi_3^*=2\alpha(\dot{q}_4+\dot{q}_2)+\beta^2(q_4+q_2)$$

$$\begin{aligned}\varphi_4^*=&-\dot{q}_1^2 s_1(1+s_2)+2\dot{q}_1\dot{q}_2 c_1 c_2-\dot{q}_2^2 s_2(s_2-1)+\\ &+2\alpha[\dot{q}_1 c_1(1+s_2)+\dot{q}_2 c_2(s_1-1)]+\beta^2(s_1+s_1 s_2-s_2).\end{aligned}$$

The matrix $\underline{\psi}$ is zero since all constraints are scleronomic. Initial conditions for $q_1 \ldots q_5$ and $\dot{q}_1 \ldots \dot{q}_5$ must satisfy all four constraint equations as well as Eq. (5.195). Consider, next, the matrix $\underline{B}^*$. In the cut hinge only a torque $\boldsymbol{Y}_6$ is acting. Hence, Eq. (5.213) reads $\underline{B}^* = -\underline{\boldsymbol{p}}\,\underline{T}\,\underline{S}^* \cdot \boldsymbol{Y}_6 = -\boldsymbol{Y}_6 \cdot \underline{\boldsymbol{p}}\,\underline{T}\,\underline{S}^*$. From Fig. 5.47 $\underline{T}\,\underline{S}^*$ is found to be $-\underline{1}_n$, whence follows $\underline{B}^* = \boldsymbol{Y}_6 \cdot \underline{\boldsymbol{p}}\,\underline{1}_n$ and since $\underline{\boldsymbol{p}}$ is a diagonal (5×5) matrix with the elements $\boldsymbol{e}_1^{(1)}$, $\boldsymbol{e}_2^{(2)}$, $\boldsymbol{e}_3^{(3)}$, $\boldsymbol{e}_1^{(4)}$, $\boldsymbol{e}_2^{(5)}$ along the diagonal $\underline{B}^* = [\boldsymbol{Y}_6 \cdot \boldsymbol{e}_1^{(1)}\ \boldsymbol{Y}_6 \cdot \boldsymbol{e}_2^{(2)}\ \boldsymbol{Y}_6 \cdot \boldsymbol{e}_3^{(3)}\ \boldsymbol{Y}_6 \cdot \boldsymbol{e}_1^{(4)}\ \boldsymbol{Y}_6 \cdot \boldsymbol{e}_2^{(5)}]^{\mathrm{T}}$. The torque $-\boldsymbol{Y}_6$ is acting on body 0 about the axis $\boldsymbol{e}_3^{(0)}$. Let q_6 be the angle of rotation of body 0 relative to body 5 about this axis with $q_6 = 0$ in the cube configuration. The torque is then $-\boldsymbol{Y}_6 = -(k_6 q_6 + d_6 \dot{q}_6)\boldsymbol{e}_3^{(0)}$. If the symmetry properties of the linkage were not known q_6 and $\dot{q}_6$ would have to be expressed in terms of the generalized coordinates $q_1 \ldots q_5$ and velocities $\dot{q}_1 \ldots \dot{q}_5$ as follows. The transformation matrix $\underline{G}_6$ defined by the equation $\underline{\boldsymbol{e}}^{(0)} = \underline{G}_6 \underline{\boldsymbol{e}}^{(5)}$ is

$$\underline{G}_6 = \begin{bmatrix} c_6 & s_6 & 0 \\ -s_6 & c_6 & 0 \\ 0 & 0 & 1 \end{bmatrix}.$$

According to Eq. (5.210) this matrix is related to the generalized coordinates $q_1 \ldots q_5$ by the equation $\underline{G}_6 = (\underline{G}_5 \underline{G}_4 \underline{G}_3 \underline{G}_2 \underline{G}_1)^{\mathrm{T}}$. Hence, q_6 is found to be the inverse sine of the element with indices (1, 2) of this matrix. The vector $\dot{q}_6 \boldsymbol{e}_3^{(0)}$ represents the angular velocity $\boldsymbol{\Omega}_6$ of body 0 relative to body 5. Eq. (5.211) relates it to the generalized coordinates and velocities:

$$\dot{q}_6 \boldsymbol{e}_3^{(0)} = -(\boldsymbol{\Omega}_1 + \boldsymbol{\Omega}_2 + \boldsymbol{\Omega}_3 + \boldsymbol{\Omega}_4 + \boldsymbol{\Omega}_5) = -(\dot{q}_1 \boldsymbol{e}_1^{(1)} + \dot{q}_2 \boldsymbol{e}_2^{(2)} + \dot{q}_3 \boldsymbol{e}_3^{(3)} + \dot{q}_4 \boldsymbol{e}_1^{(4)} + \dot{q}_5 \boldsymbol{e}_2^{(5)}).$$

These two results for q_6 and $\dot{q}_6 \boldsymbol{e}_3^{(0)}$ have to be substituted into the expression for $\boldsymbol{Y}_6$. In the present case the symmetry properties allow the formulation of much simpler relationships for q_6 and $\dot{q}_6$. The element with indices (2, 1) of the matrix $\underline{G}_5 \underline{G}_4 \underline{G}_3 \underline{G}_2 \underline{G}_1$ is $-c_4 s_3 c_2 + s_4 c_2$ or, in view of Eq. (5.220), $-s_2$, whence follows $q_6 = -q_2$. Hence, $-\boldsymbol{Y}_6 = (k_6 q_2 + d_6 \dot{q}_2)\boldsymbol{e}_3^{(0)}$. Together with Eqs. (5.214), (5.220) and (5.221) this yields

$$\underline{B}^* = -(k_6 q_2 + d_6 \dot{q}_2)\,[0 \quad s_1 \quad c_1 c_2 \quad -s_1 \quad 0]^{\mathrm{T}}.$$

Let the side length of the cube in Fig. 5.46 be l and let the bodies be homogeneous solids of density ϱ. Then, the center of mass of body 0 is located in $\underline{\boldsymbol{e}}^{(0)}$ at a point with the coordinates $l[a/4 \quad 1/2 \quad -a/4]$ where a is an abbreviation for $\tan\gamma$. The mass of the body is $\varrho l^3 a^2/6$ and its central inertia matrix in the base $\underline{\boldsymbol{e}}^{(0)}$ is

$$\varrho l^5 a^2/40 \begin{bmatrix} 1/3 + a^2/4 & a/6 & -a/12 \\ a/6 & a^2/2 & -a/6 \\ -a/12 & -a/6 & 1/3 + a^2/4 \end{bmatrix}.$$

■

Illustrative Example 5.4 Formulate the equations of motion for the three-legged table shown in Fig. 5.49a. At the end of each leg a wheel is held in a vertical position by means of a cage which is free to rotate about a vertical axis fixed in the leg (Fig. 5.49b). It is assumed that the wheels are in contact with the ground at all times and that they are rolling without slipping. The angular velocity of one of the wheels

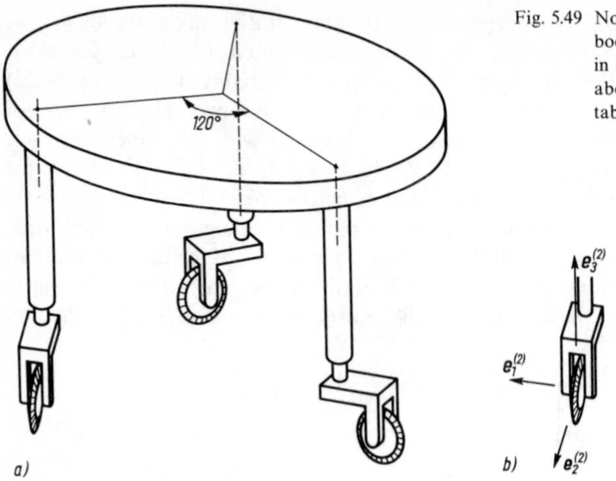

Fig. 5.49 Nonholonomic system with seven bodies (a)). Each wheel is held in a cage which is free to rotate about the vertical axis in the table leg (b))

relative to its cage is controlled to be a given time function $\Omega(t)$. This introduces a rheonomic constraint. The inertia of, both, cages and wheels is to be taken into account.

Solution. First, a reduced system is constructed which differs from the given system only in that the nonholonomic constraints at the wheel support points are missing. A schematic presentation is shown in Fig. 5.50a. The table plane (body 1) is connected with the ground (body 0) by a three-degree-of-freedom hinge which allows the table

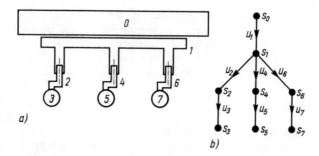

Fig. 5.50
a) The reduced system for the system of Fig. 5.49 and b) its directed graph

plane to translate in a plane fixed on body 0 and to rotate about the normal to this plane. The associated directed graph is depicted in Fig. 5.50b. Body-fixed vector bases: On bodies $i=0,1,2,4$ and 6 the base vectors $e_3^{(i)}$ are directed vertically upward. On bodies $i=2,3,4,5,6$ and 7 the base vectors $e_1^{(i)}$ are parallel to the respective wheel axes. On bodies $i=2,4$ and 6 the base vectors $e_2^{(i)}$ are pointing in the direction shown in Fig. 5.49b for one representative cage. All unspecified base vector orientations can be chosen arbitrarily. Generalized coordinates: Hinge 1: Cartesian coordinates q_{11} and q_{12} of the table center along $e_1^{(0)}$ and $e_2^{(0)}$, respectively, and a rotation angle q_{13} of body 1 relative to body 0 about $e_3^{(0)}$. Hinges $a=2,4$ and 6: Rotation angle q_{a1} of body $i^-(a)$ (a cage) relative to body $i^+(a)$ (body 1) about $e_3^{(1)}$. Hinges $a=3,5$ and 7: Rotation angles q_{a1} of body $i^-(a)$ (a wheel) relative to body $i^+(a)$ (a cage) about $e_1^{(i^-(a))}$. The coordinates q_{11}, q_{12}, q_{13} are zero for some arbitrarily defined position of body 1, and q_{31}, q_{51}, q_{71} are zero for some arbitrarily defined positions of the wheels relative to the cages. The coordinates q_{21}, q_{41}, q_{61} are zero when the base vectors $\boldsymbol{e}_2^{(2)}$, $\boldsymbol{e}_2^{(4)}$ and $\boldsymbol{e}_2^{(6)}$ are pointing radially away from the central vertical table axis. One out of the nine degrees of freedom of the system is eliminated by the rheonomic constraint that for one wheel, say body 3, the angular velocity relative to its cage (body 2) is a prescribed function of time $\Omega(t)$. This means that $\dot{q}_{31}=\Omega(t)$, $\ddot{q}_{31}=\dot{\Omega}(t)$ and $q_{31}=\int\Omega(t)\mathrm{d}t+\text{const}$ are known functions of time. Eq. (5.116) for hinge 3 reads $\boldsymbol{\Omega}_3=\boldsymbol{p}_{30}(t)=\Omega(t)\boldsymbol{e}_1^{(3)}$. This shows that the matrix $\underline{\boldsymbol{p}}$ defined by Eq. (5.126) has no entries in the third column ($n_3=0$) while the matrix $\underline{\boldsymbol{p}}_0$ in Eq. (5.129) is $[0\ 0\ \boldsymbol{p}_{30}\ 0\ 0\ 0\ 0]^{\mathrm{T}}$. With the quantities thus described and with numerical data for all body dimensions, masses and inertia components the matrices $\underline{A}$ and $\underline{B}$ from Eqs. (5.171) and (5.172) for the reduced system are well-defined functions of the eight independent coordinates $q_{11}, q_{12}, q_{13}, q_{21}, q_{41}, q_{51}, q_{61}, q_{71}$, of their first time derivatives and of time t. The equations for the complete system including the nonholonomic constraints have the form of Eq. (5.204). The matrices $\underline{H}$, $\underline{\varphi}^*$ and $\underline{\psi}$ will now be determined from the constraint equations. These equations which have to be formulated first have the general form of Eq. (5.184). It is sufficient to consider one wheel only, say body 3. For the other two the equations are found by an interchange of variables. Fig. 5.51 illustrates the bodies 1, 2 and 3. Shown are the reference base $\underline{\boldsymbol{e}}^{(0)}$ fixed in

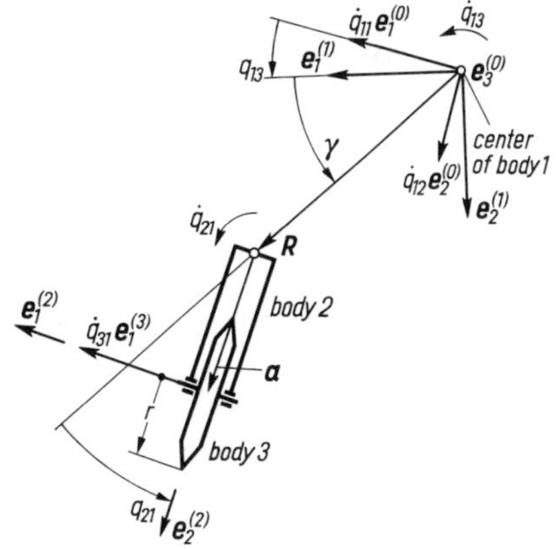

Fig. 5.51
Kinematics of one cage-and-wheel unit

inertial space, base vectors of the body-fixed bases $\underline{\boldsymbol{e}}^{(1)}$, $\underline{\boldsymbol{e}}^{(2)}$ and $\underline{\boldsymbol{e}}^{(3)}$, the angles q_{13} and q_{21} and the velocities $\dot{q}_{11}\boldsymbol{e}_1^{(0)}$, $\dot{q}_{12}\boldsymbol{e}_2^{(0)}$ and angular velocities $\dot{q}_{13}\boldsymbol{e}_3^{(0)}$, $\dot{q}_{21}\boldsymbol{e}_3^{(0)}$ and $\dot{q}_{31}\boldsymbol{e}_1^{(3)}$. The quantity $\dot{q}_{31}$ will later be replaced by $\Omega(t)$. In addition, quantities γ, $\boldsymbol{R}$ and $\boldsymbol{a}$ are indicated which define the locations of the cage axis and of the center of the wheel. The wheel radius is called r. The constraints require that the absolute velocity $\boldsymbol{v}$ of the ground contact point of the wheel have no components along $\boldsymbol{e}_1^{(2)}$ and $\boldsymbol{e}_2^{(2)}$. The figure yields for $\boldsymbol{v}$ the expression

$$\boldsymbol{v}=\dot{q}_{11}\boldsymbol{e}_1^{(0)}+\dot{q}_{12}\boldsymbol{e}_2^{(0)}+\dot{q}_{13}\boldsymbol{e}_3^{(0)}\times(\boldsymbol{R}+\boldsymbol{a})+\dot{q}_{21}\boldsymbol{e}_3^{(0)}\times\boldsymbol{a}-\dot{q}_{31}\boldsymbol{e}_1^{(3)}\times r\,\boldsymbol{e}_3^{(0)}.$$

It is a matter of simple algebra to determine the scalar products $\boldsymbol{v}\cdot\boldsymbol{e}_1^{(2)}$ and $\boldsymbol{v}\cdot\boldsymbol{e}_2^{(2)}$ which must, both, be zero. The desired constraint equations read

$$g_1=\dot q_{11}\sin(q_{13}+q_{21}+\gamma)-\dot q_{12}\cos(q_{13}+q_{21}+\gamma)-\dot q_{13}(R\cos q_{21}+a)-\dot q_{21}a=0$$
$$g_2=\dot q_{11}\cos(q_{13}+q_{21}+\gamma)+\dot q_{12}\sin(q_{13}+q_{21}+\gamma)+\dot q_{13}R\sin q_{21}+\dot q_{31}r=0.$$

To obtain the constraint equations $g_i=0$ $(i=3\ldots6)$ for the other wheels one has to replace γ by $\gamma+120^\circ$ and $\gamma+240^\circ$, respectively, and q_{21}, q_{31} by q_{41}, q_{51} and q_{61}, q_{71}, respectively. When this is done $\dot q_{31}$ is replaced by $\Omega(t)$ so that the constraint equations contain eight generalized velocities. The term $\Omega(t)r$ in g_2 represents the function a_2 of Eq. (5.184). In the (6×8) matrix $\underline{H}$ each row corresponds to one constraint equation and contains as elements the coefficients of all eight generalized velocities $\dot q_{11}$, $\dot q_{12}$, $\dot q_{13}$, $\dot q_{21}$, $\dot q_{41}$, $\dot q_{51}$, $\dot q_{61}$ and $\dot q_{71}$. The matrix $\underline{\varphi}^*$ is defined by Eq. (5.199). The terms σg_i $(i=1\ldots6)$ are already known. The remaining terms φ_i $(i=1\ldots6)$ are found from Eq. (5.194) as those parts of $\dot g_i$ which are independent of the second derivatives of the generalized coordinates. The first two elements of $\underline{\varphi}^*$ are, for example,

$$\varphi_1^*=\dot q_{11}(\dot q_{13}+\dot q_{21})\cos(q_{13}+q_{21}+\gamma)+\dot q_{12}(\dot q_{13}+\dot q_{21})\sin(q_{13}+q_{21}+\gamma)+$$
$$+\dot q_{13}\dot q_{21}R\sin q_{21}+\sigma g_1$$
$$\varphi_2^*=-\dot q_{11}(\dot q_{13}+\dot q_{21})\sin(q_{13}+q_{21}+\gamma)+\dot q_{12}(\dot q_{13}+\dot q_{21})\cos(q_{13}+q_{21}+\gamma)+$$
$$+\dot q_{13}\dot q_{21}R\cos q_{21}+\dot\Omega r+\sigma g_2.$$

The matrix $\underline{\psi}$ in Eq. (5.204) is given by Eq. (5.195) to be $[0\ \ \Omega(t)r\ \ 0\ \ 0\ \ 0\ \ 0]^{\mathrm{T}}$. ■

Illustrative Example 5.5 An anthropomorphic figure consisting of ten bodies (head, trunk, arms and legs composed of two bodies each) is in plane motion. The angle ϕ_i between a base vector $\boldsymbol{e}_1$ fixed in inertial space and a base vector $\boldsymbol{e}_1^{(i)}$ fixed on body i is used as generalized coordinate to describe the orientation of body i $(i=1\ldots10)$. All angles are zero when the figure is in a vertical position with closed feet, knees and elbows stretched. Formulate the differential equations for a phase of motion in which both feet are resting on the ground a distant L apart. The system including the labeling of bodies and indications of leg dimensions is shown in Fig. 5.52[1)].

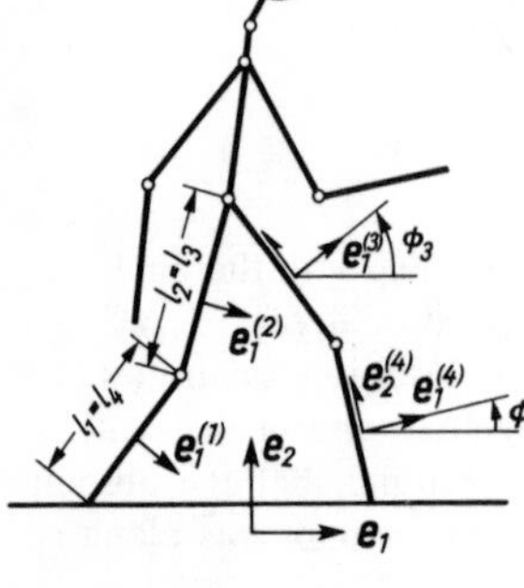

Fig. 5.52
Anthropomorphic figure in plane motion with two-feet contact to the ground

[1)] For testing purposes a FORTRAN program was written for numerical integrations of Eq. (5.173) for the anthropomorphic figure in the general case of spatial motions. Knees and elbows were assumed to be pin joints, hips and shoulders ball-and-socket joints and the neck a five-degree-of-freedom hinge allowing translational motions of the head perpendicular to the spine. Motions relative to an accelerating car were investigated.

Solution. The hinge between the front foot and the ground is cut. The resulting reduced system with tree structure has pin joints only, and one of its bodies is coupled to an external body whose motion is prescribed (the external body is the ground and it is actually at rest in inertial space). Plane motions of the reduced system are governed by Eq. (5.37). In Sec. 5.2.8 it was shown that the equations of motion of Sec. 5.2.3 are special cases of the general Eq. (5.173) which resulted from d'Alembert's principle. For Eq. (5.37) this principle has the form

$$\delta\underline{\phi}^{\mathrm{T}}\{\underline{A}\,\ddot{\underline{\phi}}+\underline{B}[\dot{\phi}_1^2\ldots\dot{\phi}_n^2]^{\mathrm{T}}-\underline{S}\,\underline{Y}-\underline{R}\}=0.$$

Constraint equations are incorporated in the usual fashion. Hence, the equations for the complete system of Fig. 5.52 read (cf. Eq. (5.204))

$$\ddot{\underline{\phi}}=\underline{A}^{-1}[\underline{Q}-\underline{H}^{\mathrm{T}}(\underline{H}\,\underline{A}^{-1}\underline{H}^{\mathrm{T}})^{-1}(\underline{H}\,\underline{A}^{-1}\underline{Q}+\underline{\varphi}^*)]$$

with $\quad \underline{Q}=-\underline{B}[\dot{\phi}_1^2\ldots\dot{\phi}_n^2]^{\mathrm{T}}+\underline{S}\,\underline{Y}+\underline{R}+\underline{B}^*.$

The matrices $\ddot{\underline{\phi}}$, $\underline{A}$, $\underline{B}$, $\underline{S}$, $\underline{Y}$ and $\underline{R}$ in this equation are the same as in Eq. (5.37). The matrix $\underline{B}^*$ accounts for eventually existing internal forces and torques (other than constraint forces and torques) in the cut hinge. The matrices $\underline{H}$ and $\underline{\varphi}^*$ are determined by the constraints in the cut hinge. In terms of the quantities defined in Fig. 5.52 these constraints are

$$f_1=\sum_{k=1}^{4} l_k\sin\phi_k=0,\qquad f_2=-\sum_{k=1}^{4} l_k\cos\phi_k+L=0.$$

The (2×10) matrix $\underline{H}$ has nonzero elements in the first four columns only. These columns are

$$\begin{bmatrix} l_1\cos\phi_1 & l_2\cos\phi_2 & l_2\cos\phi_3 & l_1\cos\phi_4 \\ l_1\sin\phi_1 & l_2\sin\phi_2 & l_2\sin\phi_3 & l_1\sin\phi_4 \end{bmatrix}.$$

The column matrix $\underline{\varphi}^*$ has two elements which, according to Eq. (5.198) are

$$\varphi_1^*=\sum_{k=1}^{4} l_k[(\beta^2-\dot{\phi}_k^2)\sin\phi_k+2\alpha\dot{\phi}_k\cos\phi_k]$$

$$\varphi_2^*=\sum_{k=1}^{4} l_k[-(\beta^2-\dot{\phi}_k^2)\cos\phi_k+2\alpha\dot{\phi}_k\sin\phi_k]+\beta^2 L.$$ ■

5.4 Concluding remarks

Of the material presented in this chapter the content of Secs. 5.2.1 and 5.2.4 was developed first. It was the subject of ref. [1] by Roberson and Wittenburg. At the same time Hooker and Margoulis [20] investigated the same problem. The two groups knew of each other's attempts. What neither one of them knew was that Fischer [21], sixty years earlier, had considered the same problem already. His book Introduction to the Mechanics of Living Mechanisms contains all results of Secs. 5.2.2 and 5.2.4, in particular the concept of augmented bodies. Fischer's approach was different, though. He did not apply concepts of graph theory and instead of working with absolute angular velocities he defined a set of Euler angles for each body, formulated the Lagrangian and went through the laborious procedure of constructing Lagrange's equations term by term writing out all vector and tensor products in scalar form. His final equations are so lengthy that he wrote them down in explicit form only for a system of two bodies with a ball-and-socket joint located on a principal axis of either body. He was, however, in the possession of the equations for general systems with tree structure and with ball-and-socket joints and actually applied these equations to the problem of gait of the human body. The physical interpretations given following Eqs. (5.29) and (5.59) can be found in his book. It was Roberson who, in 1967, came across this early work. Hooker [22] first showed how to establish scalar equations of motion for systems with tree structure and with pin joints. Roberson [23] made the first attempt at treating systems in which motions of contiguous bodies relative to one another are not purely rotational. Lilov and Wittenburg [24] with the first author playing the leading role used d'Alembert's principle for establishing equations of motion. Boland, Samin and Willems [25] had done the same, restricting their investigation to linearized equations, however. The approach via d'Alembert's principle opened the road to treating closed kinematic chains the way it is done in Sec. 5.3.

In this book only rigid body systems and only exact nonlinear equations of motion are considered. The methods developed here serve as a starting point also for investigations into systems of coupled nonrigid bodies. Leight-weight structures of spacecraft with rotating antennas and solar panels are typical examples of systems in which the flexibility of bodies must be taken into consideration. In this area contributions were made by Likins [26], Likins and Fleischer [27], Boland, Samin and Willems [25,28], Frisch [29] and others. The idea is to apply finite element methods to flexible bodies and to treat hinge coordinates the way described in this chapter. For the combination of small displacement variables describing body deformations and of hinge variables undergoing large changes the term hybrid coordinates was proposed by Likins. It should be pointed out that there have been some other attempts at treating multi-body systems [30 to 35]. None of these methods is as general as the one described in this chapter. A survey of different methods is given in ref. [35] by Renaud who himself developed a computer-oriented procedure based on Lagrange's equations for systems with tree structure and with rotational and translational degrees of freedom in the hinges.

6 Impact Problems in Holonomic Multi-Body Systems

The previous chapter was devoted to the mathematical description of continuous motions of multi-body systems. In contrast, in the present chapter the case is studied in which a system experiences discontinuous changes of velocities and angular velocities. Such phenomena occur when a multi-body system collides with a single body or with another multi-body system or when two bodies belonging to one and the same multi-body system collide with each other. Typical examples are illustrated in Fig. 6.1. The actual physical processes during impact are highly complex. In

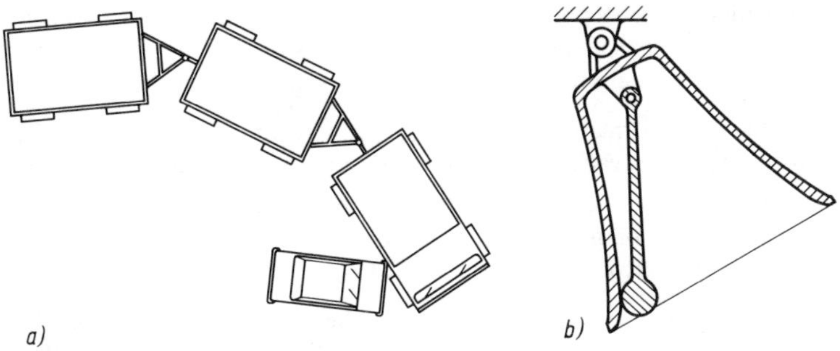

Fig. 6.1 a) Collision between two systems and b) within one system

order to render the problem amenable to mathematical treatment some simplifying assumptions must be made. Whether these are acceptable must be judged in every particular case of application of the resulting formalism. The assumptions concern the dynamic behavior of bodies under the action of impulsive forces and the phenomena in the immediate vicinity of the point of collision, in particular. They are the same which are made in the classical treatment of the collision between two rigid bodies. In addition, some assumptions are made about the nature of constraints in the hinges. These latter assumptions do not go beyond the ones made in the last chapter.

6.1 Basic assumptions

As regards the behavior of bodies it is postulated that the impact is taking place in such a short time interval Δt that in the mathematical description the idealization $\Delta t \to 0$ is possible. This implies that any propagation of waves of deformations and tensions through bodies is neglected since such processes require finite periods of time. Hence, all bodies of a system are treated as rigid bodies during impact. During the infinitesimally short time interval the positions and angular orientations of all bodies remain unchanged since all velocities and angular velocities remain finite. Springs, dampers and similar elements in the hinges of a system do not play any role since they exert forces and torques of finite magnitude whose integrals over the infinitesimally short time interval are zero. Only impulsive forces can cause discontinuous changes of velocities and angular velocities. A force $\boldsymbol{F}(t)$ is called an impulsive force if the integral $\int_{t_0}^{t_0+\Delta t} \boldsymbol{F}(t)\mathrm{d}t$ over the time interval Δt converges toward a finite quantity $\hat{\boldsymbol{F}}$ when Δt approaches zero. For this to be the case the magnitude of $\boldsymbol{F}(t)$ must tend toward infinity during Δt for the limiting case $\Delta t \to 0$. The quantity $\hat{\boldsymbol{F}}$ is called an impulse. Impulses occur at the point of collision as well as in hinges with kinematic constraints. The torque of an impulse is called an impulse couple.

After these comments on the simplifying assumptions concerning the behavior of bodies the idealizations regarding the processes at the point of collision are discussed. In a volume surrounding the two colliding body-fixed points elastic and/or plastic deformations occur. It is assumed that this volume has infinitesimally small dimensions so that the previous assumptions of rigid bodies and of an infinitesimally short time interval of impact are not violated. Two cases are distinguished:

Case (i) The deformations are ideally plastic, and the colliding bodies have zero velocity relative to each other at the point of collision immediately after impact.

Case (ii). The impulsive interaction forces between the colliding bodies are directed normal to the common tangent plane at the point of collision. This means that there is no friction between the bodies.

Case (ii) represents a whole family of cases because it must be distinguished whether the deformation of the bodies is a purely plastic compression or whether the phase of compression is followed by a phase of either total or partial decompression. Let $\hat{\boldsymbol{F}}^c$ be the impulse which is exerted on one of the colliding bodies by the other colliding body during the phase of compression. A parameter e called the coefficient of restitution is defined. It is the ratio between the impulse in the phase of decompression and $\hat{\boldsymbol{F}}^c$. Hence, the total impulse during impact is

$$\hat{\boldsymbol{F}} = (1+e)\hat{\boldsymbol{F}}^c. \tag{6.1}$$

The parameter e is zero if the compression is purely plastic and one if it is purely elastic. For partially elastic compressions it is in the range $0<e<1$.

It has been said already that springs and dampers in the hinges between bodies do not influence the finite increments of velocities and angular velocities during impact. Only kinematic constraints are important since they cause internal impulses and impulse couples in the hinges. In this chapter only holonomic constraints will be taken into consideration. They have the general form

$$f(q_1 \ldots q_s, t) = 0$$

where $q_1 \ldots q_s$ is some unspecified set of generalized coordinates. Differentiation with respect to time yields

$$\sum_{i=1}^{s} \frac{\partial f}{\partial q_i} \dot{q}_i + \frac{\partial f}{\partial t} = 0\,.$$

The derivatives $\partial f/\partial q_i\,(i=1\ldots s)$ and $\partial f/\partial t$ are, in general, functions of $q_1\ldots q_s$ and t. Since these variables are constant during impact the derivatives are constant, as well. Writing down the equation once for the instant immediately after impact and once for the instant immediately prior to impact and taking the difference one obtains

$$\sum_{i=1}^{s} \frac{\partial f}{\partial q_i} \Delta \dot{q}_i = 0 \tag{6.2}$$

where $\Delta \dot{q}_i$ $(i=1\ldots s)$ are the finite increments of the generalized velocities $\dot{q}_1\ldots\dot{q}_s$ during impact. This shows that any holonomic constraint results in a homogeneous linear equation with constant coefficients for velocity increments during impact. Eq. (6.2) is applicable only if the following conditions are fulfilled. First, there must be no play in the hinges. Second, the hinges must not be deformed by internal impulses and impulse couples which occur during impact. Finally, no loss of kinetic energy is allowed to occur in the hinges during impact. This implies, among other things, the absence of dry friction. As was said in the beginning these conditions do not go beyond the ones underlying the results of the previous chapter.

6.2 Instantaneous velocity increments

With the assumptions specified in the previous subsection the problem of impact is now reduced to the determination of two groups of mechanical quantities. One group is made up of the instantaneous finite increments of all velocities. Velocities of interest are the first time derivatives of generalized coordinates as well as angular velocities of bodies and translational velocities of body centers of mass and of other body-fixed points. The second group comprises internal impulses and impulse couples in hinges with holonomic constraints as well as the impulse which is applied to a system at the point of collision. The development of explicit solutions for the said quantities begins with the formulation of a kinematic relationship for the two colliding body-fixed points. Two kinds of collision were distinguished. For the ideally plastic impact (case (i)) the kinematic relationship is trivial. Let $\boldsymbol{v}_1$ and $\boldsymbol{v}_2$ be the absolute velocities immediately before impact of the two colliding body-fixed points and let $\Delta\boldsymbol{v}_1$ and $\Delta\boldsymbol{v}_2$ be the finite increments of these velocities as the result of the collision. The condition that the velocity of the two body-fixed points relative to each other is zero immediately after impact reads

$$(\boldsymbol{v}_1 + \Delta\boldsymbol{v}_1) - (\boldsymbol{v}_2 + \Delta\boldsymbol{v}_2) = \boldsymbol{0}\,. \tag{6.3}$$

This is the desired relationship for case (i). Next, case (ii) is considered. In Fig. 6.2 the point of collision is called P. The two colliding bodies belong either to two different systems or to one and the same multi-body system. For the present purpose it does not matter which of these two cases applies. It does not matter either to how many contiguous bodies the two colliding bodies are coupled by hinges. In Fig. 6.2 only one contiguous body is indicated for each of the two bodies. All hinges as well as the material contact between the colliding bodies at the point P are cut. The result is a set of isolated bodies which, at the moment of collision, are subject to impulses and impulse couples at the points where the cuts are made. Each impulse and impulse

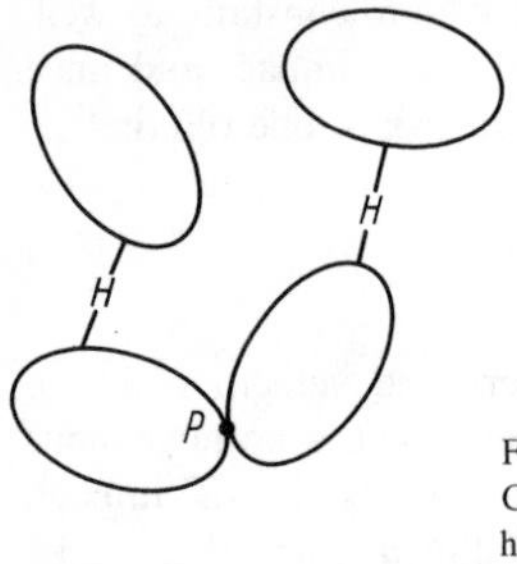

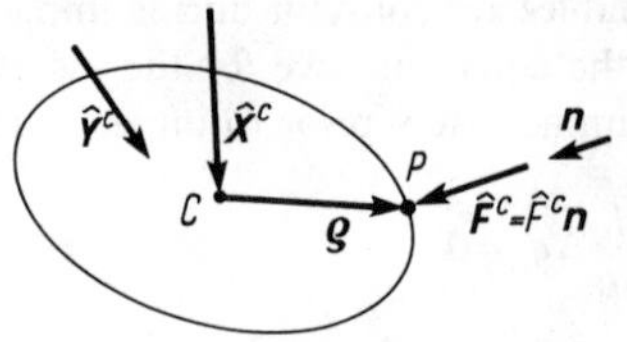

Fig. 6.3 Free-body diagram for one of the colliding bodies of Fig. 6.2

Fig. 6.2
Collision between two bodies which are coupled to other bodies by hinges (indicated by the letter "H")

couple is acting with equal magnitude and opposite direction on two bodies. In Fig. 6.3 one of the colliding bodies is shown with all impulses and impulse couples which act in the phase of compression. At the point of collision the impulse $\hat{\boldsymbol{F}}^c=\hat{F}^c\boldsymbol{n}$ is applied. The unit vector $\boldsymbol{n}$ is normal to the tangent plane at this point. The single impulse $\hat{\boldsymbol{X}}^c$ pointing through the body center of mass C and the single impulse couple $\hat{\boldsymbol{Y}}^c$ are equivalent to the impulses and impulse couples which act in the cut hinge. The three quantities $\hat{\boldsymbol{F}}^c$, $\hat{\boldsymbol{X}}^c$ and $\hat{\boldsymbol{Y}}^c$ are the integrals

$$\int_{t_0}^{t_0+\Delta t^c} \boldsymbol{F}(t)\,\mathrm{d}t\,, \qquad \int_{t_0}^{t_0+\Delta t^c} \boldsymbol{X}(t)\,\mathrm{d}t \quad \text{and} \quad \int_{t_0}^{t_0+\Delta t^c} \boldsymbol{Y}(t)\,\mathrm{d}t\,,$$

respectively, over the time interval Δt^c of the compression phase in the limit $\Delta t^c \to 0$. The quantities $\boldsymbol{F}(t)$, $\boldsymbol{X}(t)$ and $\boldsymbol{Y}(t)$ are impulsive forces and torques whose magnitudes are finite as long as Δt^c is finite. Newton's law and the law of moment of momentum for the body read

$$m\dot{\boldsymbol{v}}_C=\boldsymbol{F}^c(t)+\boldsymbol{X}^c(t), \qquad \boldsymbol{J}\cdot\dot{\boldsymbol{\omega}}+\boldsymbol{\omega}\times\boldsymbol{J}\cdot\boldsymbol{\omega}=\boldsymbol{\varrho}\times\boldsymbol{F}^c(t)+\boldsymbol{Y}^c(t). \tag{6.4}$$

Integration over Δt^c yields in the limit $\Delta t^c \to 0$

$$m\Delta\boldsymbol{v}_C^c=\hat{\boldsymbol{F}}^c+\hat{\boldsymbol{X}}^c, \qquad \boldsymbol{J}\cdot\Delta\boldsymbol{\omega}^c=\boldsymbol{\varrho}\times\hat{\boldsymbol{F}}^c+\hat{\boldsymbol{Y}}^c \tag{6.5}$$

(the term $\boldsymbol{\omega}\times\boldsymbol{J}\cdot\boldsymbol{\omega}$ remains finite so that in the limit the integral is zero). The quantity $\Delta\boldsymbol{v}_C^c$ is the finite increment of the absolute velocity $\boldsymbol{v}_C$ of the body center of mass in the compression phase. Similarly, $\Delta\boldsymbol{\omega}^c$ is the increment of the absolute angular velocity in the compression phase. Also the point of application of $\hat{\boldsymbol{F}}^c$ experiences a velocity increment. It is

$$\Delta\boldsymbol{v}_1^c=\Delta\boldsymbol{v}_C^c+\Delta\boldsymbol{\omega}^c\times\boldsymbol{\varrho}\,. \tag{6.6}$$

This kinematic relationship for rigid bodies is applicable since the velocity of deformation of the body is zero at the beginning as well as at the end of the compression phase. Similar equations are valid for the other colliding body shown in Fig. 6.2. They yield, among other things, the velocity increment $\Delta\boldsymbol{v}_2^c$ of the body-fixed point which coincides with P. At the end of the compression phase the two colliding body-fixed points have a velocity relative to each other whose component along the unit vector $\boldsymbol{n}$ is zero. This can be expressed in the form

$$[(\boldsymbol{v}_1+\Delta\boldsymbol{v}_1^c)-(\boldsymbol{v}_2+\Delta\boldsymbol{v}_2^c)]\cdot\boldsymbol{n}=0 \tag{6.7}$$

where $\boldsymbol{v}_1$ and $\boldsymbol{v}_2$ are the absolute velocities of the two points immediately before impact. The dynamic equations (6.5) as well as the kinematic equations (6.6) and (6.7) are linear equations with constant coefficients for the unknown velocity increments, impulses and impulse couples. This allows the conclusion that for all these unknowns relationships in the form of Eq. (6.1) exist. In particular, $\Delta\boldsymbol{v}_1=(1+e)\Delta\boldsymbol{v}_1^c$ and $\Delta\boldsymbol{v}_2=(1+e)\Delta\boldsymbol{v}_2^c$, are the velocity increments of the two colliding body-fixed points between the moments immediately before and immediately after impact. With this the previous equation becomes

$$\left[\left(\boldsymbol{v}_1+\frac{\Delta\boldsymbol{v}_1}{1+e}\right)-\left(\boldsymbol{v}_2+\frac{\Delta\boldsymbol{v}_2}{1+e}\right)\right]\cdot\boldsymbol{n}=0$$

or

$$[(\boldsymbol{v}_1+\Delta\boldsymbol{v}_1)-(\boldsymbol{v}_2+\Delta\boldsymbol{v}_2)]\cdot\boldsymbol{n}=-e(\boldsymbol{v}_1-\boldsymbol{v}_2)\cdot\boldsymbol{n}\,. \tag{6.8}$$

The quantity on the left hand side is the component along $\boldsymbol{n}$ of the velocity of the two colliding body-fixed points relative to each other immediately after impact. It is $(-e)$ times the same quantity immediately before impact. This is the desired kinematic relationship for the case (ii). It has the same simple form as in the classical theory of collision between two rigid bodies (cf. Routh [36]).

It is now a straightforward procedure to determine all velocity increments of interest which occur when two multi-body systems collide with each other or when two bodies of one and the same multi-body system collide with one another. In Fig. 6.4a, b these two cases are illustrated, again. The colliding bodies are labeled k and l. The collision

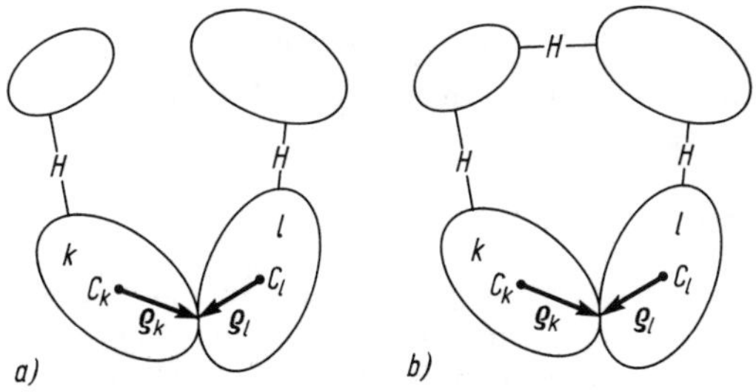

Fig. 6.4
a) Collision between two systems and b) between two bodies belonging to the same system

points on the bodies are specified by body-fixed vectors $\boldsymbol{\varrho}_k$ and $\boldsymbol{\varrho}_l$. These vectors originate from the respective body centers of mass. The interaction impulse $\hat{\boldsymbol{F}}$ which is exchanged between the two colliding bodies is an unknown quantity which must be determined along with the desired velocity increments. The impulse is applied to the two bodies with equal magnitude and opposite direction. The positions and velocities immediately before impact are known for all bodies involved in the process. Let $\Delta\boldsymbol{v}_k$ and $\Delta\boldsymbol{v}_l$ be the increments of the absolute translational velocities of the two body-fixed collision points. It was shown that velocity increments and impulses are related through linear equations with constant coefficients. These coefficients cannot be scalars since an equation of the form $\Delta\boldsymbol{v}_k=\alpha\hat{\boldsymbol{F}}$ with a scalar α would imply that

$\Delta\boldsymbol{v}_k$ has the direction of $\hat{\boldsymbol{F}}$. This is certainely not true, in general. The most general linear function of a vector is a scalar product of this vector with a tensor. An example is the term $\boldsymbol{J}\cdot\Delta\boldsymbol{\omega}^c$ in Eq. (6.5). The equations for $\Delta\boldsymbol{v}_k$ and $\Delta\boldsymbol{v}_l$ are, therefore, written in the form

$$\left.\begin{aligned} \Delta\boldsymbol{v}_k &= \boldsymbol{U}_{kk}\cdot\hat{\boldsymbol{F}} \\ \Delta\boldsymbol{v}_l &= \boldsymbol{U}_{ll}\cdot(-\hat{\boldsymbol{F}}) \end{aligned}\right\} \quad \text{collision between two multi-body systems} \tag{6.9}$$

and

$$\left.\begin{aligned} \Delta\boldsymbol{v}_k &= \boldsymbol{U}_{kk}\cdot\hat{\boldsymbol{F}} + \boldsymbol{U}_{kl}\cdot(-\hat{\boldsymbol{F}}) \\ &= (\boldsymbol{U}_{kk}-\boldsymbol{U}_{kl})\cdot\hat{\boldsymbol{F}} \\ \Delta\boldsymbol{v}_l &= \boldsymbol{U}_{lk}\cdot\hat{\boldsymbol{F}} + \boldsymbol{U}_{ll}\cdot(-\hat{\boldsymbol{F}}) \\ &= -(-\boldsymbol{U}_{lk}+\boldsymbol{U}_{ll})\cdot\hat{\boldsymbol{F}} \end{aligned}\right\} \quad \text{collision within one multi-body system.} \tag{6.10}$$

The tensors $\boldsymbol{U}_{kk}$, $\boldsymbol{U}_{kl}$, $\boldsymbol{U}_{lk}$ and $\boldsymbol{U}_{ll}$ are as yet unknown. They are constants which depend only on the state of the system (or systems) immediately before impact. How they are determined will be shown later. Let us assume, for the moment, that they are known already. Suppose two multi-body systems collide with each other (Fig. 6.4a) and the collision is ideally plastic. The latter assumption means that Eq. (6.3) applies. The indices 1 and 2 in this equation have to be replaced by k and l, respectively. With Eq. (6.9) this yields

$$(\boldsymbol{v}_k + \boldsymbol{U}_{kk}\cdot\hat{\boldsymbol{F}}) - (\boldsymbol{v}_l - \boldsymbol{U}_{ll}\cdot\hat{\boldsymbol{F}}) = \boldsymbol{0}$$

or

$$(\boldsymbol{U}_{kk}+\boldsymbol{U}_{ll})\cdot\hat{\boldsymbol{F}} = -(\boldsymbol{v}_k - \boldsymbol{v}_l)\,.$$

In order to solve this equation for $\hat{\boldsymbol{F}}$ all vectors and tensors are decomposed into scalar coordinates in some common reference frame, for example in the base $\underline{e}^{(0)}$ on body 0. This yields the matrix equation

$$(\underline{U}_{kk}+\underline{U}_{ll})\underline{\hat{F}} = -(\underline{v}_k - \underline{v}_l)$$

with a scalar (3×3) matrix $\underline{U}_{kk}+\underline{U}_{ll}$ and with scalar column matrices $\underline{\hat{F}}$, $\underline{v}_k$ and $\underline{v}_l$. The solution reads

$$\underline{\hat{F}} = -(\underline{U}_{kk}+\underline{U}_{ll})^{-1}(\underline{v}_k - \underline{v}_l) \quad \text{ideally plastic collision between two multi-body systems} \tag{6.11}$$

The inverse exists if $\underline{\hat{F}}$ has a unique solution, i. e. if the problem is physically meaningful. With this result for the interaction impulse the velocity increments $\Delta\boldsymbol{v}_k$ and $\Delta\boldsymbol{v}_l$ can be obtained from Eq. (6.9). Finally, all velocity increments of interest can be determined since they are linear functions of $\hat{\boldsymbol{F}}$. The coefficients in these functions have, of course, still to be formulated. This will be done later. If the collision between two multi-body systems is not ideally plastic but frictionless then Eq. (6.8) applies instead of Eq. (6.3). The unknown impulse $\hat{\boldsymbol{F}}$ has the direction of the unit vector $\boldsymbol{n}$ which allows writing $\hat{\boldsymbol{F}} = \hat{F}\boldsymbol{n}$. Substituting this and Eq. (6.9) into Eq. (6.8) one obtains

$$(\boldsymbol{v}_k - \boldsymbol{v}_l)\cdot\boldsymbol{n} + \hat{F}\boldsymbol{n}\cdot(\boldsymbol{U}_{kk}+\boldsymbol{U}_{ll})\cdot\boldsymbol{n} = -e(\boldsymbol{v}_k - \boldsymbol{v}_l)\cdot\boldsymbol{n}$$

whence follows for the scalar magnitude of $\hat{\boldsymbol{F}}$ the result

$$\hat{F} = -(1+e)\frac{(\boldsymbol{v}_k - \boldsymbol{v}_l)\cdot\boldsymbol{n}}{\boldsymbol{n}\cdot(\boldsymbol{U}_{kk}+\boldsymbol{U}_{ll})\cdot\boldsymbol{n}} \quad \text{frictionless collision between two multi-body systems.} \tag{6.12}$$

What has just been done for the case of collision between two multi-body systems can be repeated analogously for the case of a collision within one multi-body system. Eq. (6.9) is then replaced by Eq. (6.10). This yields the results

$$\hat{\underline{F}} = -(\underline{U}_{kk} - \underline{U}_{kl} - \underline{U}_{lk} + \underline{U}_{ll})^{-1}(\underline{v}_k - \underline{v}_l) \qquad \text{ideally plastic collision within one multi-body system} \tag{6.13}$$

$$\hat{F} = -(1+e)\frac{(\boldsymbol{v}_k - \boldsymbol{v}_l)\cdot\boldsymbol{n}}{\boldsymbol{n}\cdot(\boldsymbol{U}_{kk} - \boldsymbol{U}_{kl} - \boldsymbol{U}_{lk} + \boldsymbol{U}_{ll})\cdot\boldsymbol{n}} \qquad \text{frictionless collision within one multi-body system.} \tag{6.14}$$

We shall now determine the tensors $\boldsymbol{U}_{kk}$, $\boldsymbol{U}_{kl}$, $\boldsymbol{U}_{lk}$ and $\boldsymbol{U}_{ll}$ as well as other tensors which relate velocity increments to impulses. For the problems treated above it would suffice to consider a single multi-body system under the simultaneous action of two impulses $\hat{\boldsymbol{F}}$ and $-\hat{\boldsymbol{F}}$ on two different bodies k and l. This is the case shown in Fig. 6.4b. It includes the special case of Fig. 6.4a where only one impulse is applied to a multi-body system. However, in view of problems which will be treated later a more general situation is considered. Not to only two bodies but to all n bodies of a multi-body system simultaneously acting impulses are applied. The one on body i $(i=1\ldots n)$ is called $\hat{\boldsymbol{F}}_i$. It is applied to a point whose location is specified by the body-fixed vector $\boldsymbol{\varrho}_i$ which originates at the body center of mass C_i (Fig. 6.5). In order to be even more

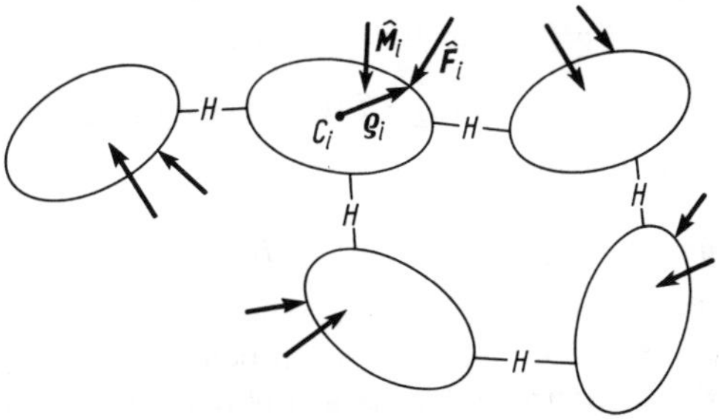

Fig. 6.5
A system with one impulse and one impulse couple applied to each body

general it is assumed that each body i is at the same time subject to an impulse couple $\hat{\boldsymbol{M}}_i$ $(i=1\ldots n)$. As motivation for the formulation of such a problem it may suffice to say that some of the impulses and impulse couples will later be interpreted as internal impulses and impulse couples in cut hinges. The equations to be developed will serve to determine also these unknown quantities. The desired relationships for velocity increments can be developed from the differential equations for continuous motions of multi-body systems. For this purpose the external forces and torques which appear in these equations are assumed to be impulsive forces and impulsive torques which act during a time interval Δt such that their integrals over Δt are the above finite impulses and impulse couples in the limit $\Delta t \to 0$. The desired equations are, therefore, obtained by an integration of the differential equations over the time interval Δt in the limiting case $\Delta t \to 0$. This method has already been used in the development of Eqs. (6.5) from Eqs. (6.4). The most general set of differential equations of motion is Eq. (5.204):

$$\ddot{\underline{q}} = \underline{A}^{-1}\{\underline{B} + \underline{B}^* - \underline{H}^{\mathrm{T}}(\underline{H}\,\underline{A}^{-1}\underline{H}^{\mathrm{T}})^{-1}[\underline{H}\,\underline{A}^{-1}(\underline{B} + \underline{B}^*) + \underline{\varphi}^*]\}\,. \tag{6.15}$$

It is valid for systems with closed chains. For systems without closed kinematic chains the matrix $\underline{H}$ does not exist, and the equation has the simpler form

$$\ddot{\underline{q}}=\underline{A}^{-1}(\underline{B}+\underline{B}^*)$$

where $\underline{B}^*$ accounts for eventually existing closed non-kinematic chains. The various matrices are defined by Eqs. (5.171), (5.172), (5.213), (5.189) and (5.201). External forces and torques on the bodies of the system appear in the matrix $\underline{B}$ only. This matrix can be written in the form

$$\underline{B}=\underline{B}_0+[\underline{p}\times\underline{T}(\underline{C}+\underline{Z})\underline{T}-\underline{k}\,\underline{T}]\cdot\underline{\boldsymbol{F}}-\underline{p}\,\underline{T}\cdot\underline{\boldsymbol{M}} \tag{6.16}$$

where $\underline{B}_0$ is an abbreviation for the sum of all terms which are independent of the external forces and torques. The coefficient matrices of $\underline{\boldsymbol{F}}$ and $\underline{\boldsymbol{M}}$ are based on the convention that the line of action of the force $\boldsymbol{F}_i$ on body i is passing through the body i center of mass (see Fig. 5.11). In the present case, however, $\boldsymbol{F}_i$ is required to act at a point which is located by the vector $\boldsymbol{\varrho}_i$ (Fig. 6.5). In order to adapt the equation for $\underline{B}$ to this a torque $\boldsymbol{\varrho}_i\times\boldsymbol{F}_i$ must be added to $\boldsymbol{M}_i$. The column matrix $[\boldsymbol{\varrho}_1\times\boldsymbol{F}_1\ldots\boldsymbol{\varrho}_n\times\boldsymbol{F}_n]^{\mathrm{T}}$ of these torques is written in the form $\underline{\boldsymbol{\varrho}}\times\underline{\boldsymbol{F}}$ where $\underline{\boldsymbol{F}}$ is the same matrix as in Eq. (6.16) and $\underline{\boldsymbol{\varrho}}$ is an $(n\times n)$ diagonal matrix with the elements

$$\boldsymbol{\varrho}_{ij}=\delta_{ij}\boldsymbol{\varrho}_i \qquad i,j=1\ldots n\,.$$

In Eq. (6.16) $\underline{\boldsymbol{M}}$ is now replaced by $\underline{\boldsymbol{M}}+\underline{\boldsymbol{\varrho}}\times\underline{\boldsymbol{F}}$. Because of the identity $\underline{p}\,\underline{T}\cdot\underline{\boldsymbol{\varrho}}\times\underline{\boldsymbol{F}}=\underline{p}\times\underline{T}\,\underline{\boldsymbol{\varrho}}\cdot\underline{\boldsymbol{F}}$ the matrix $\underline{B}$ becomes

$$\underline{B}=\underline{B}_0+\{\underline{p}\times\underline{T}[(\underline{C}+\underline{Z})\underline{T}-\underline{\boldsymbol{\varrho}}]-\underline{k}\,\underline{T}\}\cdot\underline{\boldsymbol{F}}-\underline{p}\,\underline{T}\cdot\underline{\boldsymbol{M}}\,. \tag{6.17}$$

With this Eq. (6.15) reads

$$\ddot{\underline{q}}=[\underline{A}^{-1}-\underline{A}^{-1}\underline{H}^{\mathrm{T}}(\underline{H}\,\underline{A}^{-1}\underline{H}^{\mathrm{T}})^{-1}\underline{H}\,\underline{A}^{-1}]\{\{\underline{p}\times\underline{T}[(\underline{C}+\underline{Z})\underline{T}-\underline{\boldsymbol{\varrho}}]-\underline{k}\,\underline{T}\}\cdot\underline{\boldsymbol{F}}-\underline{p}\,\underline{T}\cdot\underline{\boldsymbol{M}}\}+\underline{B}_1\,.$$

The column matrix $\underline{B}_1$ combines all terms which are independent of the external forces and torques. It is a function of the generalized coordinates of the system, of the generalized velocities and of time. It is, therefore, finite at all times, even at the moment of an impact on the system. This means that the integral of $\underline{B}_1$ over the time interval Δt is zero in the limit $\Delta t\to 0$. The coefficient matrices of $\underline{\boldsymbol{F}}$ and $\underline{\boldsymbol{M}}$ are functions of the generalized coordinates and of time, but not of the generalized velocities. They can, consequently, be treated as constants when the equations are integrated over Δt with $\Delta t\to 0$. Hence, this integration yields

$$\Delta\dot{\underline{q}}=[\underline{A}^{-1}-\underline{A}^{-1}\underline{H}^{\mathrm{T}}(\underline{H}\,\underline{A}^{-1}\underline{H}^{\mathrm{T}})^{-1}\underline{H}\,\underline{A}^{-1}]\{\{\underline{p}\times\underline{T}[(\underline{C}+\underline{Z})\underline{T}-\underline{\boldsymbol{\varrho}}]-\underline{k}\,\underline{T}\}\cdot\hat{\underline{\boldsymbol{F}}}-\underline{p}\,\underline{T}\cdot\hat{\underline{\boldsymbol{M}}}\}\,. \tag{6.18}$$

In the absence of closed kinematic chains in the system this relationship has the simpler form

$$\Delta\dot{\underline{q}}=\underline{A}^{-1}\{\{\underline{p}\times\underline{T}[(\underline{C}+\underline{Z})\underline{T}-\underline{\boldsymbol{\varrho}}]-\underline{k}\,\underline{T}\}\cdot\hat{\underline{\boldsymbol{F}}}-\underline{p}\,\underline{T}\cdot\hat{\underline{\boldsymbol{M}}}\}\,. \tag{6.19}$$

These equations establish the desired relationship between the increments of the generalized velocities and the external impulses $\hat{\boldsymbol{F}}_1\ldots\hat{\boldsymbol{F}}_n$ and impulse couples $\hat{\boldsymbol{M}}_1\ldots\hat{\boldsymbol{M}}_n$, respectively. It is now a simple matter to determine the increment of any other velocity in the system. Eq. (5.130), for instance, relates the absolute angular velocities of the bodies to $\dot{\underline{q}}$. From this follows

$$\Delta\underline{\omega} = -(\underline{p}\,\underline{T})^{\mathrm{T}}\Delta\underline{\dot{q}}\,. \tag{6.20}$$

Likewise, Eq. (5.158) yields, when integrated over Δt with $\Delta t \to 0$,

$$\Delta\underline{\dot{r}} = [\underline{p}\times\underline{T}(\underline{C}+\underline{Z})\underline{T} - \underline{k}\,\underline{T}]^{\mathrm{T}}\Delta\underline{\dot{q}}\,. \tag{6.21}$$

The column matrix on the left hand side contains the increments of the absolute velocities of the body centers of mass. Finally, the velocity increments of the points of application of the impulses can be determined. They are called $\Delta\boldsymbol{v}_i$ $(i=1\ldots n)$. The kinematic equations

$$\Delta\boldsymbol{v}_i = \Delta\dot{\boldsymbol{r}}_i - \boldsymbol{\varrho}_i\times\Delta\boldsymbol{\omega}_i \qquad i=1\ldots n \tag{6.22}$$

are combined in the matrix equation

$$\Delta\underline{\boldsymbol{v}} = \Delta\underline{\dot{r}} - \underline{\varrho}\times\Delta\underline{\omega}\,. \tag{6.23}$$

The matrix $\underline{\varrho}$ is the same diagonal matrix as in Eq. (6.18), and $\Delta\underline{\boldsymbol{v}}$ is the column matrix $[\Delta\boldsymbol{v}_1\ldots\Delta\boldsymbol{v}_n]^{\mathrm{T}}$. With $\Delta\underline{\dot{r}}$ and $\Delta\underline{\omega}$ from Eqs. (6.21) and (6.20) one gets

$$\Delta\underline{\boldsymbol{v}} = \{\underline{p}\times\underline{T}[(\underline{C}+\underline{Z})\underline{T}-\underline{\varrho}] - \underline{k}\,\underline{T}\}^{\mathrm{T}}\Delta\underline{\dot{q}} \tag{6.24}$$

or with Eq. (6.18)

$$\Delta\underline{\boldsymbol{v}} = \underline{\boldsymbol{U}}\cdot\underline{\hat{\boldsymbol{F}}} + \underline{\boldsymbol{V}}\cdot\underline{\hat{\boldsymbol{M}}} \tag{6.25}$$

where $\underline{\boldsymbol{U}}$ and $\underline{\boldsymbol{V}}$ are the tensorial matrices

$$\begin{aligned}\underline{\boldsymbol{U}} = \{&\underline{p}\times\underline{T}[(\underline{C}+\underline{Z})\underline{T}-\underline{\varrho}]-\underline{k}\,\underline{T}\}^{\mathrm{T}}[\underline{A}^{-1} - \\ &-\underline{A}^{-1}\underline{H}^{\mathrm{T}}(\underline{H}\,\underline{A}^{-1}\underline{H}^{\mathrm{T}})^{-1}\underline{H}\,\underline{A}^{-1}]\{\underline{p}\times\underline{T}[(\underline{C}+\underline{Z})\underline{T}-\underline{\varrho}]-\underline{k}\,\underline{T}\}\end{aligned} \tag{6.26}$$

$$\underline{\boldsymbol{V}} = -\{\underline{p}\times\underline{T}[(\underline{C}+\underline{Z})\underline{T}-\underline{\varrho}]-\underline{k}\,\underline{T}\}^{\mathrm{T}}[\underline{A}^{-1}-\underline{A}^{-1}\underline{H}^{\mathrm{T}}(\underline{H}\,\underline{A}^{-1}\underline{H}^{\mathrm{T}})^{-1}\underline{H}\,\underline{A}^{-1}](\underline{p}\,\underline{T}). \tag{6.27}$$

The matrix $\underline{\boldsymbol{U}}$ is conjugate symmetric, i. e. its elements have the property $\boldsymbol{U}_{ji} = \bar{\boldsymbol{U}}_{ij}$ (conjugate of $\boldsymbol{U}_{ij}$) for $i, j = 1\ldots n$. For the matrix $\underline{\boldsymbol{U}}$ itself this can be expressed in the symbolic form $\underline{\bar{\boldsymbol{U}}}^{\mathrm{T}} = \underline{\boldsymbol{U}}$. If the multi-body system under consideration has no closed kinematic chain then the term $\underline{A}^{-1}\underline{H}^{\mathrm{T}}(\underline{H}\,\underline{A}^{-1}\underline{H}^{\mathrm{T}})^{-1}\underline{H}\,\underline{A}^{-1}$ in the central matrix in the equations for $\underline{\boldsymbol{U}}$ and $\underline{\boldsymbol{V}}$ must be omitted.

Let us now return to the problem of collision between two bodies belonging to either two different systems or to one and the same multi-body system. The tensors $\boldsymbol{U}_{kk}$, $\boldsymbol{U}_{kl}$, $\boldsymbol{U}_{lk}$ and $\boldsymbol{U}_{ll}$ in Eqs. (6.11) to (6.14) still had to be determined. They were defined by Eqs. (6.9) and (6.10) as elements of the coefficient matrix in the relationship between $\Delta\underline{\boldsymbol{v}}$ and $\underline{\hat{\boldsymbol{F}}}$. Hence, they are elements of the conjugate symmetric matrix $\underline{\boldsymbol{U}}$ defined above. If two bodies of one and the same multi-body system collide all four coefficients are elements of the matrix $\underline{\boldsymbol{U}}$ for this system. If, on the other hand, two bodies k and l belonging to two different systems collide then a matrix $\underline{\boldsymbol{U}}$ has to be calculated for each system separately. One matrix furnishes $\boldsymbol{U}_{kk}$ and the other $\boldsymbol{U}_{ll}$. With these tensors the interaction impulse $\hat{\boldsymbol{F}}$ can be determined from one of the equations (6.11) to (6.14). With it also the matrix $\underline{\hat{\boldsymbol{F}}}$ in Eq. (6.18) is known. The matrix $\underline{\hat{\boldsymbol{M}}}$ is zero. The equation yields the velocity increments $\Delta\underline{\dot{q}}$. From Eqs. (6.20), (6.21) and (6.24) follow then $\Delta\underline{\omega}$, $\Delta\underline{\dot{r}}$ and $\Delta\underline{\boldsymbol{v}}$. With these results all velocity increments of interest are known.

6.3 An analogy to the law of Maxwell and Betti

The discussion of the general problem with impulses and impulse couples on all bodies will now be continued. When the expression for $\Delta\underline{\dot{q}}$ in Eq. (6.18) is substituted into Eq. (6.20) $\Delta\underline{\omega}$ turns out to be

$$\Delta\underline{\omega} = \underline{\bar{V}}^{\mathrm{T}} \cdot \underline{\hat{F}} + \underline{W} \cdot \underline{\hat{M}} .$$

The coefficient matrix of $\underline{\hat{F}}$ is the conjugate transpose of the matrix $\underline{V}$ in Eq. (6.25), i.e. its elements are $(\underline{\bar{V}}^{\mathrm{T}})_{ij} = \bar{V}_{ji}$. The tensorial matrix $\underline{W}$ is

$$\underline{W} = (\underline{p}\,\underline{T})^{\mathrm{T}} [\underline{A}^{-1} - \underline{A}^{-1} \underline{H}^{\mathrm{T}} (\underline{H}\,\underline{A}^{-1} \underline{H}^{\mathrm{T}})^{-1} \underline{H}\,\underline{A}^{-1}] (\underline{p}\,\underline{T}). \tag{6.28}$$

Like $\underline{U}$ it is conjugate symmetric. When the equations for $\Delta\underline{v}$ and $\Delta\underline{\omega}$ are combined in the form

$$\begin{bmatrix} \Delta\underline{v} \\ \Delta\underline{\omega} \end{bmatrix} = \begin{bmatrix} \underline{U} & \underline{V} \\ \underline{\bar{V}}^{\mathrm{T}} & \underline{W} \end{bmatrix} \cdot \begin{bmatrix} \underline{\hat{F}} \\ \underline{\hat{M}} \end{bmatrix} \tag{6.29}$$

the coefficient matrix on the right hand side also is conjugate symmetric. Consider the associated $(6n \times 6n)$ coordinate matrix equation which results when all vectors and tensors are replaced by their respective coordinate matrices measured in a common reference frame. This coefficient matrix is symmetric. The symmetry establishes an analogy between rigid body dynamics and elastostatics[1]. In elastostatics the following problem is considered. A linearly elastic structure which is in a state of equilibrium is subject to forces and couples at a number of points $P_1 \ldots P_n$, one force $\boldsymbol{F}_i$ and one couple $\boldsymbol{M}_i$ at each point P_i. There exists a linear relationship of the form

$$\begin{bmatrix} \underline{u} \\ \underline{\phi} \end{bmatrix} = \begin{bmatrix} \underline{A}_{11} & \underline{A}_{12} \\ \underline{A}_{21} & \underline{A}_{22} \end{bmatrix} \cdot \begin{bmatrix} \underline{F} \\ \underline{M} \end{bmatrix} . \tag{6.30}$$

The column matrices $\underline{F}$, $\underline{M}$, $\underline{u}$ and $\underline{\phi}$ are composed of n vectors each. $\underline{F} = [\boldsymbol{F}_1 \ldots \boldsymbol{F}_n]^{\mathrm{T}}$ and $\underline{M} = [\boldsymbol{M}_1 \ldots \boldsymbol{M}_n]^{\mathrm{T}}$ contain the forces and couples, respectively, $\underline{u}$ the displacements of the points $P_1 \ldots P_n$ from their locations in the unloaded state and $\underline{\phi}$ the rotation angles of the system at the points $P_1 \ldots P_n$ (small rotation angles can be treated as vectors). The law of Maxwell and Betti states that in the associated coordinate matrix equation the coefficient matrix is symmetric. This means that the tensorial coefficient matrix in Eq. (6.30) is conjugate symmetric, i.e. $\underline{A}_{11} = \underline{\bar{A}}_{11}^{\mathrm{T}}$, $\underline{A}_{21} = \underline{\bar{A}}_{12}^{\mathrm{T}}$, $\underline{A}_{22} = \underline{\bar{A}}_{22}^{\mathrm{T}}$. The equation is, thus, analogous to Eq. (6.29). Not only are the coefficient matrices in both equations conjugate symmetric. There exist also close relationships between the vectorial quantities in corresponding column matrices. This is best illustrated by writing the definitions of impulse, impulse couple, velocity increment and angular velocity increment in the form

$$\hat{\boldsymbol{F}}_i = \lim_{\Delta t \to 0} \int_{t_0}^{t_0+\Delta t} \boldsymbol{F}_i(t)\,\mathrm{d}t\,, \qquad \hat{\boldsymbol{M}}_i = \lim_{\Delta t \to 0} \int_{t_0}^{t_0+\Delta t} \boldsymbol{M}_i(t)\,\mathrm{d}t$$

[1] This analogy was first described in ref. [37]. There, the investigation was limited to systems with tree structure and to the relationship $\Delta\underline{v} = \underline{U} \cdot \underline{\hat{F}}$.

$$\Delta \boldsymbol{v}_i = \lim_{\Delta t \to 0} \left(\left. \frac{\mathrm{d}\boldsymbol{u}_i}{\mathrm{d}t} \right|_{t_0+\Delta t} - \left. \frac{\mathrm{d}\boldsymbol{u}_i}{\mathrm{d}t} \right|_{t_0} \right), \qquad \Delta \boldsymbol{\omega}_i = \lim_{\Delta t \to 0} \left(\left. \frac{\mathrm{d}\boldsymbol{\phi}_i}{\mathrm{d}t} \right|_{t_0+\Delta t} - \left. \frac{\mathrm{d}\boldsymbol{\phi}_i}{\mathrm{d}t} \right|_{t_0} \right).$$

The quantities $\boldsymbol{F}_i$, $\boldsymbol{M}_i$, $\boldsymbol{u}_i$ and $\boldsymbol{\phi}_i$ in these equations are forces, torques, displacements and rotation angles in the multi-body system which are defined exactly like the quantities of equal name in the elastic system. Note, however, the following difference between the two problems. The law of Maxwell and Betti is obtained from energy considerations. These do not provide a closed form expression for the matrix in Eq. (6.30). In fact, a closed form expression is unknown for all but the simplest types of elastic structures (trusses, for instance). In contrast, for the matrix in Eq. (6.29) a closed-form expression has been obtained in Eqs. (6.26) to (6.28) which encompasses all holonomic multi-body systems.

It will now be demonstrated that the energy considerations which lead to the law of Maxwell and Betti can be phrased in a completely analogous form so as to prove the conjugate-symmetry of the matrix in Eq. (6.29). It suffices to consider the following problem. In an arbitrary holonomic multi-body system two impulses $\hat{\boldsymbol{F}}_1$ and $\hat{\boldsymbol{F}}_2$ are applied to two arbitrarily chosen bodies labeled 1 and 2 in points P_1 and P_2, respectively (Fig. 6.6). The bodies 1 and 2 are at the same time also subject to impulse

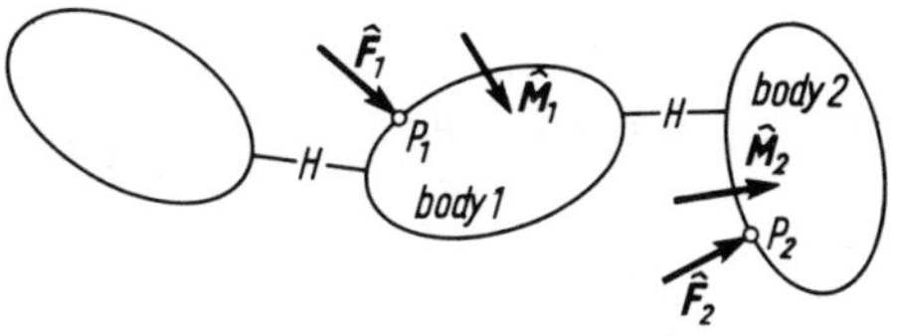

Fig. 6.6
A system under the action of two impulses and impulse couples

couples $\hat{\boldsymbol{M}}_1$ and $\hat{\boldsymbol{M}}_2$, respectively. The scalar magnitudes of the impulses and impulse couples are $\hat{F}_i$ and $\hat{M}_i$ $(i=1,2)$, respectively. The point P_i $(i=1,2)$ experiences a velocity increment. The scalar magnitude of the component of this increment in the direction of $\hat{\boldsymbol{F}}_i$ is called Δv_i. Similarly, $\Delta\omega_i$ $(i=1,2)$ is the scalar magnitude of the component along $\hat{\boldsymbol{M}}_i$ of the absolute angular velocity increment of body i. The scalar quantities are related by an equation

$$\begin{bmatrix} \Delta v_1 \\ \Delta v_2 \\ \Delta \omega_1 \\ \Delta \omega_2 \end{bmatrix} = \begin{bmatrix} a_{11} & a_{12} & a_{13} & a_{14} \\ a_{21} & a_{22} & a_{23} & a_{24} \\ a_{31} & a_{32} & a_{33} & a_{34} \\ a_{41} & a_{42} & a_{43} & a_{44} \end{bmatrix} \begin{bmatrix} \hat{F}_1 \\ \hat{F}_2 \\ \hat{M}_1 \\ \hat{M}_2 \end{bmatrix} \tag{6.31}$$

in which—it is supposed—nothing is known about the coefficient matrix. It is to be shown that the coefficients satisfy the symmetry conditions

$$a_{ij} = a_{ji} \qquad i,j=1,2,3,4\,. \tag{6.32}$$

With this proof also the conjugate-symmetry of the matrix in Eq. (6.29) is proven since Eq. (6.31) holds for arbitrary locations of the points P_1 and P_2 as well as for arbitrary directions of the impulses and impulse couples. The validity of Eq. (6.32) will be shown by a comparison of work expressions encountered in two experiments.

The first experiment is as follows. The impulse $\hat{F}_1$ and the impulse couple $\hat{M}_1$ are replaced by an impulsive force F_1 and an impulsive torque M_1, respectively, which act with finite and constant magnitudes during the finite time interval between t_0 and $t_0+\Delta t$. Immediately following, between $t_0+\Delta t$ and $t_0+2\Delta t$, are an impulsive force F_2 and an impulsive torque M_2, both of finite and constant magnitude, which replace $\hat{F}_2$ and $\hat{M}_2$. The forces and torques are chosen such that their integrals over Δt equal the given impulses and impulse couples, respectively. For the scalar magnitudes this means

$$F_i=\frac{\hat{F}_i}{\Delta t}, \qquad M_i=\frac{\hat{M}_i}{\Delta t} \qquad i=1,2.$$

In Fig. 6.7a the function $v_1(t)$ is shown. It is the scalar magnitude of the component along $\hat{F}_1$ of the absolute velocity of the point P_1. It has a certain given initial value v_{10} at $t=t_0$. In the limit $\Delta t\to 0$ its values at $t=t_0+\Delta t$ and $t=t_0+2\Delta t$ are according to Eq. (6.31) $v_{10}+a_{11}\hat{F}_1+a_{13}\hat{M}_1$ and $v_{10}+a_{11}\hat{F}_1+a_{13}\hat{M}_1+a_{12}\hat{F}_2+a_{14}\hat{M}_2$, respectively. The piecewise linear dependency of v_1 on time shown in the figure requires that the position of the system remains unchanged throughout the interval from t_0 to $t_0+2\Delta t$. The smaller Δt the more accurately this condition is fulfilled. In Fig.

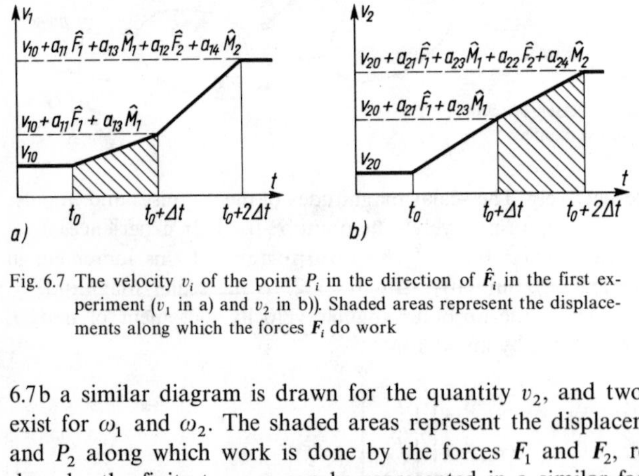

Fig. 6.7 The velocity v_i of the point P_i in the direction of $\hat{F}_i$ in the first experiment (v_1 in a) and v_2 in b)). Shaded areas represent the displacements along which the forces F_i do work

6.7b a similar diagram is drawn for the quantity v_2, and two more such diagrams exist for ω_1 and ω_2. The shaded areas represent the displacements of the points P_1 and P_2 along which work is done by the forces F_1 and F_2, respectively. The work done by the finite torques can be represented in a similar fashion in the diagrams for ω_1 and ω_2. The total work done in the time interval between t_0 and $t_0+2\Delta t$ is

$$W_2=F_1\left[v_{10}+\frac{(a_{11}\hat{F}_1+a_{13}\hat{M}_1)}{2}\right]\Delta t+M_1\left[\omega_{10}+\frac{(a_{31}\hat{F}_1+a_{33}\hat{M}_1)}{2}\right]\Delta t+$$
$$+F_2\left[v_{20}+(a_{21}\hat{F}_1+a_{23}\hat{M}_1)+\frac{(a_{22}\hat{F}_2+a_{24}\hat{M}_2)}{2}\right]\Delta t+$$
$$+M_2\left[\omega_{20}+(a_{41}\hat{F}_1+a_{43}\hat{M}_1)+\frac{(a_{42}\hat{F}_2+a_{44}\hat{M}_2)}{2}\right]\Delta t\,.$$

In the limit $\Delta t\to 0$ this work is

$$W_1 = \hat{F}_1 v_{10} + \hat{F}_2 v_{20} + \hat{M}_1 \omega_{10} + \hat{M}_2 \omega_{20} + \hat{F}_1 \frac{(a_{11}\hat{F}_1 + a_{13}\hat{M}_1)}{2} + \hat{M}_1 \frac{(a_{31}\hat{F}_1 + a_{33}\hat{M}_1)}{2} +$$

$$+ \hat{F}_2 \left[(a_{21}\hat{F}_1 + a_{23}\hat{M}_1) + \frac{(a_{22}\hat{F}_2 + a_{24}\hat{M}_2)}{2} \right] +$$

$$+ \hat{M}_2 \left[(a_{41}\hat{F}_1 + a_{43}\hat{M}_1) + \frac{(a_{42}\hat{F}_2 + a_{44}\hat{M}_2)}{2} \right].$$

In the second experiment the system is—in the same initial state at $t = t_0$—subjected to the same finite forces and torques as in the first experiment. This time, however, the bodies are loaded in reversed order, i. e. $\boldsymbol{F}_2$ and $\boldsymbol{M}_2$ are acting between t_0 and $t_0 + \Delta t$ and $\boldsymbol{F}_1$ and $\boldsymbol{M}_1$ between $t_0 + \Delta t$ and $t_0 + 2\Delta t$. The work done in this process is, in the limit $\Delta t \to 0$,

$$W_2 = \hat{F}_1 v_{10} + \hat{F}_2 v_{20} + \hat{M}_1 \omega_{10} + \hat{M}_2 \omega_{20} + \hat{F}_1 \left[(a_{12}\hat{F}_2 + a_{14}\hat{M}_2) + \frac{(a_{11}\hat{F}_1 + a_{13}\hat{M}_1)}{2} \right] +$$

$$+ \hat{M}_1 \left[(a_{32}\hat{F}_2 + a_{34}\hat{M}_2) + \frac{(a_{31}\hat{F}_1 + a_{33}\hat{M}_1)}{2} \right] +$$

$$+ \hat{F}_2 \frac{(a_{22}\hat{F}_2 + a_{24}\hat{M}_2)}{2} + \hat{M}_2 \frac{(a_{42}\hat{F}_2 + a_{44}\hat{M}_2)}{2}.$$

Because of the linearity of Eq. (6.31) the principle of superposition is applicable, whence follows the identity of W_1 and W_2. This yields

$$\hat{F}_1(a_{12}\hat{F}_2 + a_{14}\hat{M}_2) + \hat{M}_1(a_{32}\hat{F}_2 + a_{34}\hat{M}_2) = \hat{F}_2(a_{21}\hat{F}_1 + a_{23}\hat{M}_1) + \hat{M}_2(a_{41}\hat{F}_1 + a_{43}\hat{M}_1).$$

The four quantities $\hat{F}_1$, $\hat{F}_2$, $\hat{M}_1$ and $\hat{M}_2$ are independent of each other. Equating two of them to zero and doing this for all possible combinations one obtains the desired symmetry relationship of Eq. (6.32). End of proof[1].

Prior to Eq. (6.31) it was assumed that the bodies 1 and 2 are different. However, as it turns out this assumption played no role in the proof just given. It is, therefore, possible to generalize Eq. (6.29) still further so as to cover also the following case. On each body i $(i = 1 \ldots n)$ of a holonomic multi-body system one impulse couple $\hat{\boldsymbol{M}}_i$ and $\nu_i \geq 0$ impulses $\hat{\boldsymbol{F}}_{i1} \ldots \hat{\boldsymbol{F}}_{i\nu_i}$ act simultaneously. The impulses are applied to different points whose locations are specified by the body-fixed vectors $\boldsymbol{\varrho}_{i1} \ldots \boldsymbol{\varrho}_{i\nu_i}$. As an example see the body k with $\nu_k = 2$ impulses in Fig. 6.8. The absolute velocity in-

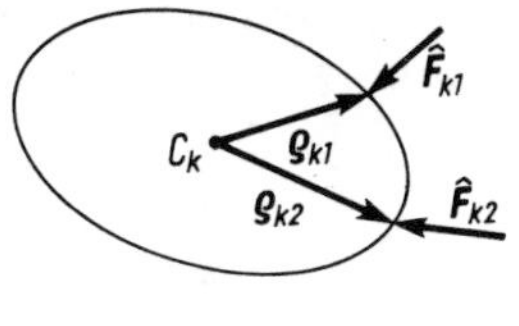

Fig. 6.8
A body with two impulses applied at different points

[1] The way in which impulses and finite impulsive forces during finite time intervals are manipulated in this proof as well as in some other places of this chapter is unsatisfactory from a mathematical point of view. For rigorous formulations the theory of distributions would have to be applied. An introduction to this theory is, however, beyond the scope of this book.

crements of the points of application of the impulses are called $\Delta\boldsymbol{v}_{i1}\ldots\Delta\boldsymbol{v}_{i\nu_i}$ $(i=1\ldots n)$. Define the column matrices

$$\underline{\hat{F}}^* = [\,\hat{\boldsymbol{F}}_{11}\ldots\hat{\boldsymbol{F}}_{1\nu_1} \quad \hat{\boldsymbol{F}}_{21}\ldots\,\hat{\boldsymbol{F}}_{2\nu_2}\,\ldots\ldots\,\hat{\boldsymbol{F}}_{n1}\ldots\,\hat{\boldsymbol{F}}_{n\nu_n}]^{\mathrm{T}}$$
$$\Delta\underline{v}^* = [\Delta\boldsymbol{v}_{11}\ldots\Delta\boldsymbol{v}_{1\nu_1} \quad \Delta\boldsymbol{v}_{21}\ldots\Delta\boldsymbol{v}_{2\nu_2}\ldots\ldots\Delta\boldsymbol{v}_{n1}\ldots\Delta\boldsymbol{v}_{n\nu_n}]^{\mathrm{T}}.$$

They have $\sigma=\sum_{i=1}^{n}\nu_i$ elements each. The matrices $\underline{\hat{M}}=[\hat{\boldsymbol{M}}_1\ldots\hat{\boldsymbol{M}}_n]^{\mathrm{T}}$ and $\Delta\underline{\omega}=[\Delta\boldsymbol{\omega}_1\ldots\Delta\boldsymbol{\omega}_n]^{\mathrm{T}}$ are defined as before. Eq. (6.32) is proof that also the coefficient matrix relating the column matrices

$$\begin{bmatrix}\underline{\hat{F}}^*\\ \underline{\hat{M}}\end{bmatrix} \text{ and } \begin{bmatrix}\Delta\underline{v}^*\\ \Delta\underline{\omega}\end{bmatrix}$$

is symmetric. An explicit expression can be developed for this matrix by a simple modification of the chain of arguments which led from Eq. (6.15) to Eq. (6.29). In Eq. (6.17) for $\underline{B}$ the matrix $\underline{F}$ is

$$\underline{F} = \begin{bmatrix}\boldsymbol{F}_{11}+\boldsymbol{F}_{12}+\cdots+\boldsymbol{F}_{1\nu_1}\\ \vdots\\ \boldsymbol{F}_{n1}+\boldsymbol{F}_{n2}+\cdots+\boldsymbol{F}_{n\nu_n}\end{bmatrix}.$$

The product $[\underline{p}\times\underline{T}(\underline{C}+\underline{Z})\underline{T}-\mathbf{k}\underline{T}]\cdot\underline{F}$—abbreviated $\underline{P}\cdot\underline{F}$—can be rewritten as a product $\underline{P}^*\cdot\underline{F}^*$ of a column matrix

$$\underline{F}^* = [\boldsymbol{F}_{11}\ldots\boldsymbol{F}_{1\nu_1} \quad \boldsymbol{F}_{21}\ldots\boldsymbol{F}_{2\nu_2}\ldots\ldots\boldsymbol{F}_{n1}\ldots\boldsymbol{F}_{n\nu_n}]^{\mathrm{T}}$$

with σ elements and an $(n\times\sigma)$ matrix $\underline{P}^*$ which has the structure

$$\underline{P}^* = \Big[\underbrace{[\quad][\quad]\cdots[\quad]}_{\substack{\nu_1 \text{ times the}\\ \text{1st column of } \underline{P}}}\cdots\cdots\cdots\cdots\underbrace{[\quad][\quad]\cdots[\quad]}_{\substack{\nu_n \text{ times the}\\ n\text{-th column of } \underline{P}}}\Big]. \tag{6.33}$$

The matrix $\underline{M}$ in Eq. (6.17) is replaced by

$$\underline{M} + \begin{bmatrix}\boldsymbol{\varrho}_{11}\times\boldsymbol{F}_{11}+\cdots+\boldsymbol{\varrho}_{1\nu_1}\times\boldsymbol{F}_{1\nu_1}\\ \boldsymbol{\varrho}_{n1}\times\boldsymbol{F}_{n1}+\cdots+\boldsymbol{\varrho}_{n\nu_n}\times\boldsymbol{F}_{n\nu_n}\end{bmatrix}.$$

This can be rewritten in the form $\underline{M}+\underline{\varrho}^*\times\underline{F}^*$ where $\underline{\varrho}^*$ is the quasi-diagonal $(n\times\sigma)$ matrix

$$\underline{\varrho}^* = \begin{bmatrix}\boldsymbol{\varrho}_{11}\cdots\boldsymbol{\varrho}_{1\nu_1} & & & \underline{0}\\ & \boldsymbol{\varrho}_{21}\cdots\boldsymbol{\varrho}_{2\nu_2} & & \\ & & \ddots & \\ \underline{0} & & & \boldsymbol{\varrho}_{n1}\cdots\boldsymbol{\varrho}_{n\nu_n}\end{bmatrix}.$$

In terms of these quantities Eq. (6.17) becomes

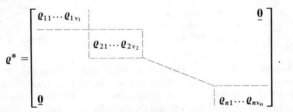

$$\underline{B}=\underline{B}_0+(\underline{\boldsymbol{P}}^*-\underline{p}\times\underline{T}\,\underline{\varrho}^*)\cdot\underline{F}^*-\underline{p}\,\underline{T}\cdot\underline{M}\,.$$

This leads to

$$\Delta\underline{\dot{q}}=[\underline{A}^{-1}-\underline{A}^{-1}\underline{H}^{\mathrm{T}}(\underline{H}\,\underline{A}^{-1}\underline{H}^{\mathrm{T}})^{-1}\underline{H}\,\underline{A}^{-1}]\,[(\underline{\boldsymbol{P}}^*-\underline{p}\times\underline{T}\,\underline{\varrho}^*)\cdot\underline{\hat{F}}^*-\underline{p}\,\underline{T}\cdot\underline{\hat{M}}] \tag{6.34}$$

instead of to Eq. (6.18). Eqs. (6.20) to (6.22) remain unchanged, whereas Eq. (6.23) must be replaced by

$$\Delta\underline{v}^*=\Delta\underline{\dot{r}}^*-\underline{\varrho}^{*\mathrm{T}}\times\Delta\underline{\omega}\,. \tag{6.35}$$

(the transposition symbol in $\underline{\varrho}^{*\mathrm{T}}$ is not required in Eq. (6.23) because the matrix $\underline{\varrho}$ is diagonal). The column matrix $\Delta\underline{\dot{r}}^*$ is

$$\Delta\underline{\dot{r}}^*=[\underbrace{\Delta\dot{r}_1\ldots\Delta\dot{r}_1}_{\nu_1\text{ times}}\;\ldots\;\underbrace{\Delta\dot{r}_n\ldots\Delta\dot{r}_n}_{\nu_n\text{ times}}]^{\mathrm{T}}.$$

From Eqs. (6.21) and (6.33) follows

$$\Delta\underline{\dot{r}}^*=\underline{\boldsymbol{P}}^{*\mathrm{T}}\Delta\underline{\dot{q}}\,.$$

Substituting this and the expressions for $\Delta\underline{\omega}$ and $\Delta\underline{\dot{q}}$ into Eq. (6.35) we obtain

$$\Delta\underline{v}^*=\underline{\boldsymbol{U}}^*\cdot\underline{\hat{F}}^*+\underline{V}^*\cdot\underline{\hat{M}}$$

with
$$\underline{\boldsymbol{U}}^*=(\underline{P}^*-\underline{p}\times\underline{T}\,\underline{\varrho}^*)^{\mathrm{T}}[\underline{A}^{-1}-\underline{A}^{-1}\underline{H}^{\mathrm{T}}(\underline{H}\,\underline{A}^{-1}\underline{H}^{\mathrm{T}})^{-1}\underline{H}\,\underline{A}^{-1}]\,(\underline{P}^*-\underline{p}\times\underline{T}\,\underline{\varrho}^*)$$
$$\underline{\boldsymbol{V}}^*=(\underline{P}^*-\underline{p}\times\underline{T}\,\underline{\varrho}^*)^{\mathrm{T}}[\underline{A}^{-1}-\underline{A}^{-1}\underline{H}^{\mathrm{T}}(\underline{H}\,\underline{A}^{-1}\underline{H}^{\mathrm{T}})^{-1}\underline{H}\,\underline{A}^{-1}]\,(\underline{p}\,\underline{T}),$$

and $\Delta\underline{\omega}$ becomes

$$\Delta\underline{\omega}=\underline{\bar{\boldsymbol{V}}}^{*\mathrm{T}}\cdot\underline{\hat{F}}^*+\underline{\boldsymbol{W}}\cdot\underline{\hat{M}}\,.$$

The matrix $\underline{\boldsymbol{W}}$ is the same as in Eq. (6.28). The formulas for $\Delta\underline{v}^*$ and $\Delta\underline{\omega}$ are combined in the equation

$$\begin{bmatrix}\Delta\underline{v}^*\\ \Delta\underline{\omega}\end{bmatrix}=\begin{bmatrix}\underline{\boldsymbol{U}}^* & \underline{\boldsymbol{V}}^*\\ \underline{\bar{\boldsymbol{V}}}^{*\mathrm{T}} & \underline{\boldsymbol{W}}\end{bmatrix}\cdot\begin{bmatrix}\underline{\hat{F}}^*\\ \underline{\hat{M}}\end{bmatrix}. \tag{6.36}$$

This is the desired generalized form of Eq. (6.29). The coefficient matrix is conjugate symmetric, again.

Illustrative Example 6.1 The point P_2 of the rigid body shown in Fig. 6.9 is constrained to move along a straight guide. At the point P_1 the body is subject to a given impulse $\hat{\boldsymbol{F}}_1$. The points P_1 and P_2 are located by the body-fixed vectors $\boldsymbol{\varrho}_1$ and $\boldsymbol{\varrho}_2$,

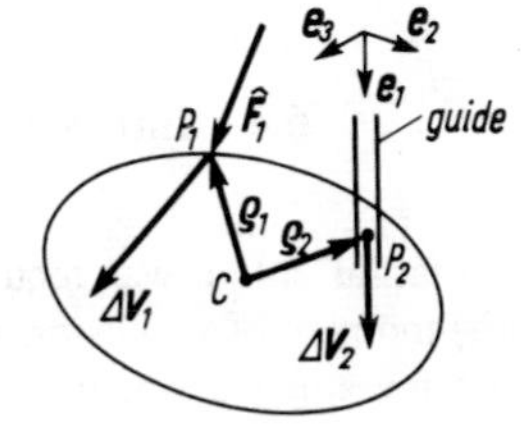

Fig. 6.9
A single constrained body under the action of one external impulse at P_1

respectively, measured from the body center of mass. Determine the reaction impulse $\hat{\boldsymbol{F}}_2$ which is exerted on the body by the guide at P_2, the velocity increments $\Delta\boldsymbol{v}_1$ and $\Delta\boldsymbol{v}_2$ of P_1 and P_2, respectively, as well as the angular velocity increment $\Delta\boldsymbol{\omega}$.

Solution. Instead of adapting Eq. (6.36) to the present simple case we develop the desired relationship from basic principles of rigid body dynamics. When the step leading from Eqs. (6.4) to Eqs. (6.5) is repeated for the present case the equations obtained are

$$m\Delta\boldsymbol{v}_C=\hat{\boldsymbol{F}}_1+\hat{\boldsymbol{F}}_2\,,\qquad \boldsymbol{J}\cdot\Delta\boldsymbol{\omega}=\hat{\boldsymbol{M}}+\boldsymbol{\varrho}_1\times\hat{\boldsymbol{F}}_1+\boldsymbol{\varrho}_2\times\hat{\boldsymbol{F}}_2\,.$$

with $\hat{\boldsymbol{M}}=\boldsymbol{0}$. As kinematic equations we have

$$\Delta\boldsymbol{v}_i=\Delta\boldsymbol{v}_C-\boldsymbol{\varrho}_i\times\Delta\boldsymbol{\omega}\qquad i=1,2\,.$$

Decomposition of all three equations in some common frame of reference yields for the coordinate matrices of $\Delta\boldsymbol{\omega}$, $\Delta\boldsymbol{v}_1$ and $\Delta\boldsymbol{v}_2$

$$\Delta\underline{\omega}=\underline{J}^{-1}(\hat{\underline{M}}+\tilde{\underline{\varrho}}_1\hat{\underline{F}}_1+\tilde{\underline{\varrho}}_2\hat{\underline{F}}_2)$$

$$\Delta\underline{v}_i=\Delta\underline{v}_C-\tilde{\underline{\varrho}}_i\Delta\underline{\omega}=\frac{\hat{\underline{F}}_1+\hat{\underline{F}}_2}{m}-\tilde{\underline{\varrho}}_i\underline{J}^{-1}(\hat{\underline{M}}+\tilde{\underline{\varrho}}_1\hat{\underline{F}}_1+\tilde{\underline{\varrho}}_2\hat{\underline{F}}_2)\qquad i=1,2$$

or in matrix form

$$\begin{bmatrix}\Delta\underline{v}_1\\ \Delta\underline{v}_2\\ \Delta\underline{\omega}\end{bmatrix}=\left[\begin{array}{c|c|c}\frac{1}{m}\underline{E}-\tilde{\underline{\varrho}}_1\underline{J}^{-1}\tilde{\underline{\varrho}}_1 & \frac{1}{m}\underline{E}-\tilde{\underline{\varrho}}_1\underline{J}^{-1}\tilde{\underline{\varrho}}_2 & -\tilde{\underline{\varrho}}_1\underline{J}^{-1}\\ \hline \frac{1}{m}\underline{E}-\tilde{\underline{\varrho}}_2\underline{J}^{-1}\tilde{\underline{\varrho}}_1 & \frac{1}{m}\underline{E}-\tilde{\underline{\varrho}}_2\underline{J}^{-1}\tilde{\underline{\varrho}}_2 & -\tilde{\underline{\varrho}}_2\underline{J}^{-1}\\ \hline \underline{J}^{-1}\tilde{\underline{\varrho}}_1 & \underline{J}^{-1}\tilde{\underline{\varrho}}_2 & \underline{J}^{-1}\end{array}\right]\begin{bmatrix}\hat{\underline{F}}_1\\ \hat{\underline{F}}_2\\ \hat{\underline{M}}\end{bmatrix}$$

($\underline{E}$ is the unit matrix). This represents the scalar form of Eq. (6.36). The coefficient matrix is symmetric.

If for decomposition of the vectors and tensors the base shown in Fig. 6.9 with the base vector $\boldsymbol{e}_1$ along the guide is used then the two column matrices are

$$[\underline{\Delta v_{11}\;\;\Delta v_{12}\;\;\Delta v_{13}\;\;\Delta v_2}\;\;0\;\;\;0\;\;\underline{\Delta\omega_1\;\;\Delta\omega_2\;\;\Delta\omega_3}]^{\mathrm{T}}$$

and $$[\hat{F}_{11}\;\;\underline{\hat{F}_{12}}\;\;\underline{\hat{F}_{13}}\;\;0\;\;\underline{\hat{F}_{22}}\;\;\underline{\hat{F}_{23}}\;\;0\;\;0\;\;0\;]^{\mathrm{T}}.$$

The underlined quantities are the unknowns. They can easily be determined from the matrix equation. ■

6.4 Internal impulses and impulse couples in hinges

So far, attention was focused on the velocity increments a system is experiencing at the moment of a collision. In the remainder of this chapter the internal hinge reactions are investigated which occur in the hinges of a multi-body system at the moment of a collision. For this purpose the bodies are isolated by cutting all hinges of the system. The internal impulses and impulse couples in a hinge labeled a are replaced by an equivalent pair of a single internal impulse $\hat{\boldsymbol{X}}_a$ and a single internal impulse couple $\hat{\boldsymbol{Y}}_a$.

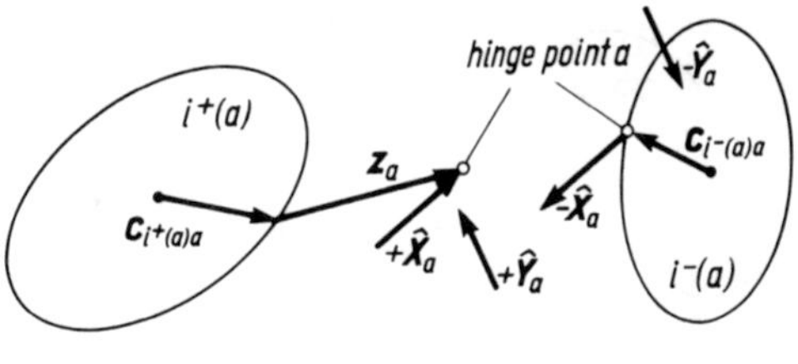

Fig. 6.10
Free-body diagrams for two contiguous bodies. At the hinge point the internal hinge impulse $\hat{\boldsymbol{X}}_a$ is applied to body $i^+(a)$ with positive and to body $i^-(a)$ with negative sign

The line of action of $\hat{\boldsymbol{X}}_a$ is passing through the hinge point of hinge a. The sign convention of Sec. 5.2.2 is adopted. It states that $+\hat{\boldsymbol{X}}_a$ and $+\hat{\boldsymbol{Y}}_a$ act on body $i^+(a)$ and $-\hat{\boldsymbol{X}}_a$ and $-\hat{\boldsymbol{Y}}_a$ on body $i^-(a)$. The problem is now reduced to the determination of $\hat{\boldsymbol{X}}_a$ and $\hat{\boldsymbol{Y}}_a$ for all hinges. Fig. 6.10 illustrates two contiguous bodies after the hinge a between them has been cut and the said impulse and impulse couple have been applied. In Chap. 5.2.9 on continuous motions internal hinge forces $\boldsymbol{X}_a$ and torques $\boldsymbol{Y}_a$ were defined in an equivalent way. The two sets of quantities are related by the equations

$$\hat{\boldsymbol{X}}_a = \lim_{\Delta t \to 0} \int_{t_0}^{t_0+\Delta t} \boldsymbol{X}_a(t)\,\mathrm{d}t\,, \qquad \hat{\boldsymbol{Y}}_a = \lim_{\Delta t \to 0} \int_{t_0}^{t_0+\Delta t} \boldsymbol{Y}_a(t)\,\mathrm{d}t$$

where Δt is the time interval of collision. These relationships will actually be used to determine the internal hinge impulses and impulse couples.

The problem is particularly simple if the multi-body system under investigation has tree structure. In this case explicit formulas have been developed for the hinge forces $\boldsymbol{X}_a$ and hinge torques $\boldsymbol{Y}_a$ $(a=1\dots n)$ for continuous motions. They have the form of Eqs. (5.180). Integration of these equations in the sense defined above immediately yields

$$\hat{\underline{\boldsymbol{X}}} = \underline{T}(\underline{m}\Delta\underline{\dot{\boldsymbol{r}}} - \hat{\underline{\boldsymbol{F}}})\,, \qquad \hat{\underline{\boldsymbol{Y}}} = \underline{T}[\underline{\boldsymbol{J}}\cdot\Delta\underline{\boldsymbol{\omega}} - \hat{\underline{\boldsymbol{M}}} - (\underline{\boldsymbol{C}} + \underline{\boldsymbol{Z}})\times\hat{\underline{\boldsymbol{X}}}]\,. \tag{6.37}$$

For $\Delta\underline{\dot{\boldsymbol{r}}}$ and $\Delta\underline{\boldsymbol{\omega}}$ Eqs. (6.21) and (6.20) have to be substituted. The equations represent already the final solution to the problem. Suppose, for instance, that two bodies k and l belonging to one and the same system with tree structure are colliding with each other. It was shown earlier how the interaction impulse $\hat{\boldsymbol{F}}$ at the point of collision and the velocity increments $\Delta\underline{\dot{\boldsymbol{r}}}$ and angular velocity increments $\Delta\underline{\boldsymbol{\omega}}$ are determined. With these results all quantities on the right hand side of Eq. (6.37) are given. The matrix $\hat{\underline{\boldsymbol{M}}}$ is zero. In the matrix $\hat{\underline{\boldsymbol{F}}}$ the k-th element is $+\hat{\boldsymbol{F}}$, the l-th element is $-\hat{\boldsymbol{F}}$, and all other elements are zero.

The problem of determining internal impulses and impulse couples is more complicated if the multi-body system under investigation has closed kinematic chains as is the case in Fig. 6.11a. No explicit formulas of the kind of Eqs. (5.180) have been developed for the internal hinge forces and torques which appear during continuous motions. The problem is, therefore, not soluable by simple integration. Instead, the solution is found as follows. From the given system a reduced system with tree structure is produced. For this purpose hinges are cut so that all closed kinematic chains are opened. Internal hinge impulses and impulse couples are, first, determined for the cut hinges alone. Once these quantities are known the internal impulses and impulse couples can be found for the hinges of the reduced system, as well. This second part of the problem is solved by Eq. (6.37). The column matrices $\underline{\hat{F}}$ and $\underline{\hat{M}}$

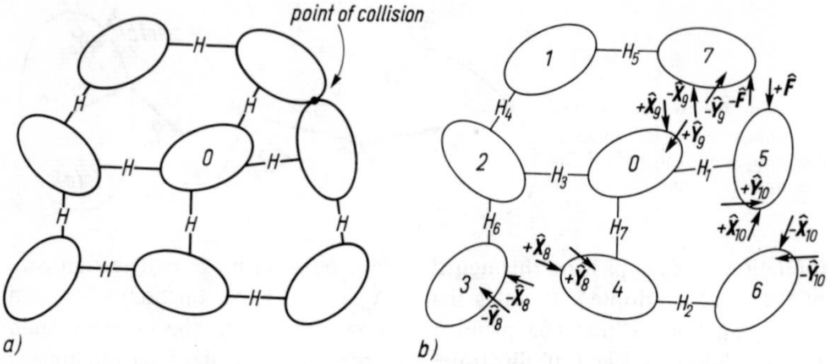

Fig. 6.11 a) A system with closed chains two bodies of which collide with one another
b) All bodies and hinges are labeled. In the cut hinges 8, 9 and 10 and at the point of collision internal impulses and impulse couples are shown. The motion of body 0 is prescribed. System structure and system graph are the same as in Fig. 5.44

are now composed of the interaction impulse $\hat{\boldsymbol{F}}$ at the point of collision ($+\hat{\boldsymbol{F}}$ and $-\hat{\boldsymbol{F}}$ if two bodies belonging to the same system collide with each other) and of the impulses and impulse couples for the cut hinges. Each of the latter quantities is applied to two contiguous bodies with opposite signs. The following discussion concerns only the first part of the problem, i. e. the determination of the internal impulses and impulse couples in the cut hinges. These hinges are labeled $n+1\ldots n+n^*$ as in Sec. 5.3, and the impulses and impulse couples are $\hat{\boldsymbol{X}}_a$ and $\hat{\boldsymbol{Y}}_a$ for $a=n+1\ldots n+n^*$. As an illustrative example Fig. 6.11b shows a reduced system for the system of Fig. 6.11a. All bodies are labeled. The two colliding bodies are subject to the known impulses $+\hat{\boldsymbol{F}}$ and $-\hat{\boldsymbol{F}}$, respectively. The velocity increments at the points of application of $+\hat{\boldsymbol{F}}$ and $-\hat{\boldsymbol{F}}$ are also known. Three hinges labeled 8, 9 and 10 are cut. The internal impulses and impulse couples in these hinges are schematically shown. Body 0 whose motion is prescribed is not affected by impulses (imagine that it has infinite inertia).

In the cut hinge a ($a=n+1\ldots n+n^*$) a certain number of holonomic constraint equations must be satisfied. Each constraint equation requires for its physical realization one internal impulse or impulse couple of a certain magnitude and direction. From this follows that each of the six coordinates which $\hat{\boldsymbol{X}}_a$ and $\hat{\boldsymbol{Y}}_a$ have together in some frame of reference can be represented as a linear combination of as many unknown quantities as there are constraint equations for hinge a. In other words this means that there are as many independent linear relationships between the six coordinates as there are degrees of freedom in hinge a. Three examples illustrate how these relationships are formulated in practice. First example: The cut hinge number 8 in Fig. 6.11a is assumed to be a ball-and-socket joint. As hinge point the geometric center of the joint is chosen. The hinge has three degrees of freedom. Hence, there are three scalar equations. They read $\hat{\boldsymbol{Y}}_8=\boldsymbol{0}$ or in more detail $\boldsymbol{n}_1\cdot\hat{\boldsymbol{Y}}_8=0$, $\boldsymbol{n}_2\cdot\hat{\boldsymbol{Y}}_8=0$ and $\boldsymbol{n}_3\cdot\hat{\boldsymbol{Y}}_8=0$ with any three non-coplanar vectors $\boldsymbol{n}_1$, $\boldsymbol{n}_2$ and $\boldsymbol{n}_3$ (for instance, the base vectors of some reference frame in which all equations are decomposed in scalar form for numerical calculations). Second Example: The cut hinge number 9 in Fig. 6.11a is assumed to be of the type shown in Fig. 5.34b. As hinge point a point on the axis is chosen. To the two degrees of freedom correspond the two equations $\boldsymbol{n}\cdot\hat{\boldsymbol{X}}_9=0$ and $\boldsymbol{n}\cdot\hat{\boldsymbol{Y}}_9=0$ in which $\boldsymbol{n}$ is a vector along the hinge axis. Third example: The cut hinge number 10 in Fig. 6.11a is assumed to be of the kind shown in Fig. 5.34e. The two contiguous bodies are in contact at a single point which is the intersection point of two guides, each guide fixed on one of the bodies (it is assumed that the collision causes the guides to be pressed against one another so that separation is prevented). Choose one of the bodies to be body $i^-(a)$ and on this body an arbitrary point as hinge point. Let the vector from the hinge point to the contact point of the guides be called $\boldsymbol{b}$. Furthermore, let $\boldsymbol{k}_1$ and $\boldsymbol{k}_2$ represent vectors which are tangent to the guides at the contact point (one vector for each of the guides). The hinge has five degrees of freedom. The corresponding five equations are $\boldsymbol{k}_1\cdot\hat{\boldsymbol{X}}_{10}=0$, $\boldsymbol{k}_2\cdot\hat{\boldsymbol{X}}_{10}=0$ and $\hat{\boldsymbol{Y}}_{10}-\boldsymbol{b}\times\hat{\boldsymbol{X}}_{10}=\boldsymbol{0}$. The last equation is equivalent to

$$\boldsymbol{n}_i\cdot(\hat{\boldsymbol{Y}}_{10}-\boldsymbol{b}\times\hat{\boldsymbol{X}}_{10})=\boldsymbol{n}_i\cdot\hat{\boldsymbol{Y}}_{10}-\boldsymbol{n}_i\times\boldsymbol{b}\cdot\hat{\boldsymbol{X}}_{10}=0 \qquad i=1,2,3$$

where $\boldsymbol{n}_1$, $\boldsymbol{n}_2$ and $\boldsymbol{n}_3$ are any non-coplanar vectors.

The total number of holonomic constraints for all cut hinges together is called ν_1 in accordance with Sec. 5.3.2. Hence, $6n^*-\nu_1$ is the total number of degrees of freedom in all cut hinges together and also the total number of scalar equations of the form just described. All these equations can be combined in the form

$$\underline{\boldsymbol{K}}\cdot\begin{bmatrix}\hat{\underline{\boldsymbol{X}}}^*\\ \hat{\underline{\boldsymbol{Y}}}^*\end{bmatrix}=\underline{0} \tag{6.38}$$

with a matrix $\underline{\boldsymbol{K}}$ of $6n^*-\nu_1$ rows and $2n^*$ columns and with the column matrix $[\hat{\boldsymbol{X}}_{n+1}\ldots\hat{\boldsymbol{X}}_{n+n^*}\,\hat{\boldsymbol{Y}}_{n+1}\ldots\hat{\boldsymbol{Y}}_{n+n^*}]^{\mathrm{T}}$ as second factor. The elements of $\underline{\boldsymbol{K}}$ are vectors as is shown by the above examples. To these equations have to be added ν_1 constraint equations. Each constraint equation has the form of Eq. (6.2). All ν_1 equations can be combined in the form

$$\underline{H}\,\Delta\dot{\underline{q}}=\underline{0} \tag{6.39}$$

in which $\underline{H}$ is the Jacobian matrix known from Eq. (5.189) (note that in the absence of nonholonomic constraints ν_2 equals ν_1). Eqs. (6.38) and (6.39) together constitute $6n^*$ scalar equations. They contain as unknowns $6n^*$ scalar coordinates of $\hat{\boldsymbol{X}}_a$ and $\hat{\boldsymbol{Y}}_a$ $(a=n+1\ldots n+n^*)$ and, in addition, a total number N of velocity increments in the matrix $\Delta\dot{\underline{q}}$. Hence, another N scalar equations for the same unknowns are needed. These are Eqs. (6.34) with $\underline{H}$ being replaced by zero (the reduced system has tree structure). Substitution into Eq. (6.39) yields

$$\underline{H}\,\underline{A}^{-1}[(\underline{\boldsymbol{P}}^*-\underline{\boldsymbol{p}}\times\underline{T}\,\underline{\boldsymbol{\varrho}}^*)\cdot\hat{\underline{\boldsymbol{F}}}^*-\underline{\boldsymbol{p}}\,\underline{T}\cdot\hat{\underline{\boldsymbol{M}}}]=\underline{0}\,. \tag{6.40}$$

The matrices $\hat{\underline{\boldsymbol{F}}}^*$ and $\hat{\underline{\boldsymbol{M}}}$ are composed of the unknowns $\hat{\boldsymbol{X}}_a$, $\hat{\boldsymbol{Y}}_a$ $(a=n+1\ldots n+n^*)$ and of the known impulse $\hat{\boldsymbol{F}}$. For the illustrative example in Fig. 6.11b, for instance, the matrices are

$$\hat{\underline{F}}^* = [0 \quad 0 \quad -\hat{X}_8 \quad +\hat{X}_8 \quad +\hat{F} \quad +\hat{X}_{10} \quad -\hat{X}_{10} \quad -\hat{F} \quad -\hat{X}_9]^{\mathrm{T}}$$
$$\hat{\underline{M}} = [0 \quad 0 \quad -\hat{Y}_8 \quad +\hat{Y}_8 \quad +\hat{Y}_{10} \quad -\hat{Y}_{10} \quad -\hat{Y}_9]^{\mathrm{T}}.$$

Eqs. (6.38) and (6.40) together determine all unknowns.

Illustrative Example 6.2 A chain consisting of eleven identical links with ball-and-socket joints is resting on a horizontal table. The links are homogeneous rods of length l, mass m and central moment of inertia $ml^2/12$ about the vertical axis. The angle between contiguous bodies is $9\pi/10$ for all pairs of contiguous bodies so that the chain is enveloping a semicircle with the center at O (Fig. 6.12a). The fourth link is hit in the center by a point mass of mass m which is moving with a velocity $\boldsymbol{v}$ normal to the direction of the link. The collision is ideally elastic. To be determined are the velocities of the centers of mass of all links and the angular velocities of all links immediately after impact. The subsequent motion of the chain is to be computed by numerical integration of the equations of motion. It is assumed that no friction occurs between the table and the chain and that no internal torques are acting in the hinges.

Solution. The system is of the kind described as *system with tree structure and ball-and-socket joints not coupled with an external body whose motion is prescribed as a function of time.* Its equations of motion were developed in Chap. 5.2.4. They consist of a single equation $\ddot{\boldsymbol{r}}_C = 1/M \sum_{j=1}^{n} \boldsymbol{F}_j$ $(n=11)$ for the system center of mass and of Eq. (5.61) describing rotational motions. For the present case of plane motions Eq. (5.61) takes the special form of Eq. (5.62). The angle ϕ_i $(i=1 \dots n)$ is measured in the plane between a base vector $\boldsymbol{e}_1^{(i)}$ fixed on body i and the base vector $\boldsymbol{e}_1$ of an inertial frame of reference. The orientation of these base vectors is chosen as shown in Fig. 6.12a. The origin of $\underline{e}$ is at the point O, $\boldsymbol{e}_1$ has the direction of $\boldsymbol{v}$ and $\boldsymbol{e}_1^{(i)}$ the direction of the longitudinal axis of body i. Under these conditions all vectors $\boldsymbol{b}_{ij}$ $(j=1 \dots n)$ on body i are parallel to $\boldsymbol{e}_1^{(i)}$ so that $\beta_{ij}=0$ for $i,j=1 \dots n$. No internal torques are acting in the hinges $(\boldsymbol{Y}_a = \boldsymbol{0})$. The external torque on body i is equal to $\boldsymbol{\varrho}_i \times \boldsymbol{F}_i$ if $\boldsymbol{\varrho}_i$ is the vector from the body i center of mass to the point of application of $\boldsymbol{F}_i$. In the phase of motion after impact no external forces are acting $(\boldsymbol{F}_i = \boldsymbol{0})$ so that the equations of motion become $\boldsymbol{r}_C = \boldsymbol{r}_C(t_0) + (t-t_0)\dot{\boldsymbol{r}}_C(t_0)$ and

$$\underline{A}\begin{bmatrix}\ddot{\phi}_1\\ \vdots \\ \ddot{\phi}_n\end{bmatrix} + \underline{B}\begin{bmatrix}\dot{\phi}_1^2\\ \vdots \\ \dot{\phi}_n^2\end{bmatrix} = \underline{0}\,. \tag{6.41}$$

The initial conditions $\boldsymbol{r}_C(t_0)$ and $\phi_i(t_0)$ for $i=1 \dots n$ are those known from Fig. 6.12a, whereas the initial conditions $\dot{\boldsymbol{r}}_C(t_0)$ and $\dot{\phi}_i(t_0)$ for $i=1 \dots n$ are the result of the first part of the problem. For this part, the equations of motion are written in the more general form

$$\underline{A}\begin{bmatrix}\ddot{\phi}_1\\ \vdots \\ \ddot{\phi}_n\end{bmatrix} + \underline{B}\begin{bmatrix}\dot{\phi}_1^2\\ \vdots \\ \dot{\phi}_n^2\end{bmatrix} = \underline{C}$$

with $\quad C_i = \sum_{k=1}^{n} (b_{ik} + \delta_{ik}\varrho_k)(-F_{k1}\sin\phi_i + F_{k2}\cos\phi_i) \qquad i=1 \dots n\,.$

The matrices $\underline{A}$ and $\underline{B}$ have the elements

$$A_{ij} = \begin{cases} K^*_{i3} & i=j \\ -M\, b_{ij} b_{ji} \cos(\phi_i - \phi_j) & i \neq j \end{cases} \quad i,j=1 \ldots n\,.$$
$$B_{ij} = -M\, b_{ij} b_{ji} \sin(\phi_i - \phi_j)$$

For the determination of the interaction impulse $\hat{\boldsymbol{F}}$ at the point of collision and of the velocity and angular velocity increments Eqs. (6.12), (6.18), (6.20) and (6.21) are available. $\hat{\boldsymbol{F}}$ was defined to be the impulse on the body labeled k. Let this be the

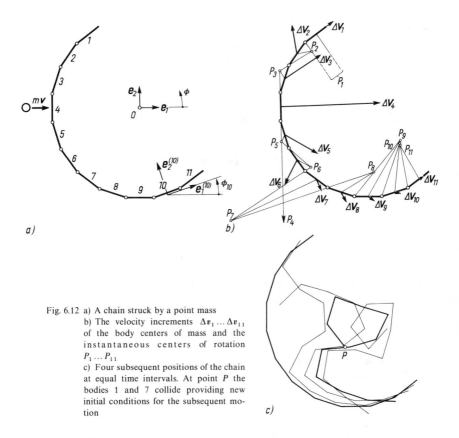

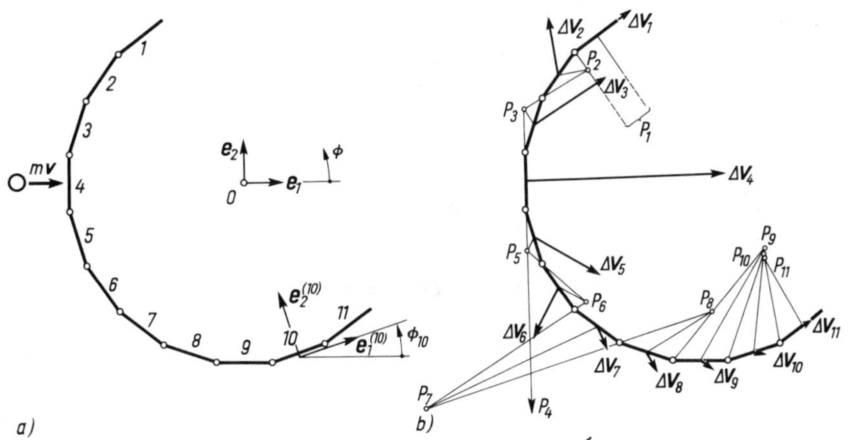

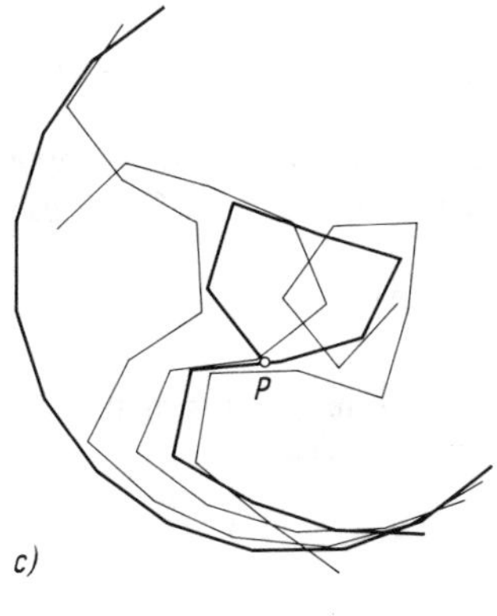

Fig. 6.12 a) A chain struck by a point mass
b) The velocity increments $\Delta \boldsymbol{v}_1 \ldots \Delta \boldsymbol{v}_{11}$ of the body centers of mass and the instantaneous centers of rotation $P_1 \ldots P_{11}$
c) Four subsequent positions of the chain at equal time intervals. At point P the bodies 1 and 7 collide providing new initial conditions for the subsequent motion

fourth link of the chain. Hence, body l is the point mass. This means $\boldsymbol{v}_k = \boldsymbol{0}$ and $\boldsymbol{v}_l = \boldsymbol{v}$. The vector $\boldsymbol{n}$ normal to the tangent plane at the point of collision is $\boldsymbol{e}_1$. The coefficient of restitution is $e=1$. The tensors $\boldsymbol{U}_{kk}$ and $\boldsymbol{U}_{ll}$ are defined by Eq. (6.9). For the point mass this equation has the simple form $\Delta \boldsymbol{v}_l = -\hat{\boldsymbol{F}}/m$. Hence, $\boldsymbol{U}_{ll} = \boldsymbol{E}/m$ with the unit tensor $\boldsymbol{E}$. Eq. (6.12), thus, becomes

$$\hat{F} = \frac{2v}{1/m + \boldsymbol{e}_1 \cdot \boldsymbol{U}_{kk} \cdot \boldsymbol{e}_1}\,. \tag{6.42}$$

The tensor $\boldsymbol{U}_{kk}$ is the element with indices (4, 4) of the matrix $\underline{\boldsymbol{U}}$ in Eq. (6.26) formulated for the chain. Because of the special character of the system this equation can be greatly simplified. The expression for $\underline{\boldsymbol{U}}$ resulted from a combination of two equations. One is the integrated equation of motion (transition from Eq. (6.15) to Eq. (6.18)), and the other is the kinematic Eq. (6.22). In the present case, integration of the equations of motion yields

$$\Delta\dot{\boldsymbol{r}}_C=\frac{1}{M}\sum_{j=1}^{n}\hat{\boldsymbol{F}}_j \quad \text{and} \quad \underline{A}\,\Delta\dot{\underline{\phi}}=\underline{\hat{C}}$$

with $$\hat{C}_i=\sum_{k=1}^{n}(b_{ik}+\delta_{ik}\,\varrho_k)\,(-\hat{F}_{k1}\sin\phi_i+\hat{F}_{k2}\cos\phi_i) \qquad i=1\ldots n\,.$$

The only impulse acting on the system is $\hat{\boldsymbol{F}}_4=\hat{F}\,\boldsymbol{e}_1$. It is applied to the center of mass of body 4 $(\varrho_4=0)$. Hence, the equations reduce to

$$\Delta\dot{r}_{C1}=\frac{\hat{F}}{M}\,,\qquad \Delta\dot{r}_{C2}=0\,,\qquad \underline{A}\,\Delta\dot{\underline{\phi}}=-\hat{F}\,\underline{Q} \tag{6.43}$$

with a column matrix $\underline{Q}$ whose elements are $Q_i=b_{i4}\sin\phi_i$ $(i=1\ldots n)$. The kinematic Eq. (6.22) is replaced by

$$\Delta\boldsymbol{v}_i=\Delta\dot{\boldsymbol{r}}_C+\Delta\dot{\boldsymbol{R}}_i+\Delta\boldsymbol{\omega}_i\times\boldsymbol{\varrho}_i \qquad i=1\ldots n$$

($\boldsymbol{R}_i$ is the vector from the system center of mass to the body i center of mass). The identity (cf. Eq. (5.54)) $\boldsymbol{R}_i=\sum_{k=1}^{n}\boldsymbol{b}_{ki}$ yields

$$\Delta\boldsymbol{v}_i=\Delta\dot{\boldsymbol{r}}_C+\sum_{k=1}^{n}\Delta\boldsymbol{\omega}_k\times(\boldsymbol{b}_{ki}+\delta_{ki}\,\boldsymbol{\varrho}_k) \qquad i=1\ldots n\,. \tag{6.44}$$

In the present case only the component Δv_{41} for the point of application of $\hat{\boldsymbol{F}}$ in the direction of $\boldsymbol{e}_1$ is needed. With $\boldsymbol{\varrho}_4=\boldsymbol{0}$ it is

$$\Delta v_{41}=\Delta\dot{r}_{C1}-\sum_{k=1}^{n}b_{k4}\sin\phi_k\,\Delta\dot{\phi}_k$$

or with the same column matrix $\underline{Q}$ as in Eq. (6.43) $\Delta v_{41}=\Delta\dot{r}_{C1}-\underline{Q}^{\mathrm{T}}\Delta\dot{\underline{\phi}}$. Substitution of Eq. (6.43) yields

$$\Delta v_{41}=\hat{F}\left(\frac{1}{M}+\underline{Q}^{\mathrm{T}}\,\underline{A}^{-1}\underline{Q}\right).$$

Comparison with Eq. (6.9) reveals that the scalar in brackets is the term $\boldsymbol{e}_1\cdot\boldsymbol{U}_{kk}\cdot\boldsymbol{e}_1$ needed for Eq. (6.42). Hence,

$$\hat{F}=\frac{2v}{1/m+1/M+\underline{Q}^{\mathrm{T}}\,\underline{A}^{-1}\underline{Q}}\,. \tag{6.45}$$

With this also $\Delta\dot{r}_{C1}=\hat{F}/M$ and $\Delta\dot{\underline{\phi}}=-\hat{F}\,\underline{A}^{-1}\,\underline{Q}$ are known. Numerical results are listed in the table below. In addition to the ratios $\Delta\dot{r}_{C1}/v$ and $\Delta\dot{\phi}_i l/v$, also $\Delta\dot{r}_{i1}/v$ and $\Delta\dot{r}_{i2}/v$ are tabulated. The quantities $\Delta\dot{r}_{i1}$ and $\Delta\dot{r}_{i2}$ are the coordinates in the base $\underline{e}$ of the velocity increments of the body centers of mass. They are calculated from Eq. (6.44) with $\varrho_i = \mathbf{0}$.

$\hat{F} = 1.323\, mv, \quad \Delta\dot{r}_{C1} = 0.1189\, v$

i	$\Delta\dot{\phi}_i l/v$	$\Delta\dot{r}_{i1}/v$	$\Delta\dot{r}_{i2}/v$
1	−0.0070	0.1132	0.0837
2	−0.3716	−0.0391	0.1958
3	0.9185	0.2474	0.1630
4	−0.0133	0.6774	0.0211
5	−0.9460	0.2209	−0.1250
6	0.3384	−0.0920	−0.1717
7	−0.0269	0.0370	−0.0831
8	0.0561	0.0377	−0.0674
9	0.0239	0.0464	−0.0287
10	0.0249	0.0425	−0.0049
11	0.0257	0.0311	0.0173

In Fig. 6.12b arrows indicate the magnitude and direction of the velocities of all body centers of mass immediately after impact. The data in the table allows identifying the location of the instantaneous centers of rotation $P_1 \dots P_n$ of the bodies. The results satisfy the condition that the line connecting the centers of rotation of any two contiguous bodies passes through the hinge point between these bodies.

If the total mass of the chain were to be concentrated in one point the interaction impulse would be $\hat{F} = 2v/(1/m + 1/M)$. The actual interaction impulse $\hat{F}$ can be represented in the form $\hat{F} = 2v/(1/m + 1/M^*)$. This equation defines an apparent system mass M^*. Comparison with Eq. (6.45) yields $1/M^* = 1/M + \underline{Q}^{\mathrm{T}} \underline{A}^{-1} \underline{Q}$. The matrix $\underline{A}$ is positive definite so that $M^* \leqslant M$. In the present case M^* equals 1.952 m. If the calculation is repeated for the case where the point mass does not strike the fourth body but the i-th body ($i = 1, 2 \dots$)—always at the center of mass and normal to the body—then the following results are obtained

i	1	2	3	4	5	6
M^*/m	1.173	1.761	1.892	1.952	1.979	1.987

The apparent system mass is always much smaller than the actual mass $M = 11$ m, and except for $i = 1$ it depends little on the location of the point of collision.

With the initial data for $\dot{r}_{C1}$ and $\dot{\phi}_i$ ($i = 1 \dots n$) numerical integrations of Eq. (6.41) yield the motion of the chain after impact. In Fig. 6.12c the chain is shown in five positions at equal intervals of time. In the fourth position the end point of body 1 collides with body 7 at the point P. This collision causes instantaneous changes of all angular velocities (not of $\dot{r}_C$) which can be calculated from equations similar to the ones used before. The details are left to the reader. With new initial conditions thus obtained the numerical integration can be continued until the next collision occurs. ■

Problems

6.1 The unknown internal impulses and impulse couples in cut hinges were obtained from Eqs. (6.38) and (6.40). The constraint equations (6.40) can be replaced by another set of constraint equations which are formulated in terms of velocity increments of the points of application of the unknown impulses and in terms of angular velocity increments of bodies. Write down these equations for the system of Fig. 6.11 assuming that the cut hinges have the same properties as in the examples given for Eq. (6.38).

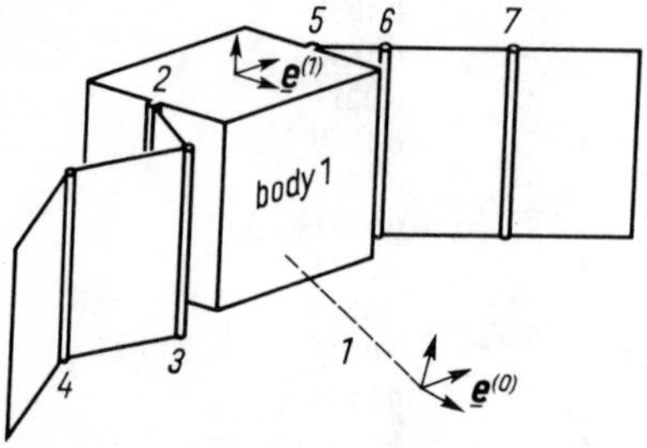

Fig. 6.13
A spacecraft with unfolding solar panels. The moment of collision in hinge 7 is depicted. Hinges 2 to 5 are still unfolding while hinge 6 has reached its final state already

6.2 The solar panels on the spacecraft in Fig. 6.13 are deployed by means of torsional springs in the pin joints. The individual bodies are (unrealistically) considered as rigid. When contiguous bodies reach their final relative orientation their motion relative to one another is suddenly stopped (ideally plastic collision). It is assumed that these collisions occur one at a time. Determine the finite angular velocity increment of the central body 1 caused by a single collision.

Answers to Problems

1.1 $\underline{E}$; 3; $\begin{bmatrix} 0 & \boldsymbol{e}_3 & -\boldsymbol{e}_2 \\ -\boldsymbol{e}_3 & 0 & \boldsymbol{e}_1 \\ \boldsymbol{e}_2 & -\boldsymbol{e}_1 & 0 \end{bmatrix}^{(r)}$; 0; $\underline{A}^{rs}$; $\underline{A}^{sr}$; $\operatorname{tr}\underline{A}^{sr}$.

1.2 $(\underline{a}\,\underline{c}\cdot\underline{b})_{ij}=\sum_k\sum_l \boldsymbol{a}_{il}c_{lk}\cdot\boldsymbol{b}_{kj}=\sum_l\sum_k \boldsymbol{a}_{il}\cdot c_{lk}\boldsymbol{b}_{kj}=(\underline{a}\cdot\underline{c}\,\underline{b})_{ij}$.

1.3 $\underline{a}^{(r)\mathrm{T}}\,\tilde{\underline{b}}^{(r)}\,\underline{A}^{rs}\,\underline{c}^{(s)}$; $\quad -\underline{b}^{(r)\mathrm{T}}\,\tilde{\underline{a}}^{(r)}\,\tilde{\underline{b}}^{(r)}\,\underline{A}^{rs}\,\underline{c}^{(s)}$; $\quad \underline{c}^{(s)\mathrm{T}}\,\underline{D}^{(s)}\,\underline{A}^{sr}\,\underline{a}^{(r)}$;
$\underline{c}^{(s)\mathrm{T}}\,\underline{A}^{sr}\,\tilde{\underline{b}}^{(r)}\,\underline{A}^{rs}\,\underline{D}^{(s)}\,\underline{c}^{(s)}$; $\quad \tilde{\underline{a}}^{(r)}\,\underline{b}^{(r)}$; $\quad \tilde{\underline{a}}^{(r)}\,\underline{A}^{rs}\,\underline{c}^{(s)}$;
$\tilde{\underline{a}}^{(r)}\,\underline{A}^{rs}\,\tilde{\underline{c}}^{(s)}\,\underline{A}^{sr}\,\underline{b}^{(r)}$; $\quad \underline{A}^{rs}\,\tilde{\underline{c}}^{(s)}\,\underline{D}^{(s)}\,\underline{A}^{sr}\,\underline{a}^{(r)}$; $\quad -\tilde{\underline{a}}^{(r)}\,\underline{A}^{rs}\,\tilde{\underline{c}}^{(s)}\,\underline{D}^{(s)}\,\underline{A}^{sr}\,\underline{b}^{(r)}$.

1.4 If $\boldsymbol{p}$ and $\boldsymbol{q}$ are any vectors for which $\boldsymbol{p}\times\boldsymbol{q}=\boldsymbol{d}$ then
$\boldsymbol{D}_{11}=\boldsymbol{b}\cdot\boldsymbol{b}\boldsymbol{E}-\boldsymbol{b}\boldsymbol{b}=\bar{\boldsymbol{D}}_{11}$; $\quad \boldsymbol{D}_{12}=(\boldsymbol{c}\cdot\boldsymbol{b}\boldsymbol{E}-\boldsymbol{c}\boldsymbol{b})+(\boldsymbol{q}\boldsymbol{p}-\boldsymbol{p}\boldsymbol{q})=\bar{\boldsymbol{D}}_{21}$;
$\boldsymbol{D}_{22}=\boldsymbol{c}\cdot\boldsymbol{c}\boldsymbol{E}-\boldsymbol{c}\boldsymbol{c}=\bar{\boldsymbol{D}}_{22}$;
$\underline{D}_{11}=\underline{b}^{\mathrm{T}}\underline{b}\,\underline{E}-\underline{b}\,\underline{b}^{\mathrm{T}}=\underline{D}_{11}^{\mathrm{T}}$; $\quad \underline{D}_{12}=\underline{c}^{\mathrm{T}}\underline{b}\,\underline{E}-\underline{c}\,\underline{b}^{\mathrm{T}}+\tilde{\underline{d}}=\underline{D}_{21}^{\mathrm{T}}$;
$\underline{D}_{22}=\underline{c}^{\mathrm{T}}\underline{c}\,\underline{E}-\underline{c}\,\underline{c}^{\mathrm{T}}=\underline{D}_{22}^{\mathrm{T}}$; $\quad$ the (6×6) matrix is symmetric.

2.1 Calculate $\underline{A}^{21}$ from Eq. (2.2) and solve Eqs. (2.15) and (2.16).

2.3 A body-fixed vector with the coordinate matrix $\underline{r}^{(2)}$ in $\underline{e}^{(2)}$ has in $\underline{e}^{(1)}$ the coordinate matrices $\underline{A}^{12}(\phi_1)\underline{r}^{(2)}$, $\underline{A}^{12}(\phi_2)\underline{A}^{12}(\phi_1)\underline{r}^{(2)}$ and $\underline{A}^{12}(\phi_3)\underline{A}^{12}(\phi_2)\underline{A}^{12}(\phi_1)\underline{r}^{(2)}$ after the first, second and third rotation, respectively, where

$$\underline{A}^{12}(\phi_1)=\begin{bmatrix}1&0&0\\0&c_1&s_1\\0&-s_1&c_1\end{bmatrix};\quad \underline{A}^{12}(\phi_2)=\begin{bmatrix}c_2&0&-s_2\\0&1&0\\s_2&0&c_2\end{bmatrix};\quad \underline{A}^{12}(\phi_3)=\begin{bmatrix}c_3&s_3&0\\-s_3&c_3&0\\0&0&1\end{bmatrix}.$$

$$\underline{A}^{21}=[\underline{A}^{12}(\phi_3)\underline{A}^{12}(\phi_2)\underline{A}^{12}(\phi_1)]^{\mathrm{T}}\quad\text{is}\quad \begin{bmatrix}0&0&-1\\0&1&0\\-1&0&0\end{bmatrix}$$

for the first two sets of angles and the unit matrix for the third.

2.4 The condition is $\boldsymbol{\omega}\perp\boldsymbol{v}_P$. Points with zero velocity lie on the instantaneous axis of rotation. It has the direction of $\boldsymbol{\omega}$ and passes through the point with radius vector $\boldsymbol{\varrho}=\boldsymbol{\omega}\times\boldsymbol{v}_P/\omega^2$ from P.

2.5 Start from $(\boldsymbol{v}_1-\boldsymbol{v}_3)\times(\boldsymbol{v}_2-\boldsymbol{v}_3)=[\boldsymbol{\omega}\times(\boldsymbol{r}_1-\boldsymbol{r}_3)]\times(\boldsymbol{v}_2-\boldsymbol{v}_3)$ which follows from Eq. (2.15). This equals $\boldsymbol{\omega}\cdot(\boldsymbol{v}_2-\boldsymbol{v}_3)(\boldsymbol{r}_1-\boldsymbol{r}_3)-\boldsymbol{\omega}(\boldsymbol{r}_1-\boldsymbol{r}_3)\cdot(\boldsymbol{v}_2-\boldsymbol{v}_3)$. The first term is zero since $\boldsymbol{v}_2-\boldsymbol{v}_3=\boldsymbol{\omega}\times(\boldsymbol{r}_2-\boldsymbol{r}_3)$. Hence

$$\boldsymbol{\omega}=\frac{(\boldsymbol{v}_1-\boldsymbol{v}_3)\times(\boldsymbol{v}_2-\boldsymbol{v}_3)}{(\boldsymbol{r}_3-\boldsymbol{r}_1)\cdot(\boldsymbol{v}_2-\boldsymbol{v}_3)}.$$

The numerator has already the desired form. In reformulating the denominator the rigid body property $(\boldsymbol{r}_\alpha - \boldsymbol{r}_\beta)^2 = \text{const}$ implying

$$(\boldsymbol{r}_\alpha - \boldsymbol{r}_\beta)\cdot(\boldsymbol{v}_\alpha - \boldsymbol{v}_\beta) = 0 \qquad \alpha, \beta = 1, 2, 3$$

is repeatedly used. With this and with the identities

$$\boldsymbol{r}_3 - \boldsymbol{r}_1 = -[(\boldsymbol{r}_1 - \boldsymbol{r}_2) + (\boldsymbol{r}_2 - \boldsymbol{r}_3)], \qquad \boldsymbol{v}_2 - \boldsymbol{v}_3 = \frac{1}{2}[(\boldsymbol{v}_2 - \boldsymbol{v}_3) + (\boldsymbol{v}_2 - \boldsymbol{v}_1) + (\boldsymbol{v}_1 - \boldsymbol{v}_3)]$$

the denominator becomes

$$\begin{aligned}(\boldsymbol{r}_3 - \boldsymbol{r}_1)\cdot(\boldsymbol{v}_2 - \boldsymbol{v}_3) &= \frac{1}{2}(\boldsymbol{r}_3 - \boldsymbol{r}_1)\cdot[(\boldsymbol{v}_2 - \boldsymbol{v}_3) + (\boldsymbol{v}_2 - \boldsymbol{v}_1)] \\ &= \frac{1}{2}\{(\boldsymbol{r}_3 - \boldsymbol{r}_1)\cdot\boldsymbol{v}_2 - [(\boldsymbol{r}_1 - \boldsymbol{r}_2) + (\boldsymbol{r}_2 - \boldsymbol{r}_3)]\cdot(-\boldsymbol{v}_3 + \boldsymbol{v}_2 - \boldsymbol{v}_1)\} \\ &= \frac{1}{2}[(\boldsymbol{r}_3 - \boldsymbol{r}_1)\cdot\boldsymbol{v}_2 + (\boldsymbol{r}_1 - \boldsymbol{r}_2)\cdot\boldsymbol{v}_3 + (\boldsymbol{r}_2 - \boldsymbol{r}_3)\cdot\boldsymbol{v}_1].\end{aligned}$$

This problem was taken from Charlamov [38].

2.6 Define $x_j = q_j^* - q_j$ $(j = 0\ldots3)$. Then $\sum_{j=0}^{3}(q_j + x_j)^2 = 1$ and

$$\frac{\partial}{\partial x_i}\left\{\sum_{j=0}^{3} x_j^2 + \lambda\left[\sum_{j=0}^{3}(q_j + x_j)^2 - 1\right]\right\} = 0 \qquad i = 0\ldots3.$$

2.7 $\underline{A}^{21} = \underline{A}^3\underline{B}^2\underline{A}^2\underline{B}^1\underline{A}^1$. The matrices $\underline{A}^3$, $\underline{A}^2$ and $\underline{A}^1$ are the same as in Eq. (2.5) and

$$\underline{B}^2 = \begin{bmatrix} 1 & 0 & 0 \\ 0 & \cos\beta & -\sin\beta \\ 0 & \sin\beta & \cos\beta \end{bmatrix}, \qquad \underline{B}^1 = \begin{bmatrix} \cos\alpha & \sin\alpha & 0 \\ -\sin\alpha & \cos\alpha & 0 \\ 0 & 0 & 1 \end{bmatrix}.$$

From $\underline{\omega} = \underline{A}^3\underline{B}^2\underline{A}^2\underline{B}^1[\dot\phi_1\ 0\ 0]^{\mathrm{T}} + \underline{A}^3\underline{B}^2[0\ \dot\phi_2\ 0]^{\mathrm{T}} + [0\ 0\ \dot\phi_3]^{\mathrm{T}}$ follows

$$[\dot\phi_1\ \dot\phi_2\ \dot\phi_3]^{\mathrm{T}} = \frac{1}{c_2 c_\alpha c_\beta}\begin{bmatrix} c_3 c_\beta & -s_3 c_\beta & 0 \\ c_2 s_3 c_\alpha + c_3(s_\alpha c_\beta + s_2 c_\alpha s_\beta) & c_2 c_3 c_\alpha - s_3(s_\alpha c_\beta + s_2 c_\alpha s_\beta) & 0 \\ -c_\alpha(s_2 c_3 + c_2 s_3 s_\beta) & c_\alpha(s_2 s_3 - c_2 c_3 s_\beta) & c_2 c_\alpha c_\beta \end{bmatrix}\begin{bmatrix}\omega_1 \\ \omega_2 \\ \omega_3\end{bmatrix}.$$

($c_i = \cos\phi_i$, $s_i = \sin\phi_i$ $(i = 1, 2, 3)$ and c_α, c_β, s_v, $s_\beta = \cos\alpha$, $\cos\beta$, $\sin\alpha$, $\sin\beta$, respectively). The case $c_2 = 0$ (gimbal lock) is critical.

3.1 $\quad \boldsymbol{L}^C = \boldsymbol{J}^C\cdot\boldsymbol{\omega}$, $\boldsymbol{L}^P = \boldsymbol{J}^P\cdot\boldsymbol{\omega}$.

3.2 The body must be an infinitely thin disc under 45° with $\boldsymbol{e}_\beta$ and containing the base vector $\boldsymbol{e}_\alpha$.

3.4 See the end of Illustrative Example 5.3.

3.5 If P is the contact point $\boldsymbol{r}_C = -b\sin\phi\,\boldsymbol{e}_1 - (R - b\cos\phi)\boldsymbol{e}_2$, $\ddot{\boldsymbol{z}}_P = -R\dot\phi^2\boldsymbol{e}_2$, $\boldsymbol{J}^P\cdot\boldsymbol{\omega} = [J^C + m(R^2 + b^2 - 2Rb\cos\phi)]\dot\phi\,\boldsymbol{e}_3$, $\dot{\boldsymbol{\omega}} = \ddot\phi\,\boldsymbol{e}_3$, $\boldsymbol{M}^P = -mgb\sin\phi\,\boldsymbol{e}_3$. This yields

$$[J^C + m(R^2 + b^2 - 2Rb\cos\phi)]\ddot\phi + mRb\dot\phi^2\sin\phi + mgb\sin\phi = 0.$$

For using the other reference points introduce reaction forces at the point of contact and eliminate these forces from the law of moment of momentum by formulating also Newton's law.

3.6 $z_P = \overrightarrow{OP_1} = c\sin\phi\, e_1$ with $c = a/\sin\alpha$, $\delta z_P = c\cos\phi\,\delta\phi\, e_1$, $\delta\pi = \delta\phi\, e_3$,

$$\ddot{z}_P = c(\ddot{\phi}\cos\phi - \dot{\phi}^2\sin\phi)e_1, \qquad \ddot{z}_C = \ddot{z}_P + \ddot{\phi}\, e_3 \times r_C - \dot{\phi}^2 r_C, \qquad M^P = a \times F_2 .$$

This yields

$$\ddot{\phi}\{J^C + m[r_C^2 + c^2\cos^2\phi - 2cr_C\cos\phi\sin(\phi+\alpha-\beta)]\} - \dot{\phi}^2 mc[c\sin\phi\cos\phi + r_C\cos(2\phi+\alpha-\beta)] = c[F_1\cos\phi + F_2\cos(\phi+\alpha)] .$$

4.1 $J_2\dot{\omega}_2 = (J_3 - J_1)\omega_3\omega_1 < 0$ for $\omega_3\omega_1 > 0$.

4.2

$$\cos\theta = \begin{cases} s_2\sqrt{J_2(D-J_3)/[D(J_2-J_3)]}\ \operatorname{sn}\tau & \text{for } D<J_2 \\ s_2\sqrt{J_2(J_1-D)/[D(J_1-J_2)]}\ \operatorname{sn}\tau & \text{for } D>J_2 . \end{cases}$$

For $D \to J_2$ the square root tends toward unity while $\operatorname{sn}\tau$ changes periodically between $+1$ and -1. This means periodic changes of θ between $\sim +90°$ and $\sim -90°$ indicating instability. For $D = J_2$ the solutions for ω_α $(\alpha = 1,2,3)$ yield $\cos\theta = +s_2\tanh\tau$, $\phi = \text{const}$, $\dot{\psi} = \sqrt{2T/J_2} = \text{const}$, $\dot{\theta} = -\dot{\tau}/\cosh\tau$. Imagine a sphere (fixed in inertial space) of radius R about the body center of mass and on this sphere the curve generated by the intersection with the axis e_2. The coordinates ψ and $\lambda = \pi/2 - \theta$ of a point of the curve are interpreted as geographic longitude and latitude, respectively (L playing the role of the polar axis). The curve is traced with velocity coordinates $R\dot{\theta} = -R\dot{\tau}/\cosh\tau$ in north-south direction and $R\dot{\psi}\sin\theta = R\dot{\psi}\sqrt{1-\tanh^2\tau} = R\dot{\psi}/\cosh\tau$ in east-west direction. The ratio of the two coordinates is constant which means that the curve is a loxodrome (curve of constant heading).

4.3 Eq. (4.65) yields

$$\begin{aligned} J_1\dot{\omega}_1 - (J_1 - J_3^*)\omega_2\omega_3 &= -\omega_2 L^r(t) \\ J_1\dot{\omega}_2 - (J_3^* - J_1)\omega_3\omega_1 &= +\omega_1 L^r(t) \\ J_3^*\dot{\omega}_3 &= -\quad M^r(t) \end{aligned}$$

with $J_3^* = J_3 - J^r$ and $L^r(t) = \int_{t_0}^{t} M^r(\tau)d\tau + L^r(t_0)$. Integrals of motion: $\omega_1^2 + \omega_2^2 = \Omega^2 = \text{const}$ and $J_3^*\omega_3 + L^r(t) = J_3\omega_3 + h = L = \text{const}$ (axial coordinate of total absolute angular momentum).

Solution $\omega_3 = \omega_3(t_0) - [L^r(t) - L^r(t_0)]/J_3^*$, $\omega_1 = \Omega\sin\alpha(t)$, $\omega_2 = \Omega\cos\alpha(t)$, $\alpha(t) = \alpha(t_0) + \int_{t_0}^{t} f(\tau)d\tau$,

$$f(t) = L/J_1 - \omega_3(t).$$

Interaction torque $M^r(t) - ah$: Eq. (4.65) is replaced by Eqs. (4.55) and (4.56) with the additional term $-ah$. This yields

$$\begin{aligned} J_1\dot{\omega}_1 - (J_1 - J_3)\omega_2\omega_3 &= -\omega_2 h \\ J_1\dot{\omega}_2 - (J_3 - J_1)\omega_3\omega_1 &= \omega_1 h \\ J_3\dot{\omega}_3 \qquad + \dot{h} &= 0 \\ J_3^r\dot{\omega}_3 \qquad + \dot{h} &= M^r(t) - ah . \end{aligned}$$

Integrals of motion: $\omega_1^2 + \omega_2^2 = \Omega^2 = \text{const}$ and $J_3\omega_3 + h = L = \text{const}$ as before.

Solution $h(t) = \phi(t) + [h(t_0) - \phi(t_0)]\exp(bt)$,

$$\omega_3 = \omega_3(t_0) + \frac{1}{J_3}\{\phi(t) + [h(t_0) - \phi(t_0)]\exp(-bt) - h(t_0)\}, \qquad b = a\frac{J_3}{J_3^*}$$

$\phi(t)$ is the particular integral of $\dot{h} + bh = M^r(t)J_3/J_3^*$, $\omega_1 = \Omega\sin\alpha(t)$, $\omega_2 = \Omega\cos\alpha(t)$ as before with the same functions $\alpha(t)$ and $f(t)$.

5.1 $s_k = s_2$: 1. s_1; 2. s_1, s_2; 3. s_3, s_5; 4. s_2, s_3, s_5; 5. s_1, s_4, s_6, s_7; 6. s_4, s_6, s_7.

$s_k = s_3$: 1. s_1, s_2; 2. s_1, s_2, s_3; 3. no vertex; 4. s_3; 5. $s_1, s_2, s_4, s_5, s_6, s_7$; 6. s_4, s_5, s_6, s_7.

5.2 A regularly labeled, unbranched chain of n vertices.

5.3 See footnote on p. 95.

5.6 Use Eqs. (5.15) and (5.24). Construct the $(3n \times n)$ coordinate matrix $\underline{C}$ associated with $\boldsymbol{C}$. Its submatrices are $S_{ia}\underline{c}_{ia}$ $(i, a = 1 \ldots n)$ where $\underline{c}_{ia}$ is the coordinate matrix of $\boldsymbol{c}_{ia}$ in $\underline{e}^{(i)}$ (FORTRAN statements are given on p. 135). The (3×1) submatrices of the product $\underline{C}\,\underline{T}$ are the coordinate matrices $\underline{d}_{ij}$ of $\boldsymbol{d}_{ij}$ in $\underline{e}^{(i)}$. The coordinate matrix in $\underline{e}^{(i)}$ of $\boldsymbol{K}_i$ is

$$\underline{K}_i = \underline{J}_i + \sum_{k=1}^{n} (\underline{d}_{ik}^{\mathrm{T}} \underline{d}_{ik} \underline{E} - \underline{d}_{ik} \underline{d}_{ik}^{\mathrm{T}}).$$

FORTRAN statements are given on p. 137.

5.7 Use Eqs. (5.53) and (5.58). The coordinate matrices $\underline{b}_{ij}$ $(i, j = 1 \ldots n)$ of $\boldsymbol{b}_{ij}$ in $\underline{e}^{(i)}$ are the submatrices of the product $-\underline{C}\,\underline{T}\,\underline{\mu}$ with $\underline{C}\,\underline{T}$ being the matrix from Problem 5.6. Eq. (5.21) yields $\underline{b}_{10} = \underline{d}_{11} + \underline{b}_{11}$ and $\underline{b}_{i0} = \underline{b}_{i1}$ $(i = 2 \ldots n)$. The coordinate matrix in $\underline{e}^{(i)}$ of $\boldsymbol{K}_i^*$ is

$$\underline{K}_i^* = \underline{J}_i + \sum_{k=1}^{n} (\underline{b}_{ik}^{\mathrm{T}} \underline{b}_{ik} \underline{E} - \underline{b}_{ik} \underline{b}_{ik}^{\mathrm{T}}).$$

5.8 If the bodies are numbered in ascending order from one end of the chain to the other then

$$K_i^* = J_i + \frac{ml^2}{4n}\{4i(n+1-i) - 3[i + (n+1-i)] + 2\} = J_i + \frac{ml^2}{4n}(-4i^2 + 4ni + 4i - 3n - 1)$$

5.9 Eq. (5.61) becomes

$$\boldsymbol{\omega} \times \boldsymbol{K}_i^* \cdot \boldsymbol{\omega} = M(\boldsymbol{b}_{ij} \times \boldsymbol{\omega}\,\boldsymbol{\omega} \cdot \boldsymbol{b}_{ji} - \boldsymbol{b}_{ij} \times \boldsymbol{b}_{ji}\,\boldsymbol{\omega} \cdot \boldsymbol{\omega}) \qquad i, j = 1, 2;\ i \neq j.$$

Scalar multiplication by $\boldsymbol{\omega}$ yields $\boldsymbol{\omega} \cdot \boldsymbol{b}_{12} \times \boldsymbol{b}_{21} = 0$ and by $\boldsymbol{b}_{ij}$ $(i \neq j)$: $\boldsymbol{b}_{12} \cdot \boldsymbol{\omega} \times \boldsymbol{K}_1^* \cdot \boldsymbol{\omega} = 0$, $\boldsymbol{b}_{21} \cdot \boldsymbol{\omega} \times \boldsymbol{K}_2^* \cdot \boldsymbol{\omega} = 0$. These equations together with Eq. (5.58) prove the statement. The orientation of the bodies relative to one another and of $\boldsymbol{\omega}$ relative to the bodies in a state of permanent rotation was investigated by Wittenburg [39].

5.11

$$2T = M\dot{\boldsymbol{r}}_C^2 + \sum_{i=1}^{n} \left[\boldsymbol{\omega}_i \cdot \boldsymbol{K}_i^* \cdot \boldsymbol{\omega}_i + M \sum_{\substack{j=1 \\ \neq i}}^{n} (\boldsymbol{\omega}_i \times \boldsymbol{b}_{ij}) \cdot (\boldsymbol{b}_{ji} \times \boldsymbol{\omega}_j) + 2\boldsymbol{\omega}_i \cdot \boldsymbol{h}_i + \sum_{k=1}^{s_i} J_{ik}\,\omega_{ik_{\mathrm{rel}}}^2 \right]$$

$$V = \frac{1}{2}\omega_0^2 \sum_{i=1}^{n} \left[3\boldsymbol{e}_3 \cdot \boldsymbol{B}_i \cdot \boldsymbol{e}_3 - M \sum_{\substack{j=1 \\ \neq i}}^{n} \boldsymbol{b}_{ij} \cdot \boldsymbol{b}_{ji} \right].$$

The quantities $\dot{\boldsymbol{r}}_C$, J_{ik} and $\omega_{ik_{\mathrm{rel}}}$ denote the absolute velocity of the satellite center of mass, the moment of inertia (about the spin axis) of the k-th rotor on body i and the angular velocity of this rotor relative to its carrier, respectively. The number of rotors on carrier i is denoted s_i. The tensors $\boldsymbol{B}_i$ $(i = 1 \ldots n)$ are defined by Eq. (5.75). For the development of these expressions see Wittenburg/Lilov [17].

5.12 The intermediate body must be counted as a regular body with finite dimensions which is coupled by a pin joint to each of its two contiguous bodies. Although the body is massless its associated augmented body has the total system mass M and, having the shape of a dumbbell, also inertia components. Eq. (5.97) can, therefore, be applied without any special considerations. The coefficient matrix $\underline{A}$ will always be positive definite.

5.13 On each of the two bodies coupled by hinge a an auxiliary body-fixed vector base can be defined such that one base vector, say the third one, is parallel to the hinge axis. Then, $\underline{G}_a=\underline{G}_a^1\underline{G}_a^2\underline{G}_a^3$ with constant matrices $\underline{G}_a^1$ and $\underline{G}_a^3$ and with

$$\underline{G}_a^2=\begin{bmatrix}\cos\phi_a & -\sin\phi_a & 0\\ \sin\phi_a & \cos\phi_a & 0\\ 0 & 0 & 1\end{bmatrix}.$$

5.14 Example: Fig. 5.34d; straight guide; body $i^-(a)$ is the pendulous body; $\boldsymbol{z}_a$ starts from a point on the guide and terminates at the pendulum suspension point. Generalized coordinates: Cartesian coordinate q_{a1} and Euler angles $q_{a2}\dots q_{a4}$. $\underline{z}_a=q_{a1}\underline{n}$ (unit vector $\boldsymbol{n}$ along the guide with constant coordinate matrix $\underline{n}$ in $\underline{\boldsymbol{e}}^{(i^+(a))}$). $\underline{G}_a=\underline{G}_a^1\underline{G}_a^2\underline{G}_a^3$ with constant matrices $\underline{G}_a^1$ and $\underline{G}_a^3$ and with $\underline{G}_a^2$ being the matrix of Eq. (2.14) with $q_{a2}\dots q_{a4}$ instead of ψ,θ,ϕ.

5.15 Body $i^+(a)$ is the pendulous body; $\boldsymbol{z}_a$ starts at the pendulum suspension point and terminates at a point on the (straight) guide. Generalized coordinates $q_{a1}\dots q_{a4}$ and matrix $\underline{G}_a$ as in solution to Problem 5.14. $\underline{z}_a=q_{a1}\underline{G}_a\underline{n}$ (unit vector $\boldsymbol{n}$ along the guide with constant coordinate matrix $\underline{n}$ in $\underline{\boldsymbol{e}}^{(i^-(a))}$). The matrix $\underline{z}_a$ depends on all four generalized coordinates and not on one only as in the case where the pendulous body is body $i^-(a)$.

5.16 Example: Fig. 5.34d. The quantities $\boldsymbol{z}_a$, $\boldsymbol{n}$ and $q_{a1}\dots q_{a4}$ are the same as in the solution to Problem 5.14. It is assumed that the matrices $\underline{G}_a^1$ and $\underline{G}_a^3$ are unit matrices which implies that q_{a2}, q_{a3} and q_{a4} are zero when the bases $\underline{\boldsymbol{e}}^{(i^+(a))}$ and $\underline{\boldsymbol{e}}^{(i^-(a))}$ are parallel. $\mathring{\boldsymbol{z}}_a=\dot{q}_{a1}\boldsymbol{n}$; $\mathring{\mathring{\boldsymbol{z}}}_a=\ddot{q}_{a1}\boldsymbol{n}$. The vectors $\boldsymbol{p}_{a1}\dots\boldsymbol{p}_{a4}$ and $\boldsymbol{w}_a$ have in $\underline{\boldsymbol{e}}^{(i^-(a))}$ the column matrices (cf. Eq. (2.28))

$$\underline{p}_{a1}=\underline{0};\qquad \underline{p}_{a2}=[\sin q_{a3}\sin q_{a4}\quad \sin q_{a3}\cos q_{a4}\quad \cos q_{a3}]^{\mathrm{T}};$$
$$\underline{p}_{a3}=[\cos q_{a4}\quad -\sin q_{a4}\quad 0]^{\mathrm{T}};\qquad \underline{p}_{a4}=[0\quad 0\quad 1]^{\mathrm{T}};$$
$$\underline{w}_a=\begin{bmatrix}\dot{q}_{a2}\dot{q}_{a3}\cos q_{a3}\sin q_{a4}+\dot{q}_{a2}\dot{q}_{a4}\sin q_{a3}\cos q_{a4}-\dot{q}_{a3}\dot{q}_{a4}\sin q_{a4}\\ \dot{q}_{a2}\dot{q}_{a3}\cos q_{a3}\cos q_{a4}-\dot{q}_{a2}\dot{q}_{a4}\sin q_{a3}\sin q_{a4}-\dot{q}_{a3}\dot{q}_{a4}\cos q_{a4}\\ -\dot{q}_{a2}\dot{q}_{a3}\sin q_{a3}\end{bmatrix}.$$

5.18 Two major simplifications: First, $\boldsymbol{z}_a\equiv\boldsymbol{0}$ $(a=1\dots n)$ if $\boldsymbol{c}_{i^-(a)a}$ and $\boldsymbol{c}_{i^+(a)a}$ are defined as in Sec. 5.2.2 and $\underline{C}_0=\underline{\boldsymbol{0}}$ if the base $\underline{\boldsymbol{e}}^{(0)}$ on body 0 is located as shown in Fig. 5.10. Second, generalized coordinates need not be introduced since $\boldsymbol{\delta\varkappa}_a$ $(a=1\dots n)$ are independent. Eq. (5.167) becomes $\delta\underline{r}^{\mathrm{T}}=-\underline{\boldsymbol{\delta\pi}}^{\mathrm{T}}\times\underline{C}\,\underline{T}$, Eq. (5.144) becomes $\ddot{\underline{r}}=(\underline{C}\,\underline{T})^{\mathrm{T}}\times\dot{\underline{\omega}}-\underline{g}+\ddot{\boldsymbol{r}}_0\underline{1}_n$ and Eq. (5.169) yields $\delta W=\sum_{a=1}^{n}\delta W_a=-\underline{\boldsymbol{\delta\varkappa}}^{\mathrm{T}}\cdot\underline{Y}$. Substitute these expressions into Eq. (5.104), then replace $\underline{\delta\pi}$ by $-\underline{\boldsymbol{\delta\varkappa}}^{\mathrm{T}}\underline{T}$ (Eq. (5.131)) and premultiply the resulting differential equation by $\underline{S}$. This yields

$$-(\underline{C}\,\underline{T})\times\underline{m}[(\underline{C}\,\underline{T})^{\mathrm{T}}\cdot\dot{\underline{\omega}}]+\underline{\boldsymbol{J}}\cdot\dot{\underline{\omega}}=-(\underline{C}\,\underline{T})\times[\underline{F}+\underline{m}(\underline{g}-\ddot{\boldsymbol{r}}_0\underline{1}_n)]+\underline{M}-\underline{V}-\underline{S}\,\underline{Y}.$$

With the help of Eqs. (5.22), (5.24) and (5.26) it is verified that this is identical with Eq. (5.34).

5.19 $\underline{A}=(\underline{k}\,\underline{T})\cdot\underline{m}(\underline{k}\,\underline{T})^{\mathrm{T}}$, $\underline{B}=-\underline{k}\cdot[\underline{T}(\underline{F}-\underline{m}\,\underline{U})+\underline{X}]$, $\underline{U}=-\underline{T}^{\mathrm{T}}\underline{s}+\ddot{\boldsymbol{r}}_0\underline{1}_n$.

5.20 Translational motions only. As generalized coordinates for hinge a $(a=1\dots n)$ cartesian coordinates q_{a1}, q_{a2}, q_{a3} in the base $\underline{\boldsymbol{e}}^{(0)}$ are used. $\underline{A}=(\underline{k}\,\underline{T})\cdot\underline{m}(\underline{k}\,\underline{T})^{\mathrm{T}}$, $\underline{B}=-\underline{k}\cdot(\underline{T}\,\underline{F}+\underline{X})$ with

$$\underline{k} = \begin{bmatrix} \underline{e}^{(0)} & & & & \\ & \underline{e}^{(0)} & & & \\ & & \underline{e}^{(0)} & & \\ \text{quasi-} & & & \underline{e}^{(0)} & \\ \text{diagonal} & & & & \underline{e}^{(0)} \end{bmatrix}, \qquad \underline{T} = \begin{bmatrix} 1 & 1 & 1 & 1 & 1 \\ 0 & 1 & 1 & 0 & 0 \\ 0 & 0 & 1 & 0 & 0 \\ 0 & 0 & 0 & 1 & 0 \\ 0 & 0 & 0 & 0 & 1 \end{bmatrix}$$

(it is assumed that all arcs in the system graph point toward s_0). Explicitly, the matrix $\underline{A}$ is

$$\underline{A} = \begin{bmatrix} (m_1+m_2+m_3+m_4+m_5)\underline{D} & (m_2+m_3)\underline{D} & m_3\underline{D} & m_4\underline{D} & m_5\underline{D} \\ & (m_2+m_3)\underline{D} & m_3\underline{D} & \underline{0} & \underline{0} \\ & & m_3\underline{D} & \underline{0} & \underline{0} \\ & & & m_4\underline{D} & \underline{0} \\ \text{symmetric} & & & & m_5\underline{D} \end{bmatrix}$$

where $\underline{D}$ and $\underline{0}$ denote (3×3) matrices all elements of which are unity and zero, respectively.

6.1 Hinge 8: $\boldsymbol{n}_i\cdot(\Delta\boldsymbol{v}_8-\Delta\boldsymbol{v}_3)=0$ ($i=1,2,3$; $\boldsymbol{n}_1$, $\boldsymbol{n}_2$, $\boldsymbol{n}_3$ are any non-coplanar vectors). Hinge 9: $\boldsymbol{n}_i\cdot\Delta\boldsymbol{v}_{72}=0$ and $\boldsymbol{n}_i\cdot\Delta\boldsymbol{\omega}_7=0$ ($i=1,2$; $\boldsymbol{n}_1$, $\boldsymbol{n}_2$ are two different vectors orthogonal to the hinge axis; $\Delta\boldsymbol{v}_{72}$ is the velocity increment of body 7 at the hinge point 9). Hinge 10:

$$(\boldsymbol{k}_1\times\boldsymbol{k}_2)\cdot[(\Delta\boldsymbol{v}_{52}+\Delta\boldsymbol{\omega}_5\times\boldsymbol{b})-(\Delta\boldsymbol{v}_6+\Delta\boldsymbol{\omega}_6\times\boldsymbol{b})]=0$$

or

$$(\boldsymbol{k}_1\times\boldsymbol{k}_2)\cdot(\Delta\boldsymbol{v}_{52}-\Delta\boldsymbol{v}_6)-[(\boldsymbol{k}_1\times\boldsymbol{k}_2)\times\boldsymbol{b}]\cdot(\Delta\boldsymbol{\omega}_5-\Delta\boldsymbol{\omega}_6)=0$$

($\boldsymbol{k}_1$, $\boldsymbol{k}_2$ and $\boldsymbol{b}$ are defined as in the example for Eq. (6.38); $\Delta\boldsymbol{v}_{52}$ and $\Delta\boldsymbol{v}_6$ are the velocity increments of body 5 and body 6, respectively, at the hinge point 10). All constraint equations are combined in the form

$$\underline{\boldsymbol{K}}^*\cdot\begin{bmatrix}\Delta\underline{\boldsymbol{v}}^*\\ \Delta\underline{\boldsymbol{\omega}}\end{bmatrix}=\underline{0}\,.$$

The elements of $\underline{\boldsymbol{K}}^*$ are vectors, and the column matrix is the same as in Eq. (6.36). Hence

$$\underline{\boldsymbol{K}}^*\cdot\begin{bmatrix}\underline{\boldsymbol{U}}^* & \underline{\boldsymbol{V}}^*\\ \underline{\boldsymbol{V}}^{*\mathrm{T}} & \underline{\boldsymbol{W}}\end{bmatrix}\cdot\begin{bmatrix}\hat{\underline{\boldsymbol{F}}}^*\\ \hat{\underline{\boldsymbol{M}}}\end{bmatrix}=\underline{0}\,.$$

This replaces Eq. (6.40).

6.2 Assume that the figure depicts the system at the moment of collision in hinge 6, and that the collision in hinge 7 has taken place already. At the moment immediately prior to the collision the system consists of six interconnected rigid bodies with 11 generalized coordinates (six in hinge 1 and one each in the hinges 2, 3, 4, 5 and 6). Eq. (6.37) reads

$$\hat{\underline{\boldsymbol{X}}}=\underline{T}\,\underline{m}\,\Delta\dot{\underline{\boldsymbol{r}}},\qquad \hat{\underline{\boldsymbol{Y}}}=\underline{T}[\underline{\boldsymbol{J}}\cdot\Delta\underline{\boldsymbol{\omega}}-(\underline{\boldsymbol{C}}+\underline{\boldsymbol{Z}})\times\hat{\underline{\boldsymbol{X}}}]$$

with

$$\Delta\dot{\underline{\boldsymbol{r}}}=[\underline{\boldsymbol{p}}\times\underline{T}(\underline{\boldsymbol{C}}+\underline{\boldsymbol{Z}})\underline{T}-\underline{\boldsymbol{k}}\,\underline{T}]^{\mathrm{T}}\Delta\dot{\underline{q}},\qquad \Delta\underline{\boldsymbol{\omega}}=-(\underline{\boldsymbol{p}}\,\underline{T})^{\mathrm{T}}\Delta\dot{\underline{q}}\,.$$

The matrices

$$\Delta\dot{\underline{q}}=[\Delta\dot{q}_{11}\ \ \ldots\ \ \Delta\dot{q}_{16}\ \ \Delta\dot{q}_{21}\ \ \Delta\dot{q}_{31}\ \ \Delta\dot{q}_{41}\ \ \Delta\dot{q}_{51}\ \ \Delta\dot{q}_{61}]^{\mathrm{T}},$$

$$\hat{\underline{\boldsymbol{X}}}=[\hat{\boldsymbol{X}}_1\ \ \hat{\boldsymbol{X}}_2\ \ \hat{\boldsymbol{X}}_3\ \ \hat{\boldsymbol{X}}_4\ \ \hat{\boldsymbol{X}}_5\ \ \hat{\boldsymbol{X}}_6]^{\mathrm{T}}\ \text{and}\ \hat{\underline{\boldsymbol{Y}}}=[\hat{\boldsymbol{Y}}_1\ \ \hat{\boldsymbol{Y}}_2\ \ \hat{\boldsymbol{Y}}_3\ \ \hat{\boldsymbol{Y}}_4\ \ \hat{\boldsymbol{Y}}_5\ \ \hat{\boldsymbol{Y}}_6]^{\mathrm{T}}$$

contain 36 unknowns altogether, namely all elements of $\Delta\dot{\underline{q}}$ except $\Delta\dot{q}_{61}$, all 15 coordinates of $\hat{\boldsymbol{X}}_2\ldots\hat{\boldsymbol{X}}_6$, eight coordinates of $\hat{\boldsymbol{Y}}_2\ldots\hat{\boldsymbol{Y}}_5$ normal to the pin joint axes and all three coordinates of $\hat{\boldsymbol{Y}}_6$. All known quantities except $\Delta\dot{q}_{61}$ are zero. Decomposition in $\underline{e}^{(1)}$ of the equations for $\hat{\underline{\boldsymbol{X}}}$ and $\hat{\underline{\boldsymbol{Y}}}$ yields 36 scalar equations from which all unknowns can be determined.

Literature References

[1] Roberson, R. E.; Wittenburg, J.: A Dynamical Formalism for an Arbitrary Number of Interconnected Rigid Bodies. With Reference to the Problem of Satellite Attitude Control. 3rd IFAC Congr. 1966, Proc. London (1968), 46D.2–46D.9

[2] Lagally, M.; Franz, W.: Vorlesungen über Vektorrechnung. Leipzig 1964

[3] Gantmacher, F. R.: Theory of Matrices. New York 1959

[4] Truesdell, C.: Die Entwicklung des Drallsatzes. Z. f. angew. Math. Mech. **44** (1964) 149–158

[5] Grammel, R.: Der Kreisel. Seine Theorie und seine Anwendungen, vol. 1. Berlin-Göttingen-Heidelberg 1950

[6] Magnus, K.: Kreisel. Theorie und Anwendungen, Berlin-Heidelberg-New York 1971

[7] Tölke, F.: Praktische Funktionenlehre, vol. 3, 4. Berlin-Heidelberg-New York 1967

[8] Magnus, K.: Der Kreisel. Eine Einführung in die Lehre vom Kreisel mit Anleitung zur Durchführung von Versuchen. Göttingen 1965

[9] Arnold, R.; Maunder, L.: Gyrodynamics and its Engineering Applications. New York-London 1961

[10] Saidov, P. I.: Theory of Gyroscopes (Russ.), vol. 1. Moscow 1965

[11] Moiseev, N. N.: Rumjancev, V. V.: Dynamic Stability of Bodies Containing Fluid. Berlin-Heidelberg-New York 1968

[12] Wittenburg, J.: Beiträge zur Dynamik von Gyrostaten. Acc. Naz. dei Lincei, Quad. **217** (1975) 217–354

[13] Wangerin, A.: Über die Bewegung miteinander verbundener Körper. Univ.-Schrift Halle 1889

[14] Volterra, V.: Sur la théorie des variations des latitudes. Acta Math. **22** (1898) 201–357

[15] Busacker, R. G.; Saaty, T. L.: Finite Graphs and Networks. An Introduction with Applications. New York 1965

[16] Belezki, V. V.: Motions of an Artificial Satellite About the Center of Mass (Russ.). Moscow 1965

[17] Wittenburg, J.; Lilov, L.: Relative Equilibrium Positions and their Stability for a Multi-Body Satellite in a Circular Orbit. Ing.-Arch. **44** (1975) 269–279

[18] Baumgarte, J.: Stabilization of Constraints and Integrals of Motion. Comput. Meth. in Appl. Mech. Eng. **1** (1972) 1–16

[19] Zurmühl, R.: Matrizen. Eine Darstellung für Ingenieure. Berlin-Göttingen-Heidelberg 1958

[20] Hooker, W. W.; Margoulis, G.: The Dynamical Attitude Equations for an n-Body Satellite, J. of Astronaut. Sci. **12** (1965) 123–128

[21] Fischer, O.: Einführung in die Mechanik lebender Mechanismen. Leipzig 1906

[22] Hooker, W. W.: A Set of r Dynamical Attitude Equations for an Arbitrary n-Body Satellite Having r Rotational Degrees of Freedom. AIAA J. **8** (1970) 1205–1207

[23] Roberson, R. E.: A Form of the Translational Dynamical Equations for Relative Motions in Systems of Many Non-Rigid Bodies. Acta Mech. **14** (1972) 297–308

[24] Lilov, L.; Wittenburg, J.: Bewegungsgleichungen für Systeme starrer Körper mit Gelenken beliebiger Eigenschaften. Z. f. angew. Math. Mech. **57** (1977) 137–152

[25] Boland, P.; Samin, J.C.; Willems, P.Y.: On the Stability of Interconnected Deformable Bodies in a Topological Tree. AIAA J. **12** (1974) 1025–1030
[26] Likins, P.W.: Dynamic Analysis of a System of Hinge-Connected Rigid Bodies With Nonrigid Appendages. JPL Tech. Rep. 32–1576, Pasadena 1974
[27] Likins, P.W.; Fleischer, G.E.: Large-Deformation Modal Coordinates for Nonrigid Vehicle Dynamics. JPL Tech. Rep. 32–1565, Pasadena 1972
[28] Boland, P.; Samin, J.C.; Willems, P.Y.: Stability Analysis of Interconnected Deformable Bodies With Closed Loop Configuration. AIAA J. **13** (1975) 864–867
[29] Frisch, H.P.: A Vector-Dyadic Development of the Equations of Motion for *n*-Coupled Flexible Bodies and Point Masses. NASA Tech. Note TN D-8047, 1975
[30] Velman, J.R.: Simulation Results for a Dual-Spin Spacecraft. Aerosp. Corp. Rep. TR-0158(3307-01)-16, El Segundo, Cal. 1967
[31] Uicker, J.J.: Dynamic Behavior of Spatial Linkages. Trans. of the ASME **68** Mech. (1968) 1–15
[32] Popov, E.P. et al.: Synthèse de la Commande des Robots Utilisant les Modèles Dynamiques des Systèmes Mécaniques. 6th IFAC Symp. Control in Space (1974) X,31–51
[33] Jerkovsky, W.: The Transformation Operator Approach to Multi-Body Dynamics. Space and Missile Systems Org. Los Angeles, Rep. SAMSO-TR-76-134 (1976), to appear in Matr. Tens. Quat.
[34] Vukobratovic, M.; Stepanenko, J.: Mathematical Models of General Anthropomorphic Systems. Math. Biosci. **17** (1973) 191–242
[35] Renaud, M.: Contribution à l'Etude de la Modélisation et de la Commande des Systèmes Mécaniques Articulés. Diss. Univ. Toulouse 1975
[36] Routh, E.J.: Dynamics of a System of Rigid Bodies (Elementary Part). New York 1960
[37] Wittenburg, J.: Stoßvorgänge in räumlichen Mechanismen. Eine Analogie zwischen Kreiseldynamik und Elastostatik. Acta Mech. **14** (1972) 309–330
[38] Charlamov, P.V.: On the Velocity Distribution in a Rigid Body. (Russ.). mech. tverdovo tela **1** (1969) 77–81
[39] Wittenburg, J.: Permanente Drehungen zweier durch ein Kugelgelenk gekoppelter, starrer Körper. Acta Mech. **19** (1974) 215–226
[40] Baumgarte, J.: Stabilisierung von Bindungen im Lagrangeschen Formalismus. To appear in Z. f. angew. Math. Mech.

Index